Spatial Modelling of the Terrestrial Environment

Editors

RICHARD E.J. KELLY

*Goddard Earth Science and Technology Center,
University of Maryland*

NICHOLAS A. DRAKE

Department of Geography, Kings College London

STUART L. BARR

School of Geography, University of Leeds

John Wiley & Sons, Ltd

Copyright © 2004   John Wiley & Sons Ltd, The Atrium, Southern Gate, Chichester,
West Sussex PO19 8SQ, England

Telephone   (+44) 1243 779777

Email (for orders and customer service enquiries): cs-books@wiley.co.uk
Visit our Home Page on www.wileyeurope.com or www.wiley.com

Cover image courtesy of NASA's Moderate Resolution Imaging Spectroradiometer (MODIS) Land Rapid Response Team.

All Rights Reserved. No part of this publication may be reproduced, stored in a retrieval system or transmitted in any form or by any means, electronic, mechanical, photocopying, recording, scanning or otherwise, except under the terms of the Copyright, Designs and Patents Act 1988 or under the terms of a licence issued by the Copyright Licensing Agency Ltd, 90 Tottenham Court Road, London W1T 4LP, UK, without the permission in writing of the Publisher. Requests to the Publisher should be addressed to the Permissions Department, John Wiley & Sons Ltd, The Atrium, Southern Gate, Chichester, West Sussex PO19 8SQ, England, or emailed to permreq@wiley.co.uk, or faxed to (+44) 1243 770620.

This publication is designed to provide accurate and authoritative information in regard to the subject matter covered. It is sold on the understanding that the Publisher is not engaged in rendering professional services. If professional advice or other expert assistance is required, the services of a competent professional should be sought.

Other Wiley Editorial Offices

John Wiley & Sons Inc., 111 River Street, Hoboken, NJ 07030, USA

Jossey-Bass, 989 Market Street, San Francisco, CA 94103-1741, USA

Wiley-VCH Verlag GmbH, Boschstr. 12, D-69469 Weinheim, Germany

John Wiley & Sons Australia Ltd, 33 Park Road, Milton, Queensland 4064, Australia

John Wiley & Sons (Asia) Pte Ltd, 2 Clementi Loop #02-01, Jin Xing Distripark, Singapore 129809

John Wiley & Sons Canada Ltd, 22 Worcester Road, Etobicoke, Ontario, Canada M9W 1L1

Wiley also publishes its books in a variety of electronic formats. Some content that appears in print may not be available in electronic books.

British Library Cataloguing in Publication Data

A catalogue record for this book is available from the British Library

ISBN 0-470-84348-9

Typeset in 10/12pt. Times and Optima by TechBooks Electronic Services, New Delhi, India
Printed and bound in Great Britain by Antony Rowe Ltd, Chippenham, Wiltshire
This book is printed on acid-free paper responsibly manufactured from sustainable forestry
in which at least two trees are planted for each one used for paper production.

For Hannah, Oscar, Rita and Geoff

Contents

List of Contributors ix
Preface xiii

1 Spatial Modelling of the Terrestrial Environment: The Coupling of Remote Sensing with Spatial Models 1
Richard E.J. Kelly, Nicholas A. Drake and Stuart L. Barr

PART I HYDROLOGICAL APPLICATIONS 7

Editorial: Spatial Modelling in Hydrology 9
Richard E.J. Kelly

2 Modelling Ice Sheet Dynamics with the Aid of Satellite-Derived Topography 13
Jonathan L. Bamber

3 Using Remote Sensing and Spatial Models to Monitor Snow Depth and Snow Water Equivalent 35
Richard E.J. Kelly, Alfred T.C. Chang, James L. Foster and Dorothy K. Hall

4 Using Coupled Land Surface and Microwave Emission Models to Address Issues in Satellite-Based Estimates of Soil Moisture 59
Eleanor J. Burke, R. Chawn Harlow and W. James Shuttleworth

5 Flood Inundation Modelling Using LiDAR and SAR Data 79
Paul D. Bates, M.S. Horritt, D. Cobby and D. Mason

viii   Contents

PART II   TERRESTRIAL SEDIMENT AND HEAT FLUX APPLICATIONS   107

    Editorial: Terrestrial Sediment and Heat Fluxes   109
    Nick Drake

6   **Remotely Sensed Topographic Data for River Channel Research:**   113
    The Identification, Explanation and Management of Error
    Stuart N. Lane, Simon C. Reid, Richard M. Westaway and D. Murray Hicks

7   **Modelling Wind Erosion and Dust Emission on Vegetated Surfaces**   137
    Gregory S. Okin and Dale A. Gillette

8   **Near Real-Time Modelling of Regional Scale Soil Erosion Using AVHRR**   157
    and METEOSAT Data: A Tool for Monitoring the Impact of Sediment
    Yield on the Biodiversity of Lake Tanganyika
    Nick Drake, Xiaoyang Zhang, Elias Symeonakis, Martin Wooster,
    Graeme Patterson and Ross Bryant

9   **Estimation of Energy Emissions, Fireline Intensity and Biomass**   175
    Consumption in Wildland Fires: A Potential Approach Using
    Remotely Sensed Fire Radiative Energy
    Martin J. Wooster, G.L.W. Perry, B. Zhukov and D. Oertel

PART III   SPATIAL MODELLING OF URBAN SYSTEM DYNAMICS   197

    Editorial: Spatial Modelling of Urban System Dynamics   199
    Stuart L. Barr

10   **Characterizing Land Use in Urban Systems via Built-Form**   201
    Connectivity Models
    Stuart Barr and Mike Barnsley

11   **Modelling the Impact of Traffic Emissions on the Urban Environment:**   227
    A New Approach Using Remotely Sensed Data
    Bernard J. Devereux, L.S. Devereux and C. Lindsay

PART IV   CURRENT CHALLENGES AND FUTURE DIRECTIONS   243

12   **Land, Water and Energy Data Assimilation**   245
    David L. Toll and Paul R. Houser

13   **Spatial Modelling of the Terrestrial Environment: Outlook**   263
    Richard E.J. Kelly, Nicholas A. Drake and Stuart L. Barr

Index   267

Contributors

Jonathan L. Bamber   School of Geographical Sciences, Bristol University, University Road, Bristol, BS8 1SS, UK

Mike Barnsley   School of Geography, University of Wales Swansea, Singleton Park, Swansea, SA2 8PP, UK

Stuart L. Barr   School of Geography, University of Leeds, Leeds, LS2 9JT, UK

Paul D. Bates   School of Geographical Sciences, University of Bristol, University Road, Bristol, BS8 1SS, UK

Adam Ross Bryant   JBA Consulting, Bowcliffe Grange, Bowcliffe Hall Estate, Bramham, West Yorkshire, LS23 6LW, UK

Eleanor J. Burke   Department of Hydrology and Water Resources, University of Arizona, Tucson, AZ 85721, USA

Alfred T.C. Chang   Hydrological Sciences Branch, Code 974/Blding 22/R169, NASA/Goddard Space Flight Center, Greenbelt, MD 20771, USA

David Cobby   ESSC, University of Reading, Whiteknights, Reading, RG9 6AL, UK

Bernard J. Devereux   Unit for Landscape Modelling, Department of Geography, University of Cambridge, Mond Building, Free School Lane, Cambridge, CB2 3RF, UK

List of Contributors

Lynn S. Devereux   Unit for Sustainable Landscapes, Department of Geography, University of Cambridge, Downing Place, Cambridge, CB2 3EN, UK

Nicholas A. Drake   Department of Geography, King's College London, Strand, WC2R 2LS, UK

James L. Foster   Hydrological Sciences Branch, Code 974/Blding 22/R169, NASA/Goddard Space Flight Center, Greenbelt, MD 20771, USA

Dale A. Gillette   Air Resources Laboratory (MD-81), Applied Modeling Research Branch, Research Triangle Park, NC 27711, USA

Dorothy K. Hall   Hydrological Sciences Branch, Code 974/Blding 22/R169, NASA/Goddard Space Flight Center, Greenbelt, MD 20771, USA

R. Chawn Harlow   Department of Hydrology and Water Resources, University of Arizona, Tucson, AZ 85721, USA

D. Murray Hicks   National Institute of Water and Atmosphere, Christchurch, New Zealand

Matt S. Horritt   School of Geographical Sciences, University of Bristol, University Road, Bristol, BS8 1SS, UK

Paul R. Houser   Hydrological Sciences Branch, NASA/Goddard Space Flight Center, Greenbelt, MD 20771, USA

Richard E.J. Kelly   Goddard Earth Science and Technology Center, University of Maryland Baltimore County, 1000 Hilltop Circle, Baltimore, Maryland 21250, USA

Stuart N. Lane   School of Geography, University of Leeds, Woodhouse Lane, Leeds, LS2 9JT, UK

Clare Lindsay   ME&P, Transport and Planning Consultants, Cambridge, UK

David Mason   ESSC, University of Reading, Whiteknights, Reading, RG9 6AL, UK

Dieter Oertel   Institute of Space Sensor Technology and Planetary Exploration, Rutherfordstr. 2, D-12489 Berlin, Germany

Gregory S. Okin   Department of Environmental Sciences, University of Virginia, Charlottesville, VA 22904-4123, USA

Graeme Patterson   Assistant Director Africa Programme, Wildlife Conservation Society, 2300 Southern Boulevard, Bronx, NY 10460, USA

George Perry Department of Geography, Kings College London, Strand, WC2R 2LS, UK

Simon C. Reid School of Geography, University of Leeds, Woodhouse Lane, Leeds, LS2 9JT, UK

W. James Shuttleworth Department of Hydrology and Water Resources, University of Arizona, Tucson, AZ 85721, USA

Elias Symeonakis Department of Geography, University of Valencia, Av. Blasco Ibanez 28, Valencia 46010, Spain

David L. Toll Hydrological Sciences Branch, NASA/Goddard Space Flight Center, Greenbelt, MD 20771, USA

Richard M. Westaway Halcrow (GIS), Swindon, SN4 0QD, UK

Martin J. Wooster Department of Geography, King's College London, Strand, WC2R 2LS, UK

Xiaoyang Zhang Department of Geography/Center for Remote Sensing, Boston University, 725 Commonwealth Avenue, Boston, MA 02215, USA

Boris Zhukov Institute of Space Sensor Technology and Planetary Exploration, Rutherfordstr. 2, D-12489 Berlin, Germany

Preface

This book contributes to the diverse and dynamic research theme of advances in remote sensing and GIS analysis. Continuing the tradition of other edited volumes, it is the product of a stimulating meeting organized jointly by members of the U.K. Remote Sensing Society's (now Photogrammetry and Remote Sensing Society) Modelling and Advanced Techniques (MAT) and Geographical Information Systems (GIS) Special Interest Groups (SIG). These two SIGs hold regular meetings and are an invaluable forum for vibrant discussions between undergraduates, graduates, professional academics and practitioners of GIS and environmental modelling. This particular meeting was held at Birkbeck College, University of London on 17 November, 2000 and attracted about 100 delegates.

The plan to organize a one-day meeting was originally hatched (perhaps during a reckless moment of bravado) at the 1999 annual conference of the Remote Sensing Society in Cardiff. Convened by Nick Drake, Stuart Barr and Richard Kelly, the one-day symposium was slightly different to previous meetings in that the theme was more narrowly focussed than before. Speakers were asked to present aspects of their research and how it relates to spatial modelling of terrestrial environments, one of the themes running through both MAT and GIS SIGs. Ten speakers presented their up-to-date research and there was plenty of scope for some vibrant discussion and exchange of ideas. From the ten presentations, six are included as chapters in this book and five new contributors were sought to enhance the existing contributions, and also add an international dimension to the volume. To provide some context for the different contributions, we have also included our own personal views about the three broad areas of applications and how spatial modelling is an important strand running through each application. These are personal views that offer our own interpretation on some very rapidly advancing GIS and modelling fields.

We hope that the book will be useful for students and researchers alike. Inevitably, there are omissions in the topics covered and responsibility for any omissions lies squarely with the editors. Furthermore, most of the topics covered in this volume are rapidly evolving in theory and application and many new research themes have recently appeared and are pushing forward rapidly at both the national and international research levels. Ultimately, while we do not claim to cater to all needs and interests, we feel that there are some important issues raised in the book that apply to the integration of remote sensing data with spatial models and trust that there are some 'nuggets' for interested academicians and practitioners in this dynamic field.

A book such as this cannot come into existence without the assistance of many people involved both with the organization of the symposium and the production of the book. For sponsoring the original meeting we are very grateful to the Photogrammetry and Remote Sensing Society, Birkbeck College, University of London, Kings College London and Leeds University. We would also like to thank the local organizing committee, especially Tessa Hilder, Chloe Hood and Matt Disney. In connection with getting the book to print and for their invaluable guidance and help in steering the editors through the publishing maze, we would like to sincerely thank John Wiley & Sons, particularly Lyn Roberts and Keily Larkins.

Richard Kelly, Nick Drake and Stuart Barr

1

Spatial Modelling of the Terrestrial Environment: The Coupling of Remote Sensing with Spatial Models

Richard E.J. Kelly, Nicholas A. Drake and Stuart L. Barr

'The creation of silicon surrogates of real-world complex systems allows us to perform controlled, repeatable experiments on the real McCoy. So, in this sense complex system theorists are much in the same position that physicists were in the time of Galileo, who was responsible for ushering in the idea of such experimentation on simple systems. It was Galileo's efforts that paved the way for Newton's development of a theory of such processes. Unfortunately, complex systems are still awaiting their Newton. But with our new found ability to create worlds for all occasions inside the computer, we can play myriad sorts of what-if games with genuine complex systems' (Casti, 1997).

1.1 Introduction

The objective of this book is to present snapshots of research that are focused on spatial modelling approaches to terrestrial environmental problems. With thirteen chapters covering three broad environmental areas, the contributions concentrate on examples of how models, particularly numerical models, can be used spatially to address a variety of practical issues. This objective seems fairly straightforward except that when we consider precise definitions, things start to get a little 'fuzzy'. For example, at first glance the *terrestrial environment* is simple enough to define since it relates to conditions, processes or events that occur on the land portion of the Earth surface. However, when we connect it to *spatial modelling*, we might need to consider land–atmosphere–ocean interactions that help to condition the purely terrestrial environment. Therefore, how should these distinctly non-terrestrial

---

Spatial Modelling of the Terrestrial Environment. Edited by R. Kelly, N. Drake, S. Barr.
© 2004 John Wiley & Sons, Ltd. ISBN: 0-470-84348-9.

components be represented? Many Earth system scientists adopt a wider and more holistic approach to spatial modelling and, with powerful computational assistance, have been able to satisfy large-scale modelling demands. Spatial models can range from simple point process models that are spatially distributed to complex spatially referenced mathematical operations performed within the context of geographical information science (GISc). And what about the data that we use to input to our spatial model? Perhaps remote sensing data are among the most obvious data types that could be harnessed for spatial modelling. In fact, if we reduce remote sensing products to their bare bones, they are probably one of the simplest forms of spatial models or spatial representation of the terrestrial (or perhaps planetary) environment. Physical remote observations from spatially discrete and often separate locations on the earth are converted to information via some algorithm or model. The derived information actually represents a spatial variable or geophysical parameter making the data implicitly spatial as an input to an environmental model. Therefore, the nature of spatial modelling of the terrestrial environment seems to have a fairly broad scope and could include many diverse scientific activities. Consequently, the objective of this book is to present a range of spatial modelling examples and to demonstrate how they can be used to inform and enhance our understanding of the terrestrial environment and ultimately pave the way to an increased accuracy of model predictions.

We have chosen to focus on the terrestrial environment primarily for reasons of self-interest. For readers seeking a more general collection, there have been several excellent recent volumes that have documented the general theme of advances in geographical information systems (GIS) and remote sensing (e.g. Foody and Curran (1994), Danson and Plummer (1998), Atkinson and Tate (1999), Tate and Atkinson (2001) and Foody and Atkinson (2002)). The fact that these volumes are highly successful is due to the high quality of the contributions, and because the editors have expertly assembled some key applications in GIS and remote sensing, both of which are dynamically interlinked and rapidly evolving sub-disciplines within geography, Earth and environmental science. For a flavour of the activity in GIS alone, the reader is directed to the comprehensive 'Big Book of GIS', now in its second edition (Longley *et al.*, 1999). We have taken a slightly different tack to these previous volumes in that we have attempted to couple the advances in remote sensing, GIS and modelling to a particular application theme, namely the terrestrial environment. In doing so, there is much that we have not covered with respect to both the terrestrial environment and the more theoretical aspects of spatial modelling. However, in understanding and perhaps predicting the terrestrial environment, the scientist often is less concerned with sub-discipline specific categorizations and more concerned with bringing these strands together to help solve a problem. We feel that not only does this book serve this purpose but also illustrates the diversity of approaches that can be taken and shows the way the field is heading.

1.2 Spatial Modelling

While the literature on general model theory is vast, the general aims of modellers usually consist of improving our understanding of a phenomenon (process, condition, etc.), simulating its behaviour to within a prescribed level of accuracy and ultimately predicting future states of the phenomenon (Kirkby *et al.*, 1993). In general, these aims are linearly related; increased environmental understanding often leads to improved simulation capability which can, but not always, result in more accurate prediction. While not wishing to become immersed in the many different modelling critiques, there are a few key aspects that are instructive for the modeller. Casti (1997) proposed four fundamental characteristics that can provide insight into the utility of a model: simplicity, clarity, bias-free and tractability. These four qualities should be self-evident for modellers. Tractability is possibly the most important property for the overall design of modern computer modelling since it relates to the ability of a computer model to undertake the task efficiently and not at a disproportionate computational cost. With increasingly complex models becoming the norm, there is sometimes a sense that the model is 'too big' for the task and that disproportionately large amounts of computer 'brawn' are needed to solve a relatively trivial problem (akin to requiring a 'sledgehammer to crack open a nut'). This is especially important when we extend models to become spatial in implementation.

A key question that modellers often need to address, then, is 'how is spatiality woven into the application?' Spatial models in GIS have generally considered the concept of data model, particularly vector and raster model representations of data (Burrough and McDonnell, 1998). More recently, research has focused on approaches that define objects or fields in geographical data modelling and most recently, Cova and Goodchild (2002) have proposed a methodology to produce fields of spatial objects. Since remote sensing data are generally rasterized, many spatial models are configured with grid cells. While this is a convenient computational device, Cracknell (1998) notes that in fact, remote sensing information rarely represents regular square fields of view. Rather, instantaneous fields of view are usually elliptical in geometry, a function of the physics of many Earth observation instruments. Thus, remote sensing data tend to be distorted to fit the modelling spatial framework that is often based on a square cell arrangement. In discussing developments of environmental modelling and the need to 'future-proof' modelling frameworks Beven (2003) observes, 'it is often inappropriate to force an environmental problem into a raster straitjacket. Treating places as flexible objects might be one way around (this) future-proofing problem.' Spatial modelling of the terrestrial environment, therefore, needs to be sensitive to this issue but as yet, most modelling approaches remain fixed in the regular grid cell or raster mode of spatial representation.

A key theme that is associated with spatial modelling is scale which Goodchild (2001) terms the 'level of detail' in a given geographical representation of the Earth's surface. Scale affects directly all four of Casti's qualities in modelling (clarity, simplicity, bias and tractability). With increased computer power and increased spatial resolution, there is a tendency towards increased spatial variation, which may or may not be required. If we also consider the gradual technological improvement of remote sensing spatial resolving power, the level of detail at which scientists can model has increased over the years. However, modellers have to use their judgement with respect to the appropriate spatial scale that is, (a) possible, and (b) required. For example, to model soil moisture spatial variability, it probably does not make sense to compute estimates at a 10×10 m spatial resolution if the scientist is interested in regional drought conditions: 500×500 m might be sufficient. Moreover, the tractability of the model might be called into question when using such a fine spatial resolution. Understanding this spatial dependence of the property to be modelled is an important aspect of scale selection but one that is sometimes neglected. Ideally, spatial dependence should determine the spatial framework of the model from the outset.

Current spatial frameworks available to environmental modellers can cater for most scales of spatial variation at the local to regional scale. Models often can be constructed using desktop computer hardware either using standalone or clustered node architectures. However, at the other end of the extreme, the regional-to-global scale (such as coupled land–ocean–atmosphere models) is less well catered for or researchers need to have access to very expensive and powerful multi-parallel processing computers. Even then, there are computational limits to what models can achieve especially with the demand for increased model complexity. The limitations can be pushed back by a coordinated infrastructure approach at the international research community level (Dickinson et al., 2002). Two approaches have led to the Program for Integrated Earth System Modelling (PRISM) in Europe and to the Earth System Modeling Framework (ESMF) in the USA (Ferraro et al., 2003). The general goals of these initiatives are similar in scope since they both aim to increase model interoperability across the Earth system science research community. In doing so, scientists will be able to compare models within a standardized and scaleable modelling framework. The aims are to enable improved model simulation and potentially identify computational technology developments (both hardware and software) required to satisfy the need for increasingly complex models over broad space and time scales. If successful, these initiatives will pave the way for the next generation of Earth system science models.

1.3 Summary of the Book

The scope of spatial modelling of the terrestrial environment is broad in its application. This is reflected by the diversity of applications and examples that are included in this book. Running through all of the contributions are three common threads that make the title of this book appropriate. First, all chapters describe mathematical operations on spatially referenced data that are a representation of a terrestrial state variable, both natural environmental and human environmental. Second, most contributions describe applications that are reliant on the use of remote sensing data as input variables to a modelling environment. Those that do not, use spatially referenced information from other authorized sources. Third, many of the contributors use geographical information systems (GIS) technology (some more formally than others) to help perform modelling procedures.

The book is divided into four Parts, which reflect the contributors' interests: hydrological applications, terrestrial sediment and heat flux applications and urban system dynamics applications. The hydrological applications Part begins with a chapter by Jonathan Bamber who discusses and demonstrates the application of radar altimetry and synthetic aperture radar (SAR) interferometry data for the estimation of ice sheet topography. Ice sheet topography is critical for numerical models that predict ice sheet dynamics and, thus, provide a means of estimating an ice sheet's mass balance. In the next chapter Richard Kelly et al. investigate how numerical hydrological models and passive microwave remote sensing observations of snow mass can be used to retrieve global snow volume. In this example, models are used both in hydrological simulation and in the retrieval algorithms that recover snow parameters such as snow depth. They also investigate spatial dependence characteristics evident in ground measurements and remote sensing observations of snow depth. Eleanor Burke et al. in Chapter 4 then describe how land surface and microwave emission models can be coupled to increase the accuracy of estimates of soil moisture from ground-based and aircraft radiometer instruments. The rationale for this work is to pave the way for accurate soil moisture retrievals from planned satellite missions designed for global soil moisture monitoring. Finally, in the last chapter of this Part, Paul Bates *et al.* demonstrate how LiDAR and SAR data can be combined with the prediction of flood inundation through their coupling to a numerical flood model.

Part II illustrates how models and remote sensing are used in terrestrial sediment and heat flux modelling. The first chapter in this Part by Stuart Lane *et al.* examines how the knowledge of errors in the derivation of digital elevation models from remote sensing data is an important component of river channel research and must be effectively managed. Understanding, quantifying and ultimately reducing errors in models not only can reflect model confidence but it can also provide insight into how the model can be improved. In Chapter 7, Greg Okin and Dale Gillette show how ground-measured wind vectors can be used to model wind erosion and dust emission. The rationale for this chapter is to explore the relative contribution of natural and land use sources of dust emission to local or regional wind erosion patterns. By quantifying micro-scale to local-scale variations in emission and the controlling processes, coarser-scale models should benefit through increased accuracy. In Chapter 8 by Nick Drake *et al.*, remote sensing data are coupled to a regional-scale soil erosion model to examine biodiversity on Lake Tanganyika. While they discuss some of the shortcomings such as the errors associated with the remote sensing products used (e.g. the normalized difference vegetation index), they show that such an approach has great promise for biodiversity monitoring of remote locations. Chapter 9 by Martin Wooster *et al.* deals with the estimation of fire radiant energy from different satellite remote sensing instruments. This information is then coupled to a model to estimate vegetation combustion, a variable that is proportional to the rate of release of pollution species. Such an application is of great interest to many regions of the world where fire monitoring and modelling are necessary civic responsibilities.

In Part III, which deals with applications of spatial modelling to urban system dynamics, Stuart Barr and Mike Barnsley demonstrate the utility of built-form connectivity models for characterizing land use in urban systems. This approach uses fine-scale digital map data, derived from remote-sensing data as input to a model. The objective is to infer urban land use from land cover parcel data. The motivation for this research is that with high spatial resolution satellite photography becoming available, such connectivity models might assist the cartographer in estimating land use in a built urban environment from space. The second 'urban' chapter is by Bernard Devereux *et al.* and discusses how a wide range of remote sensing data can be coupled to socio-economic census databases within a GIS to assist with planning policy development. The chapter, therefore, represents the powerful merging of remote sensing, spatial modelling and GIS to a direct societal application. Part IV deals with current challenges and future directions. The penultimate chapter in the book is by Dave Toll and Paul Houser. The contribution deals with land surface modelling in hydrology at large regional scales and represents an intermediate step towards the Earth System Science modelling approaches described briefly above. Large-scale atmospheric and land surface data are assembled from diverse spatial data sources (remote sensing, weather prediction models, ground surface observations, etc.) and integrated, ultimately through formal data assimilation, to generate suites of key hydrological variables that are of great interest to global water cycle modellers and climate modellers. This is a large undertaking that is being conducted at the agency level of US government (i.e. at NASA) and is helping to focus the

ESMF/PRISM initiatives. The chapter is not included in to the hydrological applications Part not because it does not deal with hydrology but because it serves as a practical example of spatial modelling of the terrestrial environment at the 'big science' scale. Furthermore, such a large-scale modelling effort could easily assimilate the examples in all preceding chapters.

In the final chapter, Nick Drake, Stuart Barr and Richard Kelly draw together some of the key strands running through the book and look towards the future of spatial modelling of the terrestrial environment. It is clear that this is a dynamic and diverse research field and that some important issues are beginning to emerge. It is also probable that these issues will affect research modelling strategies in the near future and could have an impact on the way practitioners use remote sensing, GIS and spatial modelling.

References

Atkinson, P.M. and Tate, N.J. (eds), 1999, *Advances in Remote Sensing and GIS Analysis* (Chichester: John Wiley and Sons).

Beven, K.J., 2003, On environmental models everywhere on the GRID, *Hydrological Processes*, **17**, 171–174.

Burrough, P.A. and McDonnell, R.A., 1998, *Principles of Geographical Information Systems*, 2nd edition (Oxford: Oxford University Press).

Casti, J.L., 1997, *Would-be Worlds: How Computer Simulation Is Changing the Frontiers of Science* (Chichester: John Wiley and Sons).

Cova, T.J. and Goodchild, M.F., 2002, Extending geographical representation to include fields of spatial objects, *International Journal of Geographical Information Science*, **16**, 509–532.

Cracknell, A.P., 1998, Synergy in remote sensing: what's in a pixel? *International Journal of Remote Sensing*, **19**, 2025–2047.

Danson, F.M. and Plummer, S.E. (eds), 1995, *Advances in Environmental Remote Sensing* (Chichester: John Wiley and Sons).

Dickenson, R.E., Zebiak, S.E., Anderson, J.L., Blackmon, M.L., DeLuca, C., Hogan, T.F., Iredell, M., Ji, M., Rood, R., Suarez, M.J. and Taylor, K.E., 2002, How can we advance our weather and climate models as a community? *Bulletin of the American Meteorological Society*, **83**(3), 431–434.

Ferarro, R., Sato, T., Brasseur, G., Deluca, C. and Guilyardi, E., 2003, Modeling the Earth System: critical computation technologies that enable us to predict our planet's future, *Proceedings of the International Geoscience and Remote Sensing Society*, 21–25 July, 2003, Toulouse, France.

Foody, G.M. and Atkinson, P.M. (eds), 2002, *Uncertainty in Remote Sensing and GIS* (Chichester: John Wiley and Sons).

Foody, G.M. and Curran, P.J. (eds), 1994, *Environmental Remote Sensing from Regional to Global Scales* (Chichester: John Wiley and Sons).

Goodchild, M.F., 2001, Models of scale and scales of modeling, in N.J. Tate and P.M. Atkinson (eds), *Modelling Scale in Geographical Information Science* (Chichester: John Wiley and Sons) 3–10.

Kirkby, M., Naden, P., Burt, T. and Butcher, D., 1992, *Computer Simulation in Physical Geography*, 2nd edition (Chichester: John Wiley & Sons).

Longley, P.A., Goodchild, M.F., Maquire, D.J. and Rhind, D.W. (eds), 1999, *Geographical Information Systems*, 2 volumes (Chichester: John Wiley and Sons).

Tate, N.J. and Atkinson, P.M. (eds), 2001, *Modelling Scale in Geographical Information Science* (Chichester: John Wiley and Sons).

PART I
HYDROLOGICAL APPLICATIONS

Editorial: Spatial Modelling in Hydrology

Richard E.J. Kelly

The simulation and prediction of hydrological state variables have been a significant occupation of hydrologists for more than 70 years. For example, Anderson and Burt (1985) noted that Sherman (1932) devised the unit hydrograph method to simulate location-specific river runoff with a very simple parameterization and data input. Clearly introduced before the availability of high performance and relatively inexpensive computers, this method 'was to dominate the hydrology for more than a quarter of a century, and (is) one which is still in widespread use today' (Anderson and Burt, 1985). The approach is spatial inasmuch as it represents in a 'lumped' fashion upstream catchment processes convergent at the point where runoff is simulated. While the unit hydrograph has been a very useful tool for water resource management, hydrologists have sought to 'spatialize' their methodologies of simulation (and ultimately prediction) to account for variations of catchment processes in two and three spatial dimensions. A primary need to do this has been to be able to represent systems dynamically and especially when a system is affected by an extreme event, such as a flood. Conceptually, this shift towards understanding and accounting for complex hydrological process, has manifested itself in the form of a move towards physically based models and away from simple statistically deterministic models. From the 1980s onwards, and perhaps inevitably, the availability of relatively high performance computers that could undertake vast numbers of calculations has allowed the numerical representation of distributed hydrological processes in a catchment. As a result, the propensity for models to become complex has only been a small step on. Now, complex hydrological models can be executed, in standard 'desktop' computer environments. Furthermore, the challenges facing hydrologists engaged in simulation and prediction are not so much those of technology (although this aspect is constantly pushed forward by ever demanding requirements

---

Spatial Modelling of the Terrestrial Environment. Edited by R. Kelly, N. Drake, S. Barr.
© 2004 John Wiley & Sons, Ltd. ISBN: 0-470-84348-9.

of space and time resolution), but more conceptual and theoretical in nature. For example, given the relatively small technological barriers present, how best should multi-dimensional space and time be represented, in a hydrological model digital environment? How should continuous and real hydrological processes be represented in an artificial grid cell array? Even with the ability to resolve such issues technologically, there are as many different theoretical views concerning the potential solution as there are models on this subject. As Anderson and Bates (2001) note, models are constantly evolving as new theories of hydrological processes mature. Ironically, questions about space and time demonstrate how dynamic the science of hydrology continues to be and how little it has changed from the "unit hydrograph era", when questions of space and time representation were moot.

The implementation of spatial models in hydrology has happened in different ways. Traditionally, models have consisted of some input variable (usually measured in the field, such as precipitation) passed to a transfer function (such as a mass or energy balance equation that could be stochastic or deterministic) which then estimates a state variable (such as stream flow). Models, therefore, can be used to represent hydrological processes or relationships operating within a pre-defined spatial domain. The domain might be a sub-region of a continental region or a river catchment, i.e. a basic hydrological spatial unit. While the 'lumped' spatial model is still used in some instances, many hydrological models attempt to represent explicitly mass or energy transfers discretely within the domain of interest. To do this, it has been customary to consider spatial discretization of the domain in the form of raster and/or vector representation. In the raster representation of continuous space, a model is often applied to all grid cells within a study domain. In this way, even though the model might not be explicitly 'spatial', the fact that it represents the average or dominant processes operating over a spatial area (a grid cell) implies spatiality. In other forms, distributed process models explicitly account for the spatial relationship between adjacent cells and are explicitly 'spatial' in character (for example, a runoff routing in a river catchment). Raster grid representations can take many different forms including triangles, rectangles, hexagons, etc. They can also be defined in three-dimensional space in the form of voxels, although the rectangular grid (square) is the most commonly used form. Vector representations of space consist of points, lines and polygon shapes with attributes associated with each element. The hydrological model is applied to the region bounded by polygons and represents the processes operating therein. In most respects, the implementation of models in hydrology is applied using raster representation. However, as Beven (2003) notes "... it is often inappropriate to force an environmental problem into a raster straitjacket" and the future for distributed hydrological modelling is far from clear. Nevertheless, the raster representation is a convenient form of coupling remote sensing data to models since remote-sensing data usually represent instantaneous fields of view as recorded by the observing instrument. In this Part, therefore, we examine how remote sensing data can be used with spatial models in hydrology.

Remote sensing products are coupled to hydrological models in different ways. They might be used to drive the hydrological catchment model or a land surface model (LSM) directly or they might be combined with model-estimated geophysical state variable to produce a "best estimate" via a filter. Figure E1 shows a generalized summary of the approach, which by no means covers all possibilities. Typically, remote sensing data are transformed into an estimated hydrological state variable (product) via a retrieval algorithm that can use physically based or empirical sub-models. Many algorithms are complex and require ancillary spatial data, such as topographical or meteorological data, to produce the required

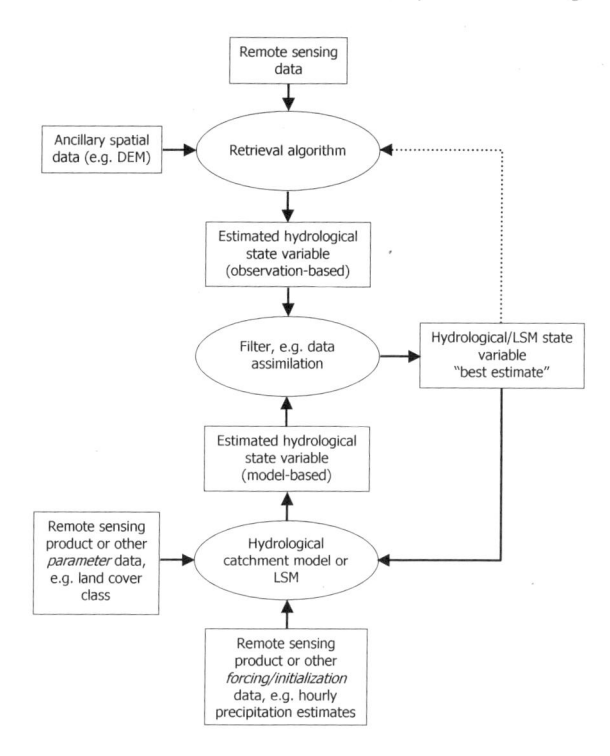

Figure E1 *Generalized view of the coupling of remote sensing data with land-based models in hydrology.*

geophysical state variable. In the early development of remote sensing applications, algorithms were often simple in scope (e.g. a simple statistical classification algorithm), but with increased understanding of the interaction between electromagnetic radiation and hydrological state variables, algorithms have generally improved, albeit often with an increase in algorithm complexity. There are always errors associated with derived remote sensing products. For operational applications, these are usually (or should be) well defined. For research purposes, however, the uncertainties may be less well defined and, therefore, can be the root cause of error propagation through the coupled model to the estimated state variable. Only by thorough model verification and testing procedures can the errors be fully specified and reduced.

The remote-sensing product (estimated hydrological state variable) can be passed directly to the hydrological model in the form of a model parameter (e.g. land cover class), as a model forcing variable (e.g. hourly precipitation) or as an initial state variable (e.g. soil moisture). It is usually only one of several parameters of forcing/initialization variables. When there is large uncertainty associated with the hydrological model estimates, it has been shown that by combining the model-based estimates with observation-based estimates, hybrid "best estimates" can be derived. This combination is often achieved through a "filter", such as data assimilation, which typically computes the best estimate of the hydrological state variable as a weighted average of the model-based and remote sensing-based state estimates. The weighting is determined from the error attributed to the model-based and remote sensing-based hydrological state variable estimates; more weight is given to the estimates with smaller associated error. A best estimate can be used to re-initialize the hydrological model at the next time step or it might also be used to initialize the remote sensing retrieval algorithm. While Toll and Houser (Chapter 12) demonstrate the use of data assimilation in LSMs, the development of data assimilation applications in hydrology, in general, is still in its infancy. For an account of its more mature application in the atmospheric sciences, Kalnay (2002) provides a good explanation of how data assimilation is used in numerical weather prediction. There are alternative filter methodologies that can be used to combine observation-based and model-based estimates and these include spatial or temporal interpolation.

In the following four chapters different specific applications are described that couple remote sensing with spatial hydrological models, in different ways. In Chapter 2, Bamber describes how the Antarctic ice sheet dynamics can be modelled, using satellite-derived topography estimates. In this example, the remote-sensing product is used as a driving parameter in physically based models of ice flow. He shows that the accuracy of the altimeter-based ice sheet elevation estimates affects the modelled estimates of ice flow, which is a key component in calculating the mass balance of Antarctica. The chapter by Kelly *et al.* (Chapter 3) analyzes and compares spatial variations of global ground-measured snow depth with satellite passive microwave estimates of snow depth. Since satellite estimates are areal in nature, while ground measurements tend to be representative of a point, the spatial scaling between these types of estimates is uncertain. For example, how far are the point measurements representative of wider areal variations in snow depth? The chapter also summarizes recent and current approaches for satellite passive microwave observations of snow depth and SWE. Burke *et al.* (Chapter 4) demonstrate how coupled land surface and microwave emission models can help with the estimation of soil moisture. The chapter describes how passive microwave soil moisture estimates can be assimilated into a state-of-the-art land surface model (LSM). They recognize the shortcomings of LSMs and suggest that improvements in model physics and improved parameterization of soil moisture heterogeneity in the pixels would improve the estimates of soil moisture and, therefore, improve the performance of the LSM. Finally, in Chapter 5, Bates *et al.* illustrate how high spatial resolution LiDAR and synthetic aperture radar (SAR) products can be used for flood inundation simulation. Interestingly, they suggest that the future of many spatially distributed modelling frameworks will rely on remote sensing data, that are better specified in terms of the accuracy of the derived hydrological variable.

References

Anderson, M.G. and Bates, P.D., 2001, Model credibility and scientific enquiry, in M.G. Anderson and P.D. Bates (eds), *Model Validation: Perspectives in Hydrological Science* (Chichester: John Wiley and Sons), 1–10.

Anderson, M.G. and Burt, T.P., 1985, Model strategies, in M.G. Anderson and T.P. Burt (eds), *Hydrological Forecasting* (Chichester: John Wiley and Sons), 1–13.

Beven, K.J., 2003, On environmental models of everywhere on the GRID, *Hydrological Processes*, **17**, 171–174.

Kalnay, E. 2002, *Atmospheric Modeling, Data Assimilation and Predictability* (Cambridge: Cambridge University Press).

Sherman, L.K., 1932, Streamflow from rainfall by unit-graph method, *Engineering News Record*, **108**, 501–505.

2
Modelling Ice Sheet Dynamics with the Aid of Satellite-Derived Topography

Jonathan L. Bamber

2.1 Introduction and Background

The Greenland and Antarctic ice sheets contain about 80% of the Earth's freshwater, and cover 10% of the Earth's land surface. They play an important role in modulating freshwater fluxes into the North Atlantic and Southern Ocean, and if they were to melt completely they would raise global sea level by around 70 m. Thus, even a relatively small imbalance in their mass budget has a significant influence on global sea level changes. The current rate of sea level rise is $\sim$2 mm a^{-1}. About half of this can be accounted for through the melting of sub-polar glaciers and thermal expansion of the oceans. The other half, however, is unaccounted for and it seems likely that the Greenland and/or Antarctic ice sheets are responsible. The uncertainty in their mass budget is, however, equivalent to the total sea level rise signal of 2 mm a^{-1}. To reduce this uncertainty, and to improve our ability to model future changes, accurate measurements of the form and flow of the ice sheets are essential.

The over-arching rationale for studying the ice sheets is often limited to their influence on sea level as there is a clear cause and effect and it is one of the key issues associated with global warming. It is important to note, however, that the ice sheets also have a fundamental impact on the climate system in other ways. In particular, they are primary sources of freshwater fluxes into the North Atlantic and Southern Oceans. Changes in these fluxes could have, for example, a profound impact on the thermohaline circulation of the North Atlantic. Such changes have been associated with rapid climate change during the last

---

Spatial Modelling of the Terrestrial Environment. Edited by R. Kelly, N. Drake, S. Barr.
© 2004 John Wiley & Sons, Ltd. ISBN: 0-470-84348-9.

glacial period (Manabe and Stouffer, 1993). The existence of the Antarctic ice sheet greatly reduces absorption of solar radiation as the albedo of dry snow is about 0.9 compared to 0.01 for open ocean. This is known as the ice-albedo positive feedback, which maintains lower temperatures through the high albedo of snow. This feedback mechanism is one of the reasons why the polar regions are believed to be more sensitive to global warming than lower latitude areas (Hougton, 1996).

The ice sheets interact and influence several components of the climate system and to model these interactions accurately, we need to be able to accurately model the behaviour of the ice sheets. Here, we present an overview of this modelling activity and how, accurate ice sheet topography has been used as an input to, and validation of, the models.

Satellite remote sensing is the tool of choice for studying an area as remote, hostile, expensive to operate in, and as large as the Antarctic ice sheet (AIS), which covers an area of some 13 M km^2 (greater than the conterminus USA). Although the Greenland ice sheet (GrIS) is about ten times smaller and more accessible, satellite remote sensing still plays a key role in observing and monitoring its behaviour. As a consequence, two space agencies have both recently initiated programmes whose primary focus is observations of the cryosphere (NASA with the Ice, Cloud and Land Elevation Satellite (ICESat) and ESA with CryoSat).

2.1.1 The Relevance of Topography

In hydrological modelling, accurate topography is probably *the* most important boundary condition. In the case of ice sheets, the link is perhaps less obvious but equally important. Ice sheet topography is an important parameter in numerical modelling for two reasons. First, it can be used to validate the ability of a model to reproduce the present-day geometry. Second, as with water, ice flows downhill (over an appropriate length scale) and the magnitude of the gravitational driving force that creates this flow is proportional to the surface slope (Figure 2.1). To accurately reproduce the dynamics of an ice sheet a model must, therefore, accurately reproduce surface slope or use it as an input boundary condition.

The gravitational driving force, τ, can be approximated by:

$$\tau = \rho g h \alpha \tag{1}$$

for small bed slopes, where ρ is ice density, g is gravity, h is ice thickness and α is

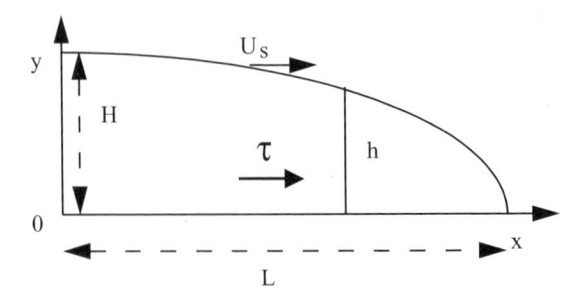

Figure 2.1 Schematic diagram of an ice sheet showing the relationship between driving stress and slope

surface slope. When τ is incorporated into the flow law of ice (known as Glen's flow law) to determine the surface velocity, U_s, we find that it is proportional to the third power of α:

$$U_S - U_{bed} = A\tau^n \frac{h}{(n+1)} \qquad (2)$$

where n is the flow law exponent, usually taken to be 3, A is a variable in the flow law of ice that determines its viscosity (Paterson, 1994) and U_{bed} is the velocity at the bed. A has an exponential relationship with temperature:

$$A = A_0 \exp-\left(\frac{q}{RT}\right). \qquad (3)$$

Equation (3) is an Arrhenius-type relationship, where A_0 is a temperature-independent 'constant', q is the activation energy for creep, R is the universal gas constant and T is temperature. The relationships between velocity and τ and between viscosity and T result in a highly non-linear system, with important consequences and challenges for numerical modelling.

Using Glen's flow law, the ice thickness, h, of an ice sheet can be shown to be related to its length, L, and position, x (see Figure 2.1) as follows:

$$h^{2+2/n} = K(L^{1+1/n} - x^{1+1/n}) \qquad (4)$$

where

$$K = \frac{2(n+2)^{1/n}}{\rho g} \left(\frac{c}{2A}\right)^{1/n} \qquad (5)$$

and c is the net accumulation at the surface (Paterson, 1994). Various assumptions have been used to derive equation (4) including a flat, horizontal bedrock and a uniform accumulation rate. From equations (4) and (5), it can be seen that, assuming a value of 3 for n, the ice thickness is insensitive to c (proportional to the eighth power), and that it varies as the eighth root of the flow law parameter A. The significance of this will be discussed later.

2.2 Remote Sensing of Topography: Methodology

Probably the most frequently used method for obtaining topography from remote sensing data is stereo-photogrammetry of airborne or satellite imagery. However, this approach, using visible sensors such as SPOT, has a number of limitations over ice sheets. First, there is rarely sufficient contrast to carry out stereo matching in the visible part of the EM spectrum. Second, cloud is ubiquitous in the polar regions and difficult to discriminate from snow. Third, to obtain absolute height measurements, ground control points are required, which are rarely available. Consequently, other approaches have proved more valuable. Microwave sensors are particularly useful in the polar regions as they are generally unaffected by clouds and can operate day or night. The two instruments that have been used most successfully for deriving topography are radar altimeters and synthetic aperture radars (SAR). The methodologies for deriving topography from these two instruments are outlined.

2.2.1 Satellite Radar Altimetry

A thorough review of the use of satellite radar altimeters (SRAs) over ice sheets can be found elsewhere (Zwally and Brenner, 2001). Presented here is a brief overview of the key issues and problems associated with the use of SRA data over ice sheets. SRAs are active microwave instruments that transmit a microwave pulse to the ground and measure its two-way travel time. With sufficiently accurate determination of orbit it is possible to measure the elevation of the sea surface with an accuracy of a few centimetres. Although SRAs were designed primarily for operation over ocean and non-ocean surfaces, such as ice sheets, certain limitations apply. In particular, the current fleet of SRAs are unable to range to surfaces that have a slope significantly greater than the antenna beamwidth of the instrument (which is typically about 0.7°). Thus their use is limited to slopes less than about 1°. Consequently, accurate height estimates can only be obtained from larger ice masses with low regional slopes, over distance of more than ~50 km, i.e., over Antarctica and Greenland.

There are four satellite missions which satisfied the dual requirement of having accurate-enough orbit determination and an orbital inclination that provided substantive coverage of the ice sheets. The first of these was Seasat, launched in 1978. This satellite had a latitudinal limit of 72° providing coverage of the southern half of Greenland and about one fifth of Antarctica (Figure 2.2). Although the mission only lasted 100 days, it clearly demonstrated

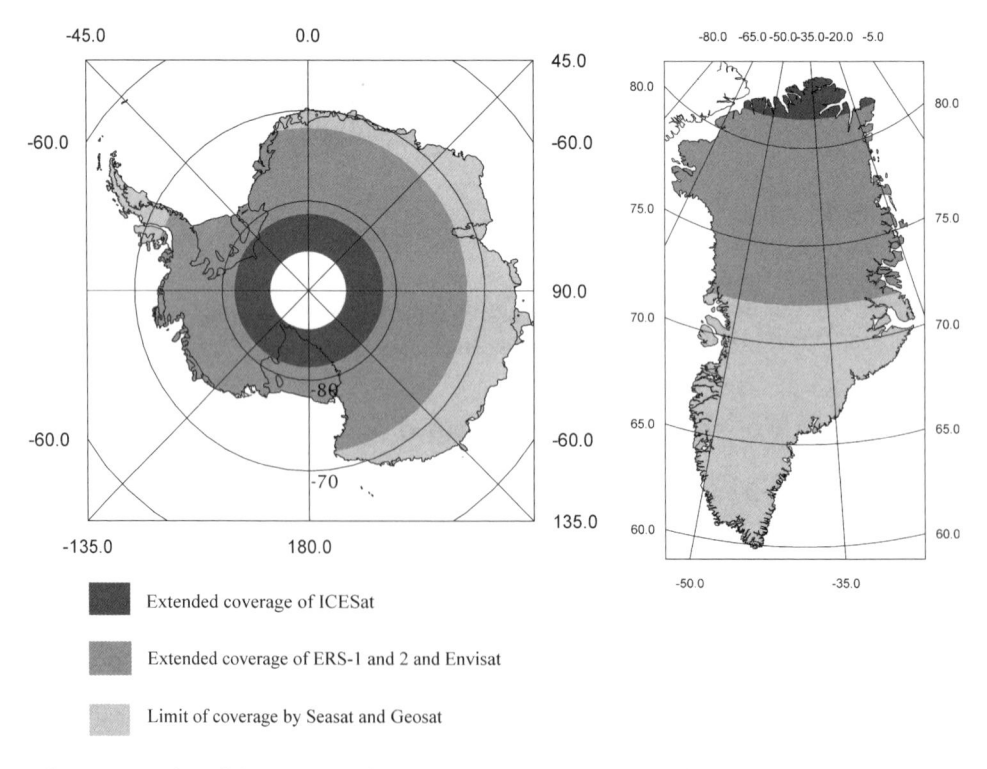

Figure 2.2 *Plot of the coverage by past, present and future satellite altimeter missions over the Greenland and Antarctic ice sheets*

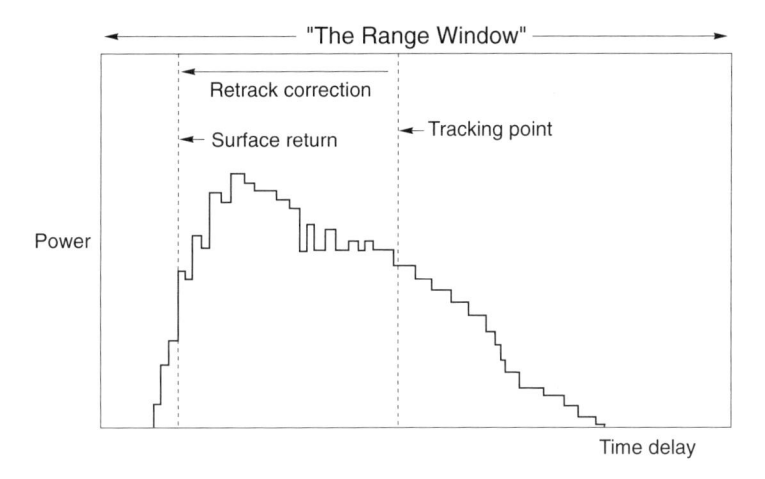

Figure 2.3 Schematic illustration of the 'range window' of a satellite radar altimeter, with a returned echo, or waveform in it. The position of the 'mean surface' and the magnitude of the retrack correction are shown

the value of SRA data for mapping the topography of the ice sheets. Geosat, which flew from 1985–1989, extended the temporal record but not the spatial coverage. In 1991 the first European remote sensing satellite, ERS-1, was placed in an orbit that provided coverage to 81.5° (Figure 2.2). In 1995, this satellite was superseded by ERS-2, which had similar characteristics. There is, thus, a continuous record of elevation change for the last decade, covering the whole of the Greenland ice sheet and four-fifths of Antarctica.

Corrections. Over non-ocean surfaces a number of corrections need to be applied due to the undulating nature of the surface. The first of these is a range-estimate refinement procedure known as waveform retracking. SRAs record the returned echo, or waveform, in a narrow range window, which for ERS-1 and 2 had a width of either 32 or 128 m (Bamber, 1994b). The narrower window was known as 'ocean mode' and the wider one as 'ice mode'. To ensure the surface return is captured in the range window the SRA needs to accurately anticipate the time delay of the returned pulse. It does this using onboard tracking software that attempts to ensure that the returned waveform is centred in the range window (Figure 2.3). This procedure works satisfactorily over the oceans, where surface slopes are extremely low. Over land surfaces, however, the onboard tracker cannot keep the waveform centred in the range window and, consequently, tends to wander about the centre point. If the topography is sufficiently rough, the echo may even move out of the range window altogether and this situation is known as 'loss of lock'. It tends to occur in steeper areas such as around rock outcrops and close to the margins of the ice sheets and results in no range estimate being obtained. Waveform retracking involves determining the offset between the centre of the range window and the position of the 'mean surface' in the echo. Different investigators have used different approaches to waveform retracking, which vary primarily in how they determine where the 'mean surface' in the returned echo is (Figure 2.3). The retrack error is typically of the order of a few metres and cannot exceed a maximum value of half the range window width (i.e., 16 or 64 m for ERS-1 or 2).

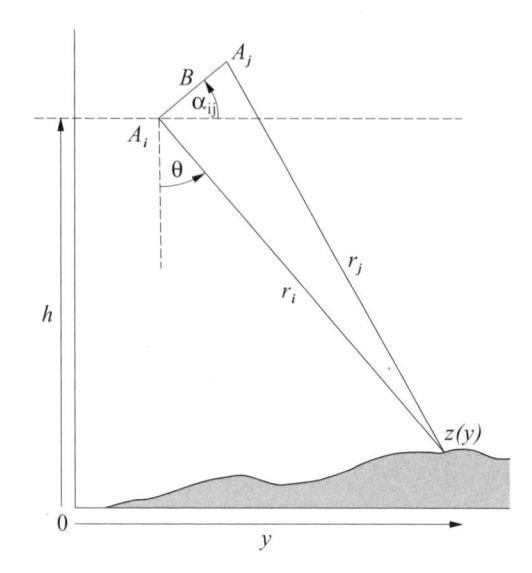

18 Spatial Modelling of the Terrestrial Environment

Figure 2.4 *Schematic diagram illustrating the geometry associated with repeat-pass SAR interferometry, where two separate measurements (A_i, and A_j) of an area of the Earth's surface are taken at different times and different locations*

The second problem that needs to be addressed is known as the slope-induced error (Brenner et al., 1983) and results from the fact that the altimeter ranges to the nearest point on the ground rather than the nadir point (Figure 2.4). For a slope of 1° the difference in range between the nadir point and the nearest point is about 120 m. The magnitude of the range difference is:

$$\Delta R = R_n(1 - \cos\theta) \qquad (6)$$

where R_n is the range to the nadir point and θ is the local surface slope. As with waveform retracking there are several methods for correcting for this error, all of which involve using slope information derived from the SRA data, either only along track (Zwally et al., 1990) or in two dimensions by combining across-track data (Bamber et al., 2001).

2.2.2 Interferometric Synthetic Aperture Radar (InSAR)

Radar interferometry is a technique that combines coherent SAR images recorded by antennas at different locations or at different times to form interferograms that permit the detection of small differences in range to a target from two points of observation (Figure 2.4). Differences of a fraction of a wavelength can be measured using the phase difference between two or more images. The phase differences produce an interference pattern or interferogram. The interference pattern is a function of: (i) the topography; (ii) any displacement of the surface that has taken place between the two image acquisitions; and (iii) the separation in space (also termed the baseline) of the SAR when the two images were acquired. A detailed description of how topography can be obtained over ice using repeat-pass interferometry is given elsewhere (Joughin et al., 1996) and only the key points are presented here.

For a given set of repeat-pass observations, from the ith and jth epochs, with baseline B_{ij} and look angle θ, the interferometric phase difference at each sample is given by:

$$\Delta\phi_{ij} = \left(\frac{4\pi}{\lambda}\right) B_{ij} \sin(\theta - \alpha_{ij}) + \left(\frac{4\pi}{\lambda}\right) \Delta\rho_{ij} = \phi_{\text{topography}} + \phi_{\text{displacement}} \quad (7)$$

where the baseline, B, is the distance separating the two points and α is the tilt of the baseline with respect to the horizontal (Figure 2.4). The first term, $\phi_{\text{topography}}$, contains phase information from the topography of the surface relative to the interferometric baseline. If the targets are displaced by $\Delta\rho_{ij}$ in the range direction between two observations, then the observed phase will include a second contribution of $(4\pi/\lambda)\Delta\rho_{ij}$ due to this displacement. The sensitivity of the measurements to topography is proportional to B. For example, if the baseline is zero, then $\phi_{\text{topography}}$ is zero and the interference pattern is dependent on displacement only. However, $\phi_{\text{displacement}}$, is independent of B. When the ith and jth observations are acquired at the same time (single-pass interferometry), only the $\phi_{\text{topography}}$ is relevant. If the ith and jth observations are separated by a time interval, the technique is generally referred to as repeat-pass interferometry. In the case where more than two observations are available, it is known as multiple pass interferometry.

InSAR can only provide relative height information and ground control points (GCPs) in the form of either high-accuracy GPS data or a coarse resolution DEM are required for three purposes: (i) to provide absolute height control; (ii) to improve the baseline estimate, which is crucial to obtain accurate results; and (iii) to separate $\phi_{\text{topography}}$ from $\phi_{\text{displacement}}$. It is also possible to achieve (iii) by using multiple-pass interferometry and assuming that $\phi_{\text{displacement}}$ is a constant term in the multiple sets of interferograms. More often, an SRA-derived DEM is used as shown in the example discussed later (Figure 2.6).

2.3 Numerical Models

Numerical ice-sheet models typically comprise three main components. These are the temporal evolution of ice thickness, which also incorporates the areal growth and contraction of the ice sheet; a reduced model of how stresses and velocities vary within an ice mass (although this is not always adequate, as discussed later); and the temporal evolution of the internal temperature field of the ice sheet. These latter two components must be linked because of the feedback between temperature and viscosity (equation (3)) and, as a consequence, the coupled system is usually referred to as a thermo-mechanical model.

There are several ancillary models which may be included such as a visco-elastic lithosphere model to account for variations in ice loading, a model for basal sliding of the ice over its underlying substrate and an ablation model to account for mass loss through surface melting. All these components are coupled together and used to simulate, through time, the ice sheet geometry, temperature and velocity fields within the ice. They can be applied either to past climatic re-constructions (e.g. Fabre *et al.*, 1997), for predictive purposes such as assessing the effect of global warming on the ice sheet (e.g. Huybrechts *et al.*, 1991) and for understanding the underlying processes and sensitivities in the fully coupled system (Marshall and Clarke, 1996).

The models solve a set of equations for the conservation of mass and energy at the surface and within the ice sheet. The equations are usually discretized onto a regular rectangular grid and solved using a finite-difference approach. Due to strong feedbacks, the non-linear flow law of ice and the exponential dependence of viscosity on temperature, the coupled system is highly non-linear. One of the key feedbacks is between internal strain heating and ice viscosity. As the ice warms, it softens, leading to increased internal strain heating and more softening. It has been suggested that in certain circumstances, this softening can lead to creep instability.

Another important feedback exists in the surface mass balance model. In Antarctica, there is almost no loss of ice at the surface, through melting. In Greenland this is not the case and an extensive ablation area exists around the margins of the ice sheet where surface melting can exceed the annual accumulation rate by more than 10 m a^{-1} (Braithwaite, 1995). Melt or ablation rates are very sensitive to surface temperature and many numerical ice sheet models use a parameterization of the surface temperature to estimate the ablation rate based on a positive degree day model (Reeh, 1991). With this approach the number of days above 0° C are estimated and a degree day factor is used to convert this to melt rate. The surface temperature is often estimated as a function of latitude, longitude and elevation (x,y,z). If melt takes place, then elevation is reduced, leading to higher temperatures and greater melting.

Fast-flow features such as outlet glaciers or ice streams (see, for example Figure 2.8) are often moving via two different mechanisms: internal ice deformation, and basal sliding, which can occur if the ice/bed interface is lubricated by meltwater. Where basal sliding does take place, it can be responsible for more than 80% of the total motion (Engelhardt and Kamb, 1998). Reliable modelling of basal sliding behaviour is, therefore, also a key component of ice sheet models. The basal sliding law, however, is poorly known and poorly constrained and invariably incorporates tuning parameters to account for unknowns such as bed roughness and pore water pressure.

Other key tuneable parameters exist in the model, in particular related to the ice rheology. This is because various unmeasurable factors affect the rheology and consequently the ice flow. One of the most important factors is the anisotropic nature of the flow law for ice, which results from the preferential orientation of ice crystals that have been subjected to quasi-uniform stresses over an extended period of time (Azuma, 1994). To overcome this problem, Some models employ a variable enhancement factor (between about 0.1 and 10) to the flow law, which represents the uncertainty in the *in situ* ice rheology. The various tuning factors employed, which are a common feature of most physical models of naturally occurring systems, have little physical meaning and exist solely to ensure the model provides a reasonable fit to present-day boundary conditions such as the surface elevation of the ice sheet and its lateral extent.

Typically, simulations of present-day conditions of the ice sheet are only validated against geometric boundary conditions such as the shape of the ice sheet (e.g. Fabre *et al.*, 1997). This is mainly because it is the most easily reproduced and best-known boundary condition. For dynamic simulations it is, however, not the most meaningful or relevant. Furthermore, the geometry of an ice mass is relatively insensitive to key parameters such as the ice viscosity. In fact, for parts of the AIS and GrIS, a perfectly plastic model of ice flow produces a reasonable fit to the surface profile (Vaughan and Bamber, 1998). Until recently, however, there were very few other data sets available, particularly at the whole ice sheet scale, to validate the models against. Consequently, derived quantities such as balance velocities (discussed in the next section) are valuable for both validating ice sheet models (Bamber *et al.*, 2000a) and for constraining them.

2.4 Results and Analysis

The primary result of the processing of SRA data for topographic information has been the production of a number of digital elevation models (DEMs) for the ice sheets. These are useful in their own right for modelling studies but have also been used to generate a suite of derived products and datasets, which have also proved valuable for constraining and validating ice sheet models. Here, we present the key primary and secondary datasets resulting from the production and analysis of satellite-derived topography.

2.4.1 Topography from SRA

A number of groups have used combinations of Seasat, Geosat and ERS SRA data to derive digital elevation models (DEMs) of both Antarctica and Greenland (Bamber, 1994a; Bamber *et al.*, 2001; Ekholm, 1996; Remy *et al.*, 1989; Zwally *et al.*, 1983).

SRA data from ERS covers almost the entirety of the GrIS (Figure 2.2) barring the steeper margins of the ice sheet. To produce a full coverage DEM of the whole ice sheet and surrounding bedrock, SRA data from ERS and Geosat were supplemented with airborne stereo-photogrammetry, airborne LiDAR data, InSAR-derived topography and, where no other adequate data were available, digitized cartography (Bamber *et al.*, 2001). The result was a 1 km posting DEM with an RMSE over the ice sheet of between 2 and 14 m. Over the bedrock the accuracy ranged from 20 to 200 m, dependent on the data source available. Figure 2.5a is a planimetric shaded relief plot of the DEM. Certain large-scale characteristics of flow are clearly visible. Major drainage basins can be identified and, conversely, ice-divides (where only vertical motion takes places) can be seen as bright 'ridges' ranging from relatively broad features for much of the northern half of the ice sheet to narrow features with a sharp appearance in the south. Much of the southern half of the main basin in the north east has a 'mottled' appearance, reflecting a 'disturbed' pattern of flow in this area, most probably associated with a rougher bedrock topography in this area and the existence of a fast-flow feature in the basin (Layberry and Bamber, 2001).

In Antarctica, the coverage by sensors other than SRAs is more limited. A 5-km spacing DEM for the whole ice sheet was produced from ERS geodetic phase data and merged with sparse and relatively poor accuracy terrestrial measurements beyond the latitudinal limit of the satellite (81.5°) (Bamber and Bindschadler, 1997). A planimetric view of this DEM is shown in Figure 2.5b. As for the GrIS, the main flow units can be identified. Extended regions of floating ice, known as ice shelves, appear as homogenous, light grey areas with low contrast. The boundary between the SRA and terrestrial data is also clearly visible and the poorer spatial resolution of the latter is manifested by the smoother appearance of the central disk. The surface expression of the sub-glacial lake at Vostok, which is the subject of considerable scientific attention, is very clear.

More recently, ERS SRA data were combined with other limited datasets, chiefly comprising digitized cartography around coastal and mountainous areas, to produce a DEM of

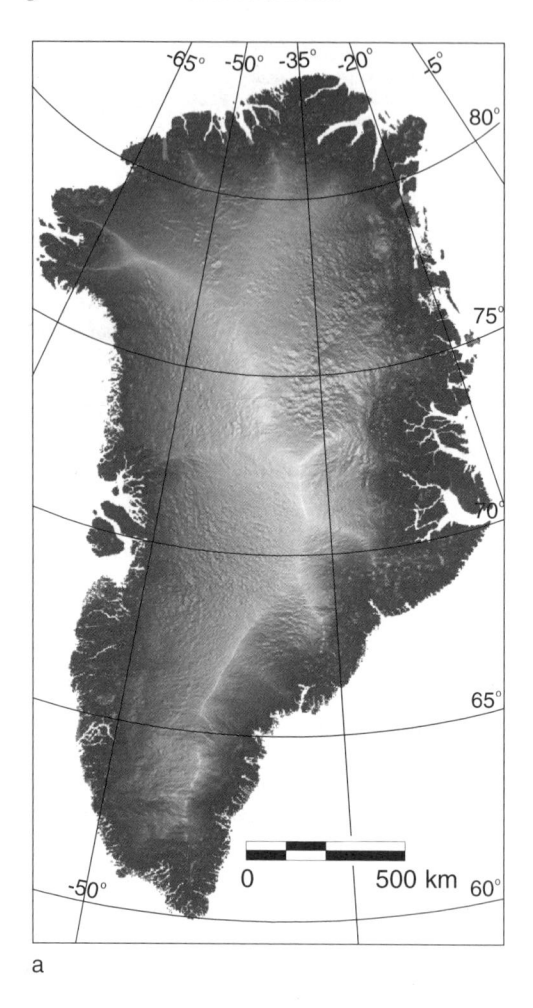

22   Spatial Modelling of the Terrestrial Environment

Figure 2.5 (a) A planimetric shaded-relief plot of a 1-km digital elevation model of the Greenland ice sheet, (b) the same as for (a) but for a 5-km DEM of the Antarctic ice sheet

Antarctica with horizontal postings at 200 m, although over the ice sheet the true resolution of the data is about 4 km (Liu *et al.*, 1999).

2.4.2 Topography from InSAR

InSAR has the capability to provide much higher-resolution topography (~25 m) than is possible with conventional SRAs. No ice sheet scale models run at resolutions as fine as this but force budget models (discussed later) or thermo-mechanical models of individual glaciers or basins can benefit from DEMs at a finer resolution than is possible from SRA data. An example of an InSAR-derived DEM for the north-east Greenland ice stream (NEGIS) is shown in Figure 2.6. It has been draped over a considerably coarser ice sheet DEM derived from SRA data (Joughin *et al.*, 2001). The higher resolution of the InSAR data provides information on flow features that are related to longitudinal stress gradients

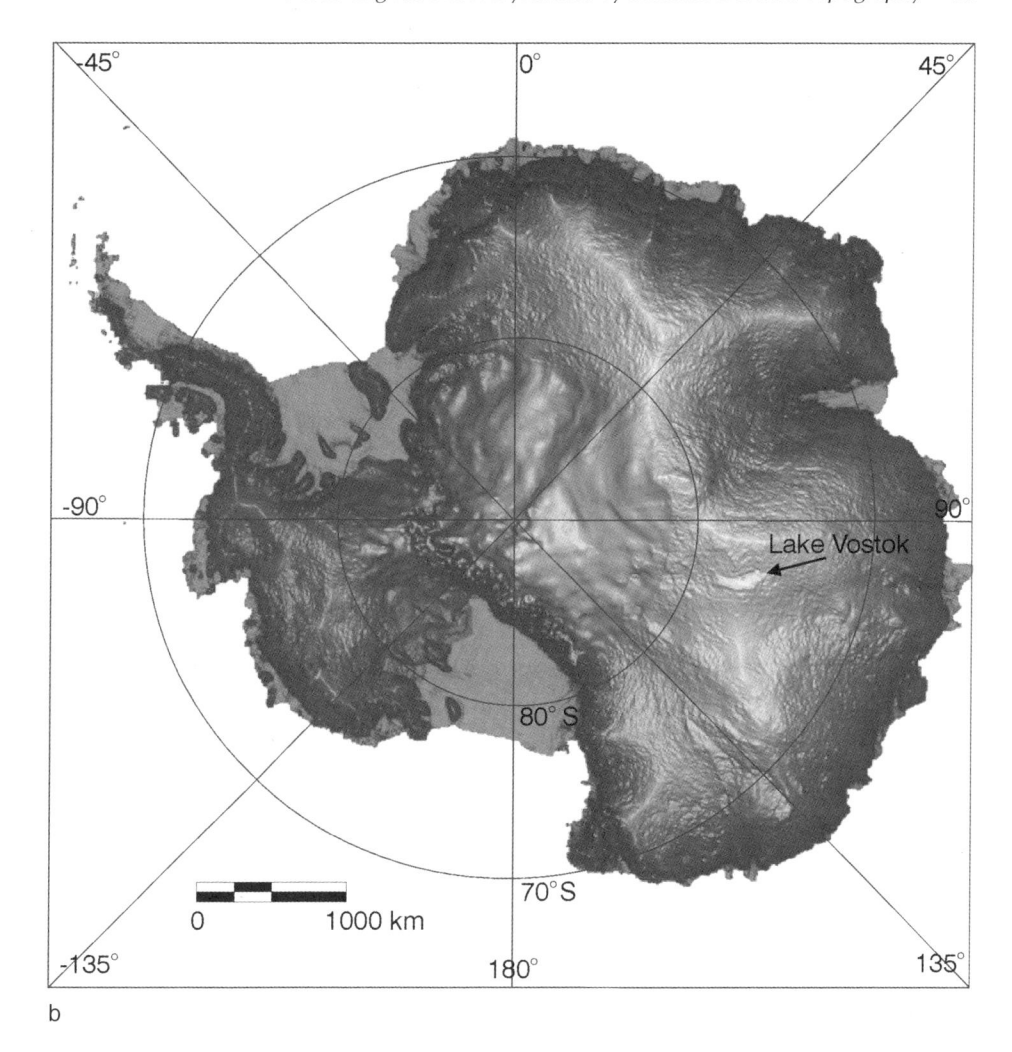

Figure 2.5 (cont.)

in the ice and basal topography. The InSAR DEM was used in a force budget calculation for the Northeast Greenland ice stream solving for basal shear stress and thus provides insights into basal conditions (Joughin *et al.*, 2001). It should be noted that InSAR produces relative elevations only and in the example discussed here, the absolute height control was provided by an SRA-derived DEM. Radar altimetry and InSAR are, therefore, complementary with neither methodology proving sufficient alone.

2.4.3 Derived Datasets

Accurate DEMs of the ice sheets have been used in a range of applications related to modelling the behaviour of the ice sheets. As mentioned, ice divides are the equivalent of watersheds in hydrology and demarcate the boundary between independent flow units, or

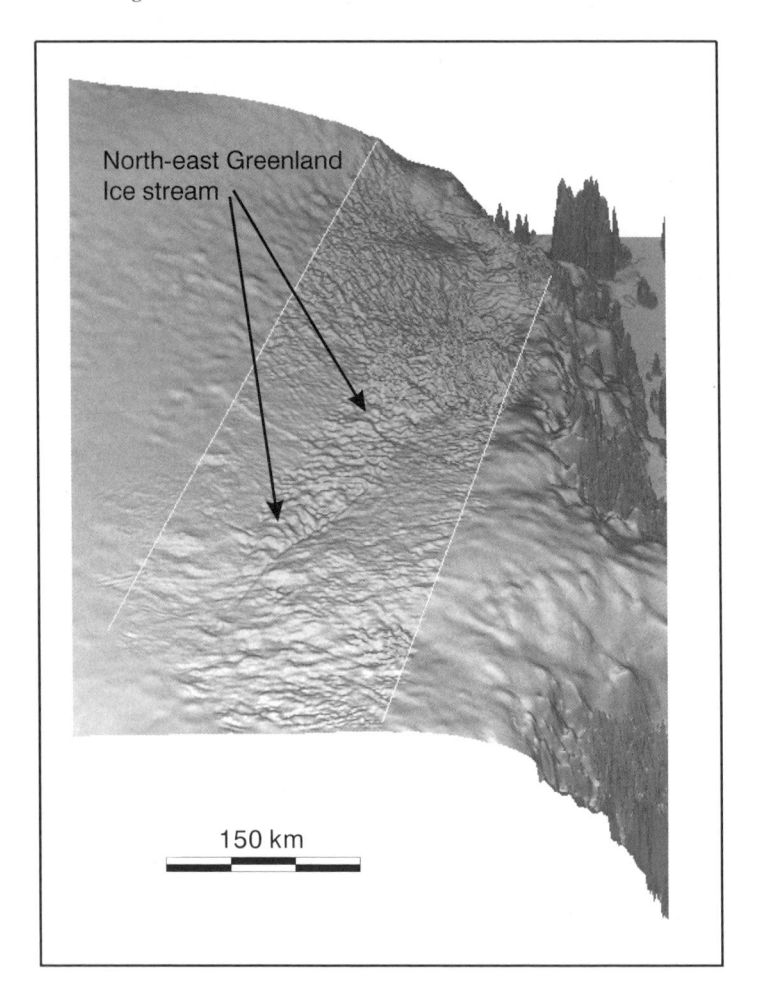

24 *Spatial Modelling of the Terrestrial Environment*

Figure 2.6 *A shaded relief plot of an InSAR-derived DEM of part of the Greenland ice sheet covering the north-east Greenland ice stream. The coverage of the InSAR data is roughly between the two white lines and the DEM was generated with a 500 m posting*

drainage basins. Due to the size of the ice sheets, their non-linear, integrative response to different forcing fields and their long-response time, different basins may be behaving in entirely different ways at any one time. In southern Greenland, for example, one basin, east of the main divide, appears to be losing mass, while to the west, the basin is gaining mass by almost the same significant amount (Thomas *et al.*, 2000). This was a surprising and puzzling result, with no clear explanation. It highlights the importance of a regional interpretation of mass balance signals. The key to this regional approach is the accurate delineation of ice divides.

This has been undertaken for both Greenland and Antarctica using a GIS-based approach by combining slope and aspect, derived from the DEMs (Hardy *et al.*, 2000; Vaughan *et al.*, 1999). The drainage basin masks were used to estimate the area of each basin and the volume

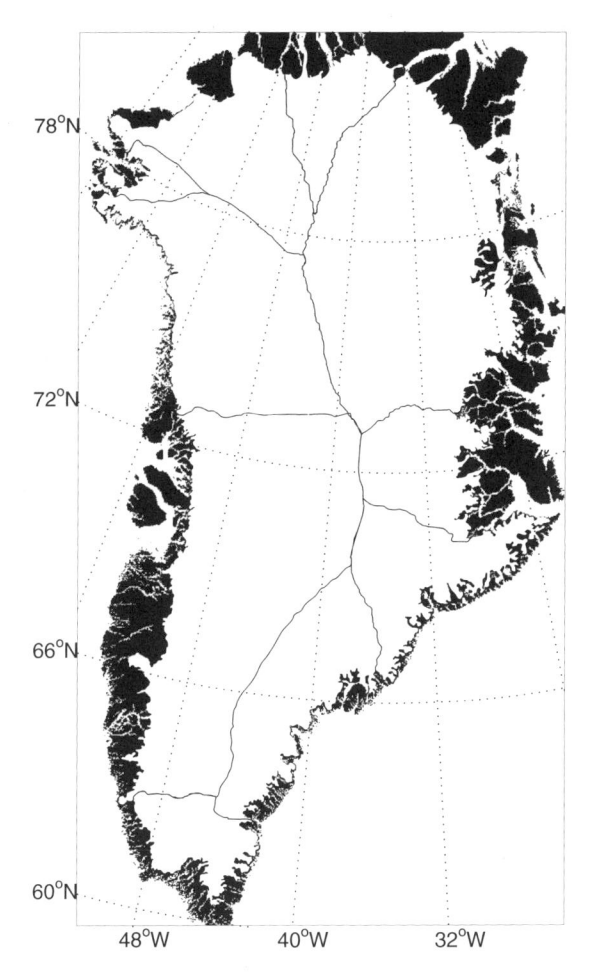

Figure 2.7 Ice divides and major drainage basins for the Greenland ice sheet obtained from slope and aspect data derived from a 2.5-km, satellite radar altimeter DEM of the ice sheet

of ice deposited in the basin based on estimates of accumulation rates. Such analyses are a key step in determining the mass balance of a basin and its relative contribution to the total mass turn-over of the whole ice sheet. Figure 2.7 shows the major ice divide locations derived from a 2.5 km resolution DEM of the GrIS.

Iceberg Fluxes. Almost all the discharge of ice from Antarctica takes places through iceberg calving at the margins of ice shelves and floating glacier tongues located around the coast of the continent. If the geoid, densities of sea water and glacier ice are known (and in general they are), it is possible, assuming hydrostatic equilibrium, to estimate the ice thickness from an estimate of its elevation above the geoid (Bamber and Bentley, 1994). If the calving rate or velocity of the floating ice is also known, the total flux of ice leaving a portion of ice shelf can be estimated. Such an inversion of surface elevations has been carried out using an SRA-derived DEM of Antarctica to estimate distributions of icebergs and model their trajectories in the Southern Ocean (Gladstone et al., 2001). The same DEM has also been used to study, in detail, the size distribution of icebergs calving from the different regions around the coast of the AIS.

Balance Velocities. As mentioned earlier, over an ice sheet, the gravitational driving force is a function of the surface slope and when smoothed over an appropriate distance (typically twenty times the ice thickness) it can be assumed that the ice flows downhill. It is, therefore, possible to trace particle paths down flow lines, from an ice divide to the coast. If the net accumulation is integrated along these flow lines, then the flux at any point down the flow line can be estimated. The depth-averaged velocity at that point is simply the flux divided by the ice thickness. This estimate of velocity is what is required to keep the ice mass in steady state (i.e., where net accumulation exactly matches the flux of ice leaving the ice mass). Hence the name, balance velocity.

Balance velocities, U_b, have been calculated for the entirety of the grounded portions of the AIS and GrIS using satellite-derived topography, combined with terrestrial measurements of accumulation and ice thickness (Bamber et al., 2000b; Bamber et al., 2000c). To do this, a finite-difference model was used to integrate ice flux along particle paths (Budd and Warner, 1996). The result for Antarctica is shown in Figure 2.8b and provides the most detailed and comprehensive picture of flow over the whole ice sheet. Tributaries of fast flow extend deep into the interior of the ice sheet, several reaching more than 1000 km inland. The balance velocities have been validated using both *in situ* GPS velocity measurements and other satellite estimates of velocity and have been demonstrated to provide an excellent representation of the pattern of flow in areas of steady state. The pattern of flow shown in Figure 2.8 differs substantially from the output of the current generation of numerical ice sheet models, which produce a less complex, and less extensive pattern of fast flow features for both Antarctica and Greenland (Bamber et al., 2000a; Bamber et al., 2000c). For the latter ice sheet, the NEGIS is completely absent in the modelled velocity field, despite being by the far the largest fast flow feature on the ice sheet. This raises the question: how reliable is the current generation of numerical models for predicting future changes in flow if they are so poor at reproducing the present-day pattern?

2.4.4 Validation of the Models

To investigate this question further, balance velocities for the GrIS were compared with diagnostic velocities (also referred to as 'dynamics' velocities or U_d) generated by a fully coupled, three-dimensional, thermomechanical model, incorporating basal sliding (Bamber et al., 2000a). Diagnostic velocities are the local solutions to Glen's flow law. They require a prescribed temperature field and, in this case, use the observed surface topography to calculate the driving stress, τ. Four simulations of the model were performed, switching on and off the thermodynamics and/or basal sliding components of the model. Figure 2.9 is a plot of the normalized difference, $(U_b - U_d)/(U_b + U_d)$, between U_b and U_d for the four experiments. Values greater than zero represent areas where U_b is larger than U_d and vice versa. A value of ± 0.333 indicates one velocity is a factor of 2 greater than the other, ± 0.5 is a factor 3 and 0.6666 is a factor 5 different. The figure caption describes the details of each simulation. Crudely, the experiments incorporated more physical processes

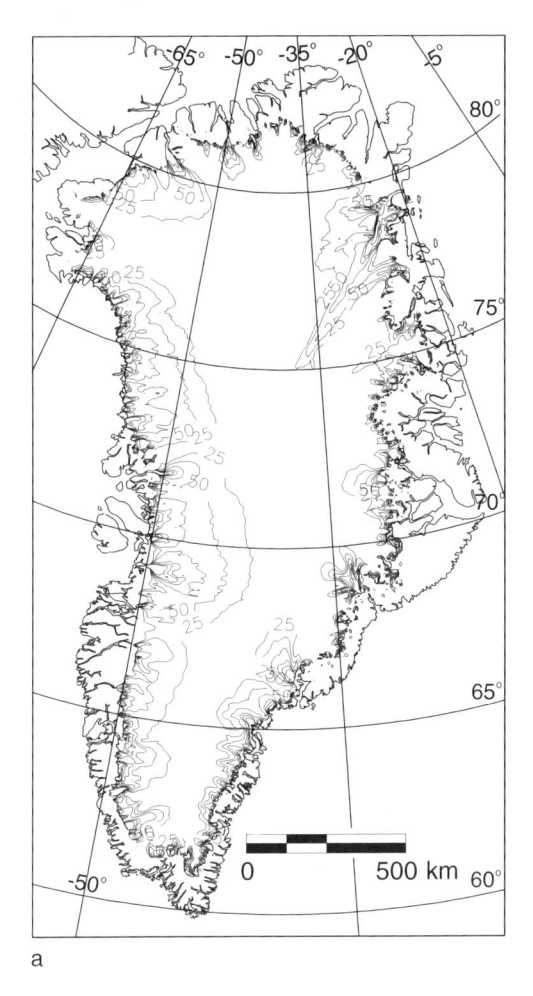

Figure 2.8 (a) Contour plot of balance velocities estimated for the Greenland ice sheet, and (b) the same as (a) but for the Antarctic ice sheet. The contour intervals plotted are 25, 50, 100, 200, 300 and 500 m a^{-1}.

moving from simulation 1 to 4. The simplest configuration was for simulation 1, where the ice was assumed to be isothermal and there was no basal sliding. This means that the thermodynamic component had been switched off and the ice viscosity had a constant value with depth. This is a gross simplification and known to be a poor representation of the englacial dynamics. In reality, most of the ice deformation takes place in the bottom 10% of the column as the ice is warmest here and subject to the highest stress. The difference in viscosity of ice between the surface and bed can be as much as 10^3. It might be expected, therefore, that this most basic simulation, lacking two crucial physical processes, would provide the poorest agreement with U_b.

In the isothermal cases (experiments 1 and 2) U_d generally decreases with respect to U_b moving from the centre towards the margins. This means the ice is too stiff inland and

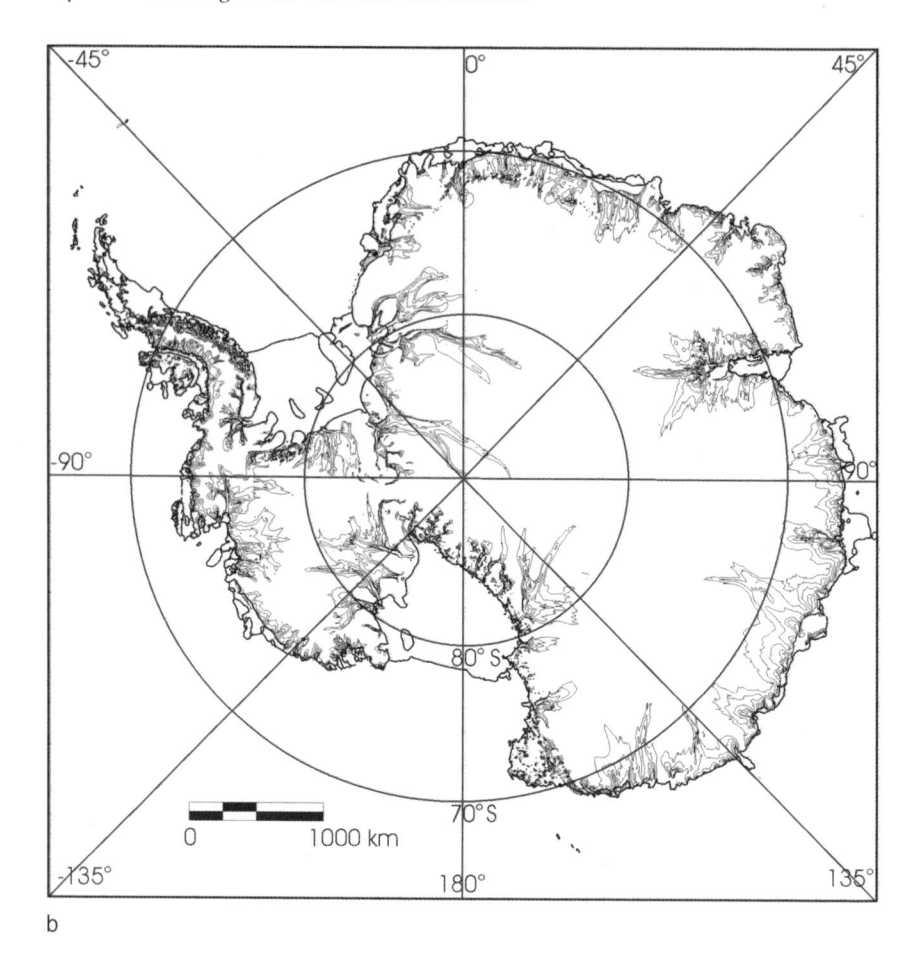

28 *Spatial Modelling of the Terrestrial Environment*

b

***Figure 2.8** (cont.)*

too soft downstream in the model. The introduction of thermodynamics in experiments 3 and 4 appears to partially remove this effect and the agreement in central Greenland is relatively good, suggesting that the ice rheology and thermodynamics (which are coupled) have been reasonably well prescribed for the summit region and northern half of the ice sheet. However, the incorporation of thermodynamics appears to have created a slight difference between the northern and southern halves of the ice sheet with the model generally under-estimating velocities, especially toward the margins, for northern Greenland (ice too stiff) and over-estimating them in the south (ice too soft). Furthermore, the thermodynamic simulations have generated several extensive regions of slow flow (white areas) and show as much variability as the isothermal cases. It should be noted that as a consequence of uncertainties in the *in situ* value for the ice viscosity, it is reasonable to scale U_d by up to a factor 2 to obtain better agreement. Thus, what is important to note in Figure 2.9 is not the absolute difference but gradients in the difference as these cannot be 'corrected' by a global scaling factor applied to U_d. Areas of high variability in greyscale indicate, therefore, a poor representation of the pattern of flow in the model. In general, there is

Figure 2.9 Normalized difference plot of the balance velocities (U_b) minus the dynamics velocities (U_d) for four different simulations of a thermo-mechanical model: (a) isothermal, with no basal sliding, (b) isothermal with basal sliding, (c) thermodynamics included, no basal sliding, (d) thermodynamics included, with basal sliding

relatively poor agreement near the margins in all the simulations and it is not clear if this is due to inaccurate modelling of the thermodynamics, a flawed relationship between ice rheology and temperature or possibly, but least likely, substantial errors in ice thickness, which would affect the accuracy of U_b.

The overestimation of velocities on the south-eastern coast does not appear to be due to basal sliding in the model as U_d is too high in all the runs, including the non-sliding cases (1 and 3). In this area the ice appears to be too soft in the model in all the experiments. On the eastern side from about 72° to 80° N the converse is the case when the thermodynamics are incorporated, suggesting that the modelled temperatures in this region are too cold.

Limitations in model resolution (20 × 20 km) also appear to be important. In Figure 2.8a, an extensive fast-flow feature can be seen in the north-east, reaching almost as far as the ice divide and splitting into two branches about half-way along (also shown in Figure 2.6). In none of the model simulations is this feature (the largest of its kind on the ice sheet) generated. This may be due to limitations in the thermodynamics (Bamber *et al.*, 2000a) but is most probably due to the coarse resolution of the model. Model resolution is not only dependent on computing power but also a key approximation used in many ice sheet models, known as the shallow ice approximation. This assumes that longitudinal stresses in the ice are averaged out and that, consequently, can be ignored over a distance of 10–20 times the ice thickness. Models that use this approximation are often termed zeroth-order models. Increasing the model resolution to a value below this scale-length may require the incorporation of vertically-averaged longitudinal stresses (a first-order model) or a full solution to the stress field (second- or higher-order models), all of which greatly increases the run-time and complexity of the model. To date, no higher-order model exists that can be used on the whole of the AIS or GrIS, yet the inter-comparison discussed above clearly demonstrates the limitations of a zeroth-order model, which appears to be unable to replicate the largest flow-feature on the GrIS.

This inter-comparison has highlighted a number of problems inherent in the current generation of numerical models. The satellite-derived data are, however, also being used to help reduce the uncertainty in some of the tuneable parameters in the models, such as the ice rheology and rates of basal sliding.

2.4.5 Use as Boundary Conditions and Inputs

SRA-derived DEMs provide the most accurate geometric boundary condition for the ice sheets (Bamber and Huybrechts, 1996). Furthermore, this geometric boundary condition can be used, directly, to determine the local gravitational driving force (equation (1)). It would seem desirable, therefore, to use accurate surface DEMs as boundary conditions in models that are attempting to reproduce present-day and future behaviour of the ice sheets. This is, however, not straightforward to achieve as the models need to be run for somewhere between 20 and 100 K years (depending on the model) to reach thermodynamic equilibrium. This normally requires a freely evolving ice sheet profile, which, invariably, does not match the actual present-day geometry. If the geometry is forced to remain fixed, while allowing the thermodynamics to evolve, then they will not be in equilibrium with each other when the fully coupled system is run forward in time.

As a consequence, to date, accurate surface DEMs have proved more useful for validation than as a boundary condition with the types of models described above. There is, however, another approach to modelling known as the force-budget approach (Vanderveen and Whillans, 1989). As the name suggests, this approach attempts to balance the various stresses being applied at the surface, bed and sides of a glacier/ice mass using measured surface strain rates and slopes. This type of model does not simulate the evolution of the ice mass through time but is useful in investigating the relationship between basal and surface stresses and can provide an insight into the basal stress regime and flow behaviour of the ice (Whillans et al., 1989). Here, accurate velocity data are crucial inputs to the model. Surface topography can be used as an input, if known well enough or it can be one of the parameters that is solved for.

As mentioned earlier, one of the key unknowns in thermo-mechanical models is the *in situ* rheology of the ice. It is also apparent that the surface profile of an ice sheet must depend, in part, on the ice rheology. Softer ice, for example, would produce a shallower profile and vice versa. The mathematical expression of the relationship between thickness and rheology is given in equations (4) and (5). From these it can be seen that if the accumulation rate, c, is known reasonably well, then it may be possible to derive information about the *in situ* ice rheology from an accurate representation of the surface and bed topography. This was attempted for an area in east Antarctica, where an ice thickness profile, obtained from airborne radio echo sounding data, was combined with surface topography from ERS-1 radar altimetry (Testut et al., 1999). Variations in flow properties along the profile were deduced by comparing the calculated shear stress, τ, with an estimate of the strain rate. Estimates of both n and A were made for different sections along the profile. Although it is clear that surface topography contains information about the ice rheology, the difficulty with this type of analysis is that any rheological signal present is vertically averaged through the column of ice and the viscosity from the bed to the surface can vary by as much as 10^3. Some knowledge of the englacial temperature field is, therefore, desirable but rarely available from experimental measurements.

Another use for high-resolution topography is as an input dataset for modelling the surface mass balance of an ice mass. As mentioned earlier, this is often undertaken using a positive degree day model, with the temperature parameterized as a function of position and elevation, with the latter being the most sensitive parameter due to the effect of the adiabatic lapse rate (the change in temperature as a function of altitude through the atmosphere). Improvements in both the spatial resolution and vertical accuracy of DEMs of the GrIS have had a substantial impact on estimates of mass loss by ablation. In a comparison of the use of two DEMs (an older one derived chiefly from terrestrial measurements and one derived from SRA data), a difference of 20% in ablation was obtained (van de Wal and Ekholm, 1996). In another study resolution was found to influence ablation due to the smoothing effect on valleys where greater ablation takes place due to their lower elevation (Reeh and Starzer, 1996). Although the mean elevation may be the same, due to the binary behaviour of the positive degree day model (melting can only take place if the temperature is above $0°$ C), the amount of melt calculated may not be. Thus the desirable DEM resolution must be sufficient to resolve glaciated valleys, typically of about 5–10 km in width. A comparison between a 2.5 km DEM and a more recent 1 km DEM of Greenland indicated that there was no difference in ablation estimated using these two models which suggests that resolution-induced errors in estimating ablation are no longer important for the GrIS at least (Bamber et al., 2001). Vertical accuracy in some marginal areas may still be, however.

2.5 Conclusion

Two quite different but complementary approaches to obtaining accurate topography over the large, homogeneous expanses of the Antarctic and Greenland ice sheets have been presented. Some examples of the datasets generated by the two methods and the application of these data to problems in spatial modelling of the ice sheets have been discussed. Accurate topography has been shown to be valuable both directly and indirectly for validating and calibrating several types of numerical ice sheet models. In particular, DEMs have been used to constrain thermo-mechanical models and test the reliability of the ice motion they predict. This latter application resulted from using the DEM directly in a simple finite-difference model, which traced and integrated the ice flux down a flow line to produce, with the benefit of surface accumulation data, an independent, global estimate of the velocity field. Ice sheet scale DEMs have also been used as the key input dataset for modelling the surface mass balance of the Greenland ice sheet, which is crucial for accurately determining its net mass balance and hence its contribution to sea level rise.

Limitations still exist in the current generation of satellite-derived DEMs, however. This is particularly true for the Antarctic ice sheet where there are still no accurate data for the central region south of $81.5°$ (the latitudinal limit of ERS-1 and -2); one-fifth of the ice sheet is still relatively poorly mapped. There are also areas around the coast of Greenland and along the TransAntarctic mountains where there is sparse radar altimeter and/or InSAR coverage. In January 2003, the first of a series of satellite missions with the primary objective of studying the cryosphere was launched. ICESat is a NASA-funded satellite that carries a dual-frequency laser. The inclination of the orbit, combined with the pointing capability of the laser, should provide valuable data over those parts of Antarctica and Greenland not reached by the current suite of sensors. Further into the future, the European

Space Agency plans to launch CRYOSat. This will be an 'imaging' radar altimeter system that will, like ICESat, provide valuable data for the marginal steeper sloping areas of the ice sheets but also for smaller ice masses such as the ice caps of the Russian and Eurasian Arctic. High accuracy and high spatial resolution topography over glacierized terrain looks set to continue to provide valuable information on the form and flow of land ice on the surface of the Earth.

References

Azuma, N., 1994, A flow law for anisotropic ice and its application to ice sheets, *Earth and Planetary Science Letters*, **128**, 601.

Bamber, J.L., 1994a, A digital elevation model of the Antarctic ice sheet derived from ERS-1 altimeter data and comparison with terrestrial measurements, *Annals of Glaciology*, **20**, 48–54.

Bamber, J.L., 1994b, Ice sheet altimeter processing scheme, *International Journal of Remote Sensing*, **14**, 925–938.

Bamber, J.L. and Bentley, C.R., 1994, A comparison of satellite altimetry and ice thickness measurements of the Ross ice shelf, Antarctica, *Annals of Glaciology*, **20**, 357–364.

Bamber, J.L. and Bindschadler, R.A., 1997, An improved elevation data set for climate and ice sheet modelling: validation with satellite imagery, *Annals of Glaciology*, **25**, 439–444.

Bamber, J.L., Ekholm, S. and Krabill, W.B., 2001, A new, high-resolution digital elevation model of Greenland fully validated with airborne laser altimeter data, *Journal of Geophysical Research*, **106**, 6733–6745.

Bamber, J.L., Hardy, R.J., Huybrechts, P. and Joughin, I., 2000a, A comparison of balance velocities, measured velocities and thermomechanically modelled velocities for the Greenland ice sheet, *Annals of Glaciology*, **30**, 211–216.

Bamber, J.L., Hardy, R.J. and Joughin, I., 2000b, An analysis of balance velocities over the Greenland ice sheet and comparison with synthetic aperture radar interferometry, *Journal of Glaciology*, **46**, 67–74.

Bamber, J.L. and Huybrechts, P., 1996, Geometric boundary conditions for modelling the velocity field of the Antarctic ice sheet, *Annals of Glaciology*, **23**, 364–373.

Bamber, J.L., Vaughan, D.G. and Joughin, I., 2000c, Widespread complex flow in the interior of the Antarctic ice sheet, *Science*, **287**, 1248–1250.

Braithwaite, R.J., 1995, Positive degree-day factors for ablation on the Greenland ice-sheet studied by energy-balance modeling, *Journal of Glaciology*, **41**, 153–160.

Brenner, A.C., Bindschadler, R.A., Thomas, R.H. and Zwally, H.J., 1983, Slope-induced errors in radar altimetry over continental ice sheets, *Journal of Geophysical Research*, **88**, 1617–1623.

Budd, W.F. and Warner, R.C., 1996, A computer scheme for rapid calculations of balance-flux distributions, *Annals of Glaciology*, **23**, 21–27.

Ekholm, S., 1996, A full coverage, high-resolution, topographic model of Greenland computed from a variety of digital elevation data, *Journal of Geophysical Research-Solid Earth*, **101**, 21961–21972.

Engelhardt, H. and Kamb, B., 1998, Basal sliding of Ice Stream B, West Antarctica, *Journal of Glaciology*, **44**, 223–230.

Fabre, A., Ritz, C. and Ramstein, G., 1997, Modelling of last glacial maximum ice sheets using different accumulation parameterizations, *Annals of Glaciology*, **24**, 223–228.

Gladstone, R.M., Bigg, G.R. and Nicholls, K.W., 2001, Iceberg trajectory modeling and meltwater injection in the Southern Ocean, *Journal of Geophysical Research*, **106**, 19,903–19,916.

Hardy, R.J., Bamber, J.L. and Orford, S., 2000, The delineation of major drainage basins on the Greenland ice sheet using a combined numerical modelling and GIS approach, *Hydrological Processes*, **14**, 1931–1941.

Houghton, J.T., Filho, L.G.M., Callander, B.A., Harris, N., Kattenberg, A. and Maskell, K., 1996, *Climate Change 1995: The Science of Climate Change* (Cambridge: Cambridge University Press).

Huybrechts, P., Letreguilly, A. and Reeh, N., 1991, The Greenland ice sheet and greenhouse warming, *Palaeogeography, Palaeoclimatology, Palaeoecology*, **89**, 399–412.

Joughin, I., Fahnestock, M., MacAyeal, D., Bamber, J.L. and Gogineni, P., 2001, Observation and analysis of ice flow in the largest Greenland ice stream, *Journal of Geophysical Research*, **106**, 34021–34034.

Joughin, I., Winebrenner, D., Fahnestock, M., Kwok, R. and Krabill, W., 1996, Measurement of ice-sheet topography using satellite radar interferometry, *Journal of Glaciology*, **42**, 10–22.

Layberry, R.L. and Bamber, J.L., 2001, A new ice thickness and bed data set for the Greenland ice sheet 2: relationship between dynamics and basal topography, *Journal of Geophysical Research*, **106**, 33781–33788.

Liu, H.X., Jezek, K.C. and Li, B.Y., 1999, Development of an Antarctic digital elevation model by integrating cartographic and remotely sensed data: a geographic information system based approach, *Journal of Geophysical Research*, **104**, 23199–23213.

Manabe, S. and Stouffer, R.J., 1993, Century-scale effects of increased atmospheric CO^2 on the ocean-atmosphere system, *Nature*, **364**, 215–218.

Marshall, S.J. and Clarke, G.K.C., 1996, Sensitivity tests of coupled ice-sheet/ice-stream dynamics in the EISMINT experimental ice block, *Annals of Glaciology*, **23**, 336–347.

Paterson, W.S.B., 1994, *The Physics of Glaciers* (Oxford: Pergamon).

Reeh, N., 1991, Parameterization of melt rate and surface temperature on the Greenland ice sheet., *Polarforschung*, **59**, 113–128.

Reeh, N., and Starzer, W., 1996, Spatial resolution of ice-sheet topography: influence on Greenland mass-balance modelling, in O.B. Olsen (ed.), *Mass Balance and Related Topics of the Greenland Ice Sheet* (Copenhagen: GEUS), 85–94.

Remy, F., Mazzega, P., Houry, S., Brossier, C. and Minster, J.F., 1989, Mapping of the topography of continental ice by inversion of satellite-altimeter data, *Journal of Glaciology*, **35**, 98–107.

Testut, L., Tabacco, I.E., Bianchi, C. and Remy, F., 1999, Influence of geometrical boundary conditions on the estimation of rheological parameters, *Proceedings of EISMINT/EPICA Symposium on Ice Sheet Modelling and Deep Ice Drilling*, International Glaciological Society, Den Haag, The Netherlands, 102–106.

Thomas, R., Akins, B., Csatho, T., Fahnestock, M., Gogineni, S.P., Kim, C. and Sonntag, J., 2000, Mass balance of the Greenland ice sheet at high elevations, *Science*, **289**, 426–428.

van de Wal, R. and Ekholm, S., 1996, On elevation models as input for mass-balance calculations of the Greenland ice sheet, in K. Hutter (ed.), *International Symposium on Ice Sheet Modelling*, International Glaciological Society, Chamonix, France, 181–186.

Vanderveen, C.J. and Whillans, I.M., 1989, Force Budget .1. Theory and numerical methods, *Journal of Glaciology*, **35**, 53–60.

Vaughan, D.G. and Bamber, J.L., 1998, Identifying areas of low-profile ice sheet and outcrop damming in the Antarctic ice sheet by ERS-1 satellite altimetry, *Annals of Glaciology*, **27**, 1–6.

Vaughan, D.G., Bamber, J.L., Giovinetto, M., Russell, J. and Cooper, A.P.R., 1999, Reassessment of net surface mass balance in Antarctica, *Journal of Climate*, **12**, 933–946.

Whillans, I.M., Chen, Y.H., Vanderveen, C.J. and Hughes, T.J., 1989, Force Budget .3. Application to 3-dimensional flow of Byrd Glacier, Antarctica, *Journal of Glaciology*, **35**, 68–80.

Zwally, H.J., Bindschaler, R.A., Brenner, A.C., Martin, T.V. and Thomas, R.H., 1983, Surface elevation contours of Greenland and Antarctic ice sheets, *Journal of Geophysical Research*, **88**, 1589–1596.

Zwally, H.J., and Brenner, A.C., 2001, Ice sheet dynamics and mass balance, in L.-L. Fu, A. Cazenave (eds), *Satellite Altimetry and the Earth Sciences* (New York: Academic Press), 351–370.

Zwally, H.J., Brenner, A., Major, J.A., Martin, T.V. and Bindschadler, R.A., 1990, *Satellite Radar Altimetry over Ice, Volume 1 – Processing and Corrections of Seasat Data over Greenland*, NASA Reference Publication 1: 1233.

3
Using Remote Sensing and Spatial Models to Monitor Snow Depth and Snow Water Equivalent

Richard E.J. Kelly, Alfred T.C. Chang, James L. Foster and Dorothy K. Hall

3.1 Introduction

Snow cover is an important hydrological parameter in the global water cycle. By influencing directly the dynamics of the global water cycle, snow cover has an important control on climate through its effect on energy budgets at the surface and lower atmospheric levels (Cohen, 1994). Therefore, for climate change studies, our ability to estimate global coverage and volumetric storage of water in seasonal and permanent snow packs impacts directly on the ability to predict changes in climate from year to year and over longer periods. At a more local scale it affects the ability to budget effectively for water supply. With the continued growth in both direct and indirect evidence of climate change, plus the increasing stresses placed on the water cycle by climate change, there is a pressing need to quantify accurately at different space and time scales, the various components of the hydrological cycle.

Earth observation has been used to monitor continental scale seasonal snow cover area for 25 years and much of this effort has focused on the use of visible and infrared sensors (e.g. Hall *et al.*, 2002a). Research suggests that snow cover extent in the Northern Hemisphere has decreased by 10% since 1966 when visible/infrared sensors were first available for use (Robinson, 1999). However, little information is available on changes of snow water equivalent (SWE) at hemispheric scale over a similar time period. The instruments capable of estimating SWE have been available for a shorter period and, more importantly, the methodologies available to estimate successfully global SWE are still in an evolutionary phase. Therefore, in hydroclimatology and climate change studies, the representation of snow is often parameterized implicitly in climate models, perhaps as a sub-component of the model, and often at very coarse spatial scales (Foster et al., 1996) that generalize SWE characteristics at grid scales of several degrees latitude/longitude in size. Satellite passive microwave estimates of snow have spatial resolutions an order of magnitude better than this. Therefore, there is an increasing need to refine methods that can provide an accurate estimation of global SWE from passive microwave sensors at spatial scales that reflect snow spatial distribution characteristics better.

Regionally calibrated approaches to SWE and snow depth estimation have been shown to work reasonably well in relatively homogeneous snow-covered areas where terrain effects are well understood (Tait, 1997; Goodison and Walker, 1994). However, in general, there is some doubt whether single regional approaches are applicable at the global scale where snow pack, land cover and terrain characteristics are more heterogeneous in nature. Thus, while the development of SWE and snow depth estimation methods has advanced in the regional domain, improvements at the global scale to estimate SWE or snow depth have been slower. Furthermore, where global studies have sought to estimate global snow volume (e.g. Chang et al., 1987), the approach taken often has been static and formulated using 'average' seasonal snow pack conditions at parameterization. While these global approaches yield reasonable snow volume retrievals when integrated over large spatial and seasonal scales, local/regional instantaneous estimates can be subject to 30–50% errors or more (Hallikainen and Jolma, 1992), although there is large uncertainty associated with these errors. Nevertheless, the predicted errors are large, even for climate model inputs, suggesting that snow volume estimates are often unreliable for catchment-based studies. It is apparent, therefore, that approaches for the estimation of snow volume from passive microwave data need to be advanced into more spatially and temporally dynamic methodologies that represent snow pack processes better and should, therefore, reduce the errors of estimates. This chapter explores the possibilities and practicalities of using hydrological models of snow pack properties and radiative transfer models of microwave emissions from snow to assist with the estimation of snow volume on a global scale. We begin by discussing the character of snow depth and SWE spatial distribution before describing the current approaches of snow volume estimation from passive microwave instruments. The issue of the determination of errors linked to snow volume estimates is also addressed towards the end of the chapter; these errors are increasingly important but probably even less straightforward to derive. The chapter then concludes with some remarks about future directions.

For clarity, in this chapter we use the term snow cover as a synonym for snow cover area extent. Mostly, however, the chapter is concerned with the estimation of SWE (mm) or snow depth (cm) per unit area. While these two are related through the snow density, they are different and can have different seasonal characteristics.

3.2 Modelling Spatial Variation of Snow Depth/SWE Using *in Situ* Snow Measurements

Walsh (1984) created a generalized map representation of the spatial distribution of global snow cover. Qualitative maps are useful for providing generalized representations of the location and climatological persistence of snow cover area. More recently, efforts are underway to build quantitative maps of global snow cover occurrence using satellite-derived estimates of snow cover (e.g. Frei and Robinson, 1999, Hall et al., 2002b). However, there is far less information available about the scale of spatial variation of snow depth or SWE. We know that snow depth or SWE varies in a snowfield for a variety of reasons (topography, vegetation, meteorology) but how can this 'spatial dependency' be defined and how might it vary through space? Field experiments that measure snow pack properties are often conducted at local scales of a few kilometres and the data are usually gridded and interpolated to produce maps of snow depth or SWE. Snow maps are then used directly in hydrological models or to test snow depth or SWE estimates from aircraft instruments. Few studies have formally quantified the spatial variability of snow depth or SWE at large regional or global scales. Brasnett (1999) used global snow depth data in the operational analysis of snow depth at the Canadian Meteorological Centre and showed how its inclusion in the analysis, through spatial interpolation, improved the overall analysis of snow. However, quantitative information about the spatial variability of these data was not reported. It is important, therefore, to determine the characteristic scales of spatial variation of snow depth or SWE in continental snow packs before we attempt to monitor snow from space.

It is known that large spatial and temporal variations exist in global and local snow cover extent and volume (Frei and Robinson, 1999) and characterization of these variations is important for effective climate prediction. Spatial scales of snow cover variation were identified by McKay and Gray (1981) who characterized snow cover distribution in terms of regional variations (up to 10^6 km^2), local variations (10^2 to 10^5 km^2) and micro-scale variations (10 to 10^2 km^2). Regional scales of snow cover distribution are controlled by latitude, elevation and orographic effects, local-scale distributions are controlled by local topographic effects such as slope and aspect and by land cover type, and micro-scale variations tend to be influenced by local transport factors such as wind redistribution (McKay and Gray, 1981). This description provided by these authors is a good starting point for understanding the nature of SWE and snow depth variation.

Snow depth and SWE can be measured directly on the ground using measurements at a point or over a limited area of a few metres with a snow pillow. In general, point measurements of snow depth or SWE produce high quality data with small location and magnitude errors. However, the spatial representativeness of these points is uncertain at larger distance scales. The only way to attempt the characterization of snow depth or SWE spatial variability necessarily relies on point measurements, usually made at official meteorological station networks and volunteer networks. Very few datasets that characterize snow cover distribution over all spatial scales are available to confirm the McKay and Gray classification. Data tend to characterize the local to regional scales with micro-scale variability snow depth or SWE observations available only at specific locations and often over short periods of time. Figure 3.1 shows three spatial scales of operational or routine snow depth measurement networks. At each site, accumulated snow depth is measured with a graduated ruler and SWE is calculated from the measured average snow depth and snow density. If no data are recorded, generally it is assumed that there was no snow present. Figure 3.1a shows World Meteorological Organization (WMO) station sites in the Global Telecommunications System network in the northern hemisphere that reported snow depth greater than 2 cm during the 2000–2001 winter season. These data are freely available from the National Climate Data Center in the USA. The maximum spatial density of measurements is 1 site per 40 km^2 although the average density is 1 site per 160 000 km^2. Figure 3.1b shows the

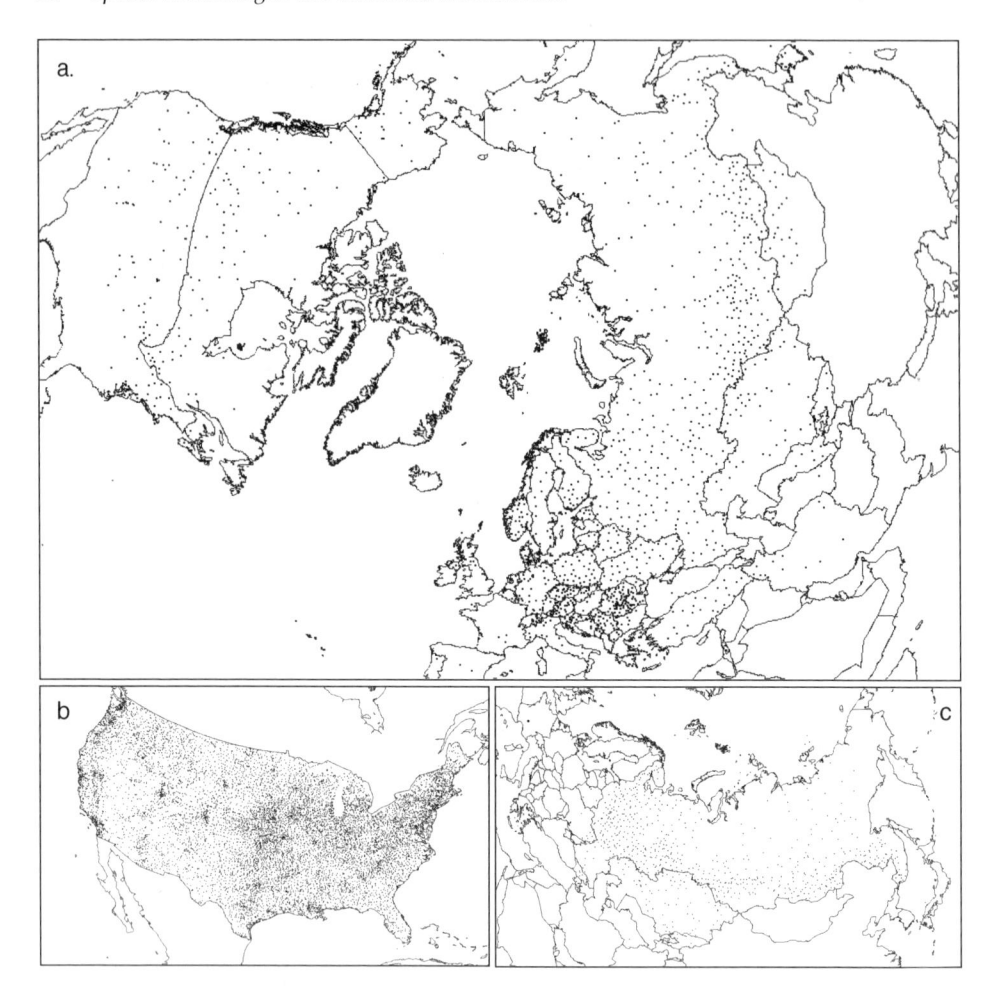

38 *Spatial Modelling of the Terrestrial Environment*

Figure 3.1 *Maps of three different snow depth measurement networks: (a) represents the number of stations in the WMO GTS network on 2 February 2001 (maximum snow extent); (b) represents the total COOP station network; (c) represents the FSU network*

active cooperative station network (COOP) for the contiguous states in the USA. These data are available from the National Oceanic and Atmospheric Administration and have a maximum and mean density of 1 site per 0.26 km^2 and 1 site per 1600 km^2, respectively. Figure 3.1c shows the station locations of snow depth observations from the former Soviet Union (FSU) for 10 February, 1989. The FSU data are available from the National Snow and Ice Data Center and have a maximum and mean density of 1 site per 0.4 km^2 and 1 site per 14 000 km^2, respectively. Using the average spatial densities (rather than the maxima), these area densities translate to scale lengths of 1 station per 40 × 40 km, 118 × 118 km and 400 × 400 km for the USA, FSU and WMO datasets, respectively. Therefore, based on McKay and Gray's specification, potentially the datasets are the most representative at local to regional scales. (It is tempting to consider that the COOP data provide a good opportunity to investigate highly local-scale or even micro-scale spatial variability. However, the actual number of stations regularly reporting snow depth or SWE is rather less than this average figure suggests.) To further investigate the spatial variability of snow depth or SWE, the analysis of snow depth variograms was undertaken.

Intuitively, there is an inherent spatial dependence of snow depth or SWE in a snow field because locations closer together tend to be more similar than those further apart (see Tobler's first Law of Geography). This concept can be used to encapsulate the micro-scale of SWE or snow depth variation. 'Spatial similarity' of snow depth also can be present over greater distances on account of local or regional controls on snow accumulation (in mountainous terrain or along the track of a snow storm). In other words, spatial autocorrelation of snow depth is probably present at different scales from micro-scales through to broad regional scales. Quantification of the spatial dependence of a variable may be expressed by the semi-variogram. In statistics, observations of a selected property are often modelled by a random variable and the spatial set of random variables covering the region of interest is known as a random function (Isaaks and Srivastava, 1989). In geostatistics, a sample of a spatially varying property is commonly represented as a regionalized variable, that is, as a realization of a random function. The semi-variance (γ) may be defined as half the expected squared difference between the random functions $Z(\mathbf{x})$ and $Z(\mathbf{x} + \mathbf{h})$ at a particular lag $\mathbf{h}$. The variogram, defined as a parameter of the random function model, is then the function that relates semi-variance to lag:

$$\gamma(\mathbf{h}) = \frac{1}{2}E[\{Z(\mathbf{x}) - Z(\mathbf{x} + \mathbf{h})\}^2] \tag{1}$$

The sample variogram $\gamma(\mathbf{h})$ can be estimated for $p(\mathbf{h})$ pairs of observations or realizations, $\{z(x_l + \mathbf{h}), l = 1, 2, \ldots, p(\mathbf{h})\}$ by:

$$\hat{\gamma}(\mathbf{h}) = \frac{1}{2}p(\mathbf{h}) \sum_{l=1}^{p(\mathbf{h})} \{z(x_l) - z(x_l + \mathbf{h})\}^2. \tag{2}$$

As Oliver (2001) states: "[the variogram] provides an unbiased description of the scale and pattern of spatial variation". It is a useful tool, therefore, for analysing the scale(s) of variation characterized by datasets of measurements of snow depth or snow water equivalent.

Variograms were calculated for measured point snow depth data (cm) for three different datasets to investigate spatial dependencies present at different sampling scales. The three dataset frameworks described in Figure 3.1 (the WMO, COOP and FSU snow depth datasets) were used to explore the characteristics of snow depth spatial variability in each dataset. Ideally, datasets containing consistent measurements of SWE should be used, but globally, they are not available routinely so snow depth records are used. Daily samples of snow depth from each dataset were selected for northern hemisphere mid-winters (early February) from the respective archives. Although the same time periods are not represented in each case, the data provide some sense of spatial variability characterized at each scale of measurement.

The point snow depth data were projected to the Equal Area Scaleable Earth Grid (EASE-grid) (Armstrong and Brodzik, 1995) and the variograms computed for 2 February, 2001 (WMO), 10 February, 1990 (FSU), 12 February, 1989 (COOP) and 12 February, 1994 (COOP). The WMO data cover the entire northern hemispehre while the FSU data cover

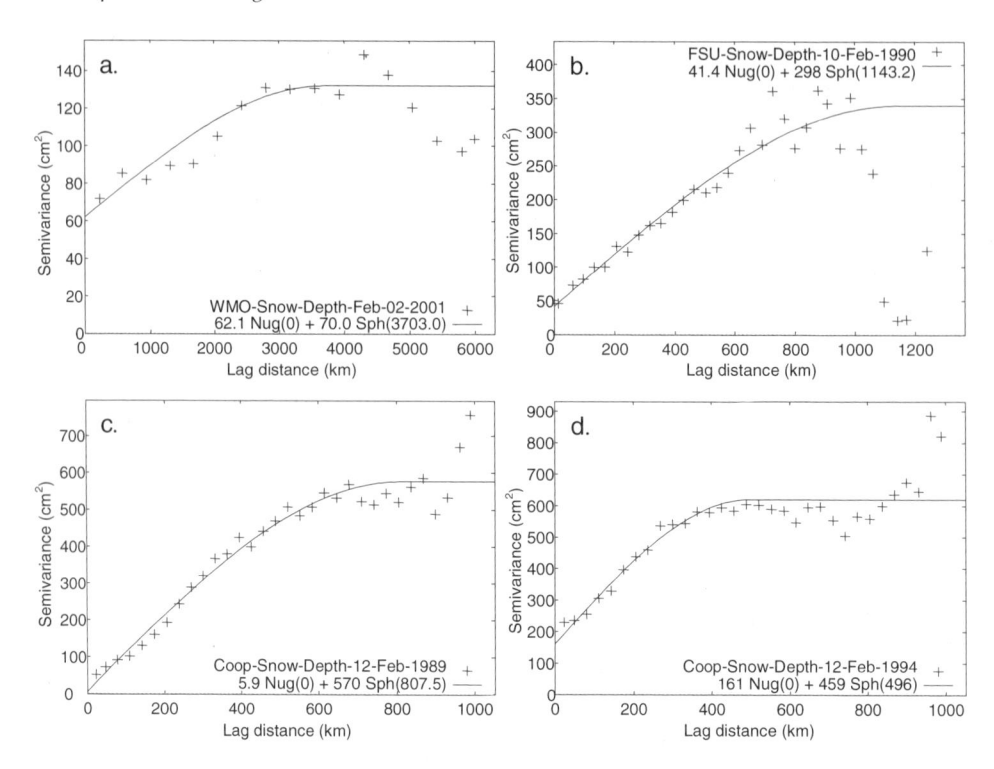

40 *Spatial Modelling of the Terrestrial Environment*

Figure 3.2 *Variograms of point snow depth variation at three different scales of spatial measurement: (a) WMO GTS snow depth for 2 February, 2001; (b) FSU snow depth for 10 February, 1990; (c) COOP snow depth for 12 February, 1989; (d) COOP snow depth for 12 February, 1994*

the region shown in Figure 3.1c. The COOP data were spatially cropped and only data from the upper mid-west USA, particularly from South Dakota, North Dakota and Wisconsin were used. The COOP data were refined in this manner to determine their utility for characterizing micro-scale snow depth variations. Figure 3.2 shows the four experimental variograms with spherical models fitted using the weighted least squares criterion. The lag separations chosen were the minimum distances between points in each of the datasets. The form of the variograms (in particular, that a bounded model provided a good fit in each case) suggests that a stationary random function model provides an adequate model of the variation in each case. In particular, trend models were not required. The key elements of the variogram of interest in this work are the sill, range and nugget variance. The sill variance determines the amount of variation in the variable (snow depth) and the range expresses the scale of variation in the variable. The nugget variance (intercept of the model on the ordinate) represents unresolved variation in the data that cannot be explained by the model. It can also be attributed to measurement error, or it is caused by uncertainty in estimating the variogram at short lags or uncertainty in fitting the model at short lags (Atkinson, 2001). Thus, for a well-structured variogram, at lag distances less than the range, spatial dependency is present while at lag distances greater than the range there is no spatial dependency. The reader is directed to other work (e.g. Oliver, 2001; Atkinson, 2001; Isaaks and Srivastava, 1989) that fully describes these parameters in formal and applied terms.

Figure 3.2a shows the snow depth variogram for the WMO global dataset. The average snow depth was 10 cm with a standard deviation of 10 cm. The number of snow depth reporting sites in this dataset is 1262. The range of the variogram is approximately 3700 km and the nugget and the sill variances are 62 cm^2. There is structure evident in the experimental variogram although the nugget variance is 50% of the total sill variance suggesting that the structure is not strong. From the interpretation of the variogram of these data, spatial dependence is evident at lag distances less than 3700 km. However, at local scales of variation (less than 1000 km lag distances) the sample design cannot effectively represent the spatial variation of snow depth since the variance of the model is very similar to the nugget variance at these smaller lag distances. Thus, while interpolation of the point data is feasible given the apparent presence of spatial dependence, it should be undertaken only at grid supports of greater than 1000 × 1000 km^2 (i.e., at a regional scale); the data should not be used to represent local scale snow depth varitions.

Figure 3.2b, shows the modelled variogram for the FSU snow depth data. The average snow depth was 20 cm with a standard deviation of 14 cm and a population of 117 stations. The variogram structure has good definition and exhibits a reasonably well-defined sill at a range of approximately 1140 km. The nugget variance is 41 cm^2, which although large, is much less than the sill variance suggesting that a distinct structure is present. Spatial dependence can be represented at the local scale of variation with these data since strong spatial dependence is exhibited at distances less than 1100 km. Micro-scale variations, however, are not represented by this dataset and so interpolation of the data should be restricted to grid support defined at the local scale.

For the North American COOP data, 283 stations comprised the 1989 data with a mean snow depth of 16 cm and standard deviation of 19 cm. For the 1994 data there were 281 stations with an average depth of 43 cm and a standard deviation of 23 cm. Figures 3.2c and d, representing 12 February, 1989 and 1994 data, respectively, show the variograms that exhibit perhaps the most distinct variogram structures. Clearly defined sills are present in both variograms with ranges located at approximately 807 and 496 km respectively for the 1989 and 1994 data. The sill variance for the 1989 data (Figure 3.2c) is 570 cm^2 while for the 1994 data the sill variance is 459 cm^2. The nugget variance for the 1989 data is 6 cm, suggesting that very little measurement error affects the data at the micro-scale variation, and so reasonable representation of the data at this scale is possible. However, for the 1994 data the nugget variance is 161 cm^2, which is larger than the 1989 data and indicates that small, micro-scale variations are less well represented by the variogram. The 1994 dataset represents effectively snow depth variability only at the local scale.

In the above examples, it is suggested that the 'standard' regional datasets are suited to the characterization of local to regional variations of snow depth (or SWE if available). While the global dataset showed some evidence of spatial variation at the regional scale, the spatial structure evident from the variogram was weak and could not be relied upon to provide a robust characterization of the spatial dependence of snow depth. At the micro-scale level of variation, the COOP and FSU data showed signs of representing snow depth variations but only when the average snow depth was shallow, when the snow depth was deeper, confidence in the representation at this scale was limited. While climate modellers are interested in snow variations at the local to regional scale, water resource managers are interested in variations at the local scale and often at the micro-scale of variation. Hence, although the 'standard' available datasets, such as those described above, might be useful for local scale applications, it is clear that they are not so useful for the characterization of micro-scale variations. Furthermore, the data are not available in real or even near-real time so their utility, potentially, is best suited to 'off-line' applications such as climate studies or the validation or testing of estimates of snow depth from alternative methodologies. For micro-scale applications or real or near-real time applications, the direct measurement of snow depth or SWE in these regional datasets is less useful. Instead, hydrologists should seek direct snow parameter measurements that are made at specific experimental catchments around the globe (e.g. Marks *et al.*, 2001).

3.3 Estimating Snow Depth and SWE Using Physically Based Models

The previous discussion concentrated on modelling the spatial dependency of snow depth represented by measurements made at discrete meteorological stations. The basis of the analysis is the purely spatial relationship of snow depth from one site to another. Several studies have demonstrated that other terrain and landcover parameters can be incorporated into statistical models to characterize snow depth or SWE spatial variability. For example, terrain variables (elevation, aspect, slope) or meteorological variables can be used to assist in the spatial modelling of snow depth or SWE (e.g. Kelly and Atkinson, 1999; Judson and Doeksen, 2000; Elder *et al.*, 1998; Carroll and Cressie, 1997). However, these approaches tend to represent snow depth or SWE at a discrete point or as a point process that represents a small homogeneous snow plot (e.g. 1 ha or less). The scaling up of these models to larger-scale applications is uncertain since the relationships in the models are statistically based and may or may not be applicable at larger spatial scales. To effectively scale up point snow data an understanding of the snow physical properties and the processes that govern their evolution in vertical and horizontal space is required. To achieve this understanding, ideally, spatially dense micro-scale measurements are needed that cover a reasonably wide area. While this need can be costly to implement, pre-existing catchments and experimental networks in place have led to some important developments in the field of snow hydrology modelling and its use in representing micro-scale to local scale snow processes.

Snow hydrology models tend to be predicated on energy balance dynamics and are usually driven at a point scale or plot scale by a suite of meteorological variables representative of micro-scale energy conditions. It is not the purpose here to provide a comprehensive review of snow hydrology models. Papers by Dozier (1987) and Bales and Harrington (1995) give full summary accounts of snow hydrology methodologies. A recent paper by Davis *et al.* (2001) provides an excellent summary of the issues concerning implementation and validation of snow model estimates. Using examples from the SNTHERM model (Jordan, 1991), the paper categorizes snow models into one-dimensional process models and two-/three-dimensional models. The two-dimensional spatial snow model generally is considered an implementation of the one-dimensional model applied to individual discrete spatial domains. One-dimensional models have been validated successfully using discrete measurements of bulk snow properties (such as snow depth and SWE), and also with detailed measurements of snow properties (such as snow density and temperature profiles, snow grain size and spectral reflectance). Snow models at the point process scale are mature and can represent at least snow depth and/or SWE variables very effectively.

Extension of the one-dimensional model to two or three dimensions has also progressed. Much of the development in this field has been through the use of instrumented catchments that have a high spatial density of meteorological instruments providing fine resolution inputs to energy balance models of snow. More recently, models have started to expand the scale to local and regional-sized supports and to examine ways that aggregate meteorological input influence snow distribution (e.g. Luce *et al.*, 1998). As Davis *et al.* note: "These efforts begin to provide frameworks for using ground-based observations as validating data in modelling exercises using large spatial domains" (Davis *et al.*, 2001, p. 278). Validation of the spatially distributed models requires data on snow extent, SWE, snow surface temperature, snow surface wetness and texture, all of which are measurable quantities. However, uncertainties associated with the spatial heterogeneity of these variables make the validation of distributed models a challenging task.

In summary, snow hydrology models appear to be mature in their ability to represent snow depth or SWE at micro-scales of spatial variation. Given our proven understanding of the physical processes controlling snow pack development, it seems reasonable to expect that application of these models to increasing support scales (e.g. local scales) should be possible. For local scale representation of snow depth or SWE, snow hydrology models should provide accurate and timely information where appropriate meteorological variables are collected in a timely fashion for input to the models. However, for global applications and in many regions of the world, real-time or near real-time meteorological data are not available routinely or at a frequent enough spatio-temporal resolution. Therefore, local or regional scale generalized climate model data will need to be used as input for which added uncertainty is associated (Slater *et al.*, 2000).

3.4 Remote Sensing Estimates of Snow Depth/SWE: Recent Approaches and Limits to Accuracy

Remote sensing of SWE or snow depth is possible using the detection of naturally upwelling microwave radiation from the earth's surface. Progress in recovering snow depth or SWE from remote sensing instruments has been made through the available 'instruments of opportunity' such as the Scanning Multichannel Microwave Radiometer (SMMR) and the Special Sensor Microwave Imager (SSM/I). Table 3.1 gives summary information about these instruments plus the new Advanced Microwave Scanning Radiometer EOS (AMSR-E) aboard NASA's Aqua platform. These instruments are passive microwave systems that measure naturally upwelling microwave radiation over relatively large instantaneous field of views (IFOV) (see Table 3.1 for details of the spatial resolution of different sensors). Neither SSM/I nor SMMR instruments were designed explicitly for snow studies but both have been found to be effective for snow applications (e.g. Chang *et al.*, 1987; Hallikainen and Jolma, 1992; Tait, 1997; Foster *et al.*, 1997; Armstrong and Brodzik, 2001). However, the scale of algorithm implementation has been firmly fixed at the regional to global scales of snow spatial variation.

The spatial resolution of observation for these microwave instruments varies with the electromagnetic wavelength or frequency of the observation, as summarized in Table 3.1. For the SSM/I, for example, at high frequencies (shorter wavelengths) such as 85 GHz, the spatial resolution is of the order of 13–15 km while the lower frequencies such as 19 GHz,

Table 3.1 Comparison of Aqua AMSR-E (Chang and Rango, 2000), SSM/I (Hollinger et al., 1990) and SMMR (Gloersen and Barath, 1977) sensor characteristics

AMSR-E	Centre frequency (GHz)	6.9	10.7	18.7	23.8	36.5	89.0
(launched 2002)	Band width (MHz)	350	100	200	400	1000	3000
	Sensitivity (K)	0.3	0.6	0.6	0.6	0.6	1.1
	IFOV (km^2)	76 × 44	49 × 28	28 × 16	31 × 18	14 × 8	6 × 4
SSM/I	Centre frequency (GHz)			19.35	22.235	37.0	85.5
(launched 1987)	Band width (MHz)			240	240	900	1400
	Sensitivity (K)			0.8	0.8	0.6	1.1
	IFOV (km^2)			69 × 43	60 × 40	37 × 29	15 × 13
SMMR	Centre frequency (GHz)	6.63	10.69	18.0	21.0	37.0	
(1979–1987)	Band width (MHz)	250	250	250	250	250	
	Sensitivity (K)	0.9	0.9	1.2	1.5	1.5	
	IFOV (km^2)	171 × 157	111 × 94	68 × 67	60 × 56	35 × 34	

the resolution is of the order of 43–69 km. The spatial scale of observation, therefore, is firmly fixed in the domain of the local to regional scale of snow variation. These IFOV dimensions, or footprints, may be considered as averaging cells in that the brightness temperature of an observation measured at the satellite is an average thermal signal for the whole area. (In fact, these spatial dimensions represent the approximate 3 dB beamwidth at the specific frequencies and do not account for side-lobe areas that also contribute to the final thermal measurement.)

Two recent papers have summarized methods to retrieve SWE or snow depth from satellite instruments (Derksen and LeDrew, 2000; König et al., 2001) and only a summary is provided here. In theory, snow acts as a scatterer of upwelling microwave radiation and at certain frequencies (or wavelengths), the scattering component dominates the overall signal if the scattering of radiation is greater than the absorption of radiation by a target. When the snow is thick, the scattering is strong and can be detected at microwave frequencies greater than 25 GHz. By comparing brightness temperatures detected at the antenna at high frequencies (potentially scattering dominated) with those brightness temperatures detected at frequencies less than 25 GHz (absorption dominated), it is possible to identify scattering surfaces. Generally, the strength of scattering is proportional to the SWE and it is this relationship that forms the basis for estimating the water equivalent or thickness of a snow pack. This has been the foundation of satellite passive microwave retrievals of SWE or snow depth.

Using SSM/I or SMMR instruments, the difference between low (19 GHz) and high (37 or 85 GHz) frequency brightness temperatures can be used to detect the presence of snow. Several diagnostic tests have been developed to screen for false snow targets such as rainfall, cold deserts and frozen ground (Grody and Basist, 1996). In addition, a dataset that masks land and sea is helpful to ensure that false snow targets are identified. Furthermore, a dataset that maps the 'climatological (im)possibility' of snow accumulation has been assembled by Dewey and Heim (1981, 1983). Using passive microwave data and the diagnostic tests of Grody and Basist (1996) along with the ancillary datasets that filter out regions where snow occurrence is unlikely, the presence of snow at any location in the world can be estimated. The advantage of the microwave approach is that snow can be mapped at locations where cloud cover obscures the snow, a perennial problem for visible/infrared wavelength sensors.

If snow is dry and uniform in density and stratigraphy, then detection of snow is straightforward. However, if the snow is stratigraphically complex, shallow or wet, its detection is more of a challenge. Armstrong and Brodzik (2001) showed that early season underestimation of snow cover area is a problem for passive microwave mapping. By comparing passive microwave estimates with visible/infrared global snow maps from the NOAA interactive multi-sensor snow and ice mapping system (IMS) product described by Ramsay (1998), the underestimation was clearly evident (Hall *et al.*, 2002a). This area of underestimation is sufficient to produce small but significant seasonal differences in snow-covered area between passive microwave and optical datasets. Figure 3.3 shows an example of this early

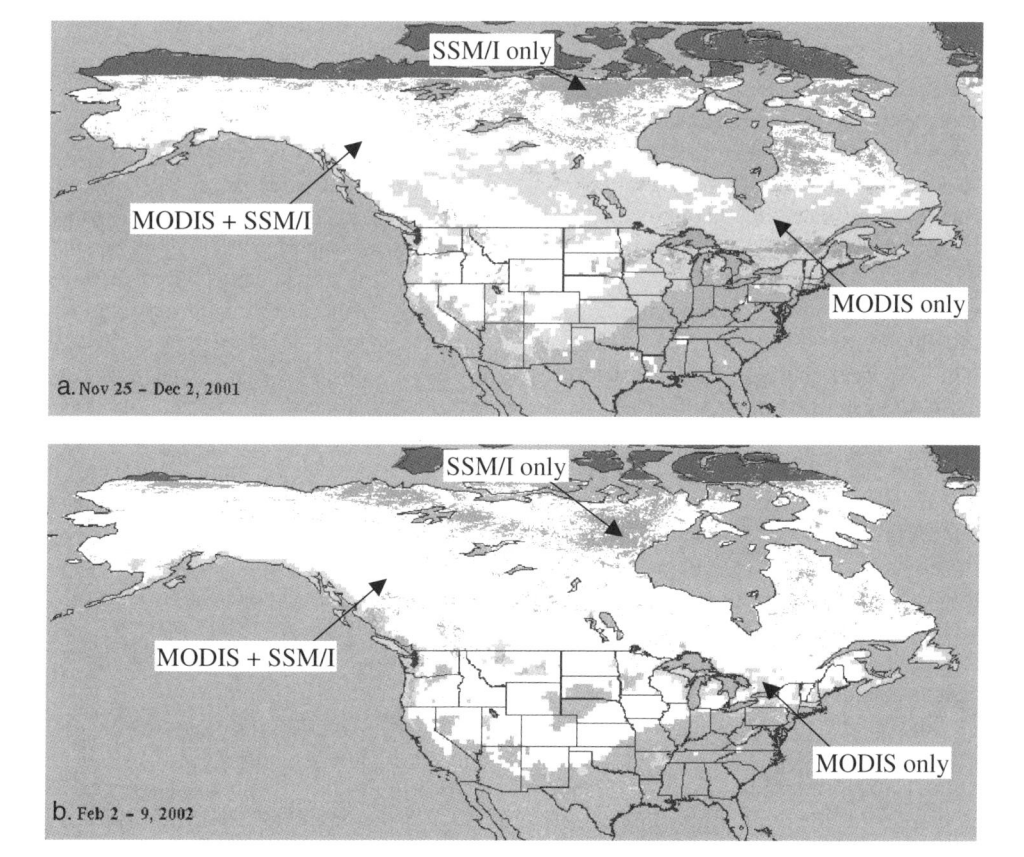

Figure 3.3 Comparison of 'static' SSM/I maximum snow area estimates and MODIS snow area estimates for 8-day periods in (a) 25 November – 2 December, 2001 and (b) 2 February – 9 February, 2002

season discrepancy between mapping approaches for SSM/I and MODIS snow mapping products in the winter of 2001–2002. In Figure 3.3a, the early season shows significant underestimation of the snow extent by the SSM/I data compared with the MODIS product. In Figure 3.3b, for mid-winter, the differences between estimates are smaller as the snow pack has thickened and a stronger scattering signal is observed by the SSM/I. Improvements in this mapping capability are expected with the development of the detection algorithms to AMSR-E data with its increased spatial resolution.

When snow is detected, it is necessary to estimate the SWE or snow depth and it is important to recognize that the estimation of SWE is the goal of most snow hydrological applications. This raises an important issue because the microwave response to a snow pack is sensitive to the bulk snow water equivalent. If, on the other hand, the snow pack can be characterized by a single homogeneous layer that has a constant density, the microwave response is sensitive to the snow thickness (assuming no free liquid water is present). The problem, however, is that most snow packs exhibit some form of layering so the density is variable both spatially and through time. The implication, for example, for two snow packs of identical thickness but with different stratigraphies (and bulk density) is that the microwave response will be different for each (Ulaby and Stiles, 1980). It seems reasonable to surmise, therefore, that passive microwave retrieval schemes should be focused on estimating SWE. However, the availability of reliable high quality ground-based measurements of SWE for validation is very poor, especially over large regions. Conversely, there are relatively abundant datasets of snow depth available for studies over selected larger areas. This fact produces a methodological dilemma for hydrologists: should they attempt to estimate SWE or snow depth? If they develop algorithms to estimate SWE, how are they to be validated over large regions? Alternatively, if the scientists opt to estimate snow depth, for which there are more spatially intensive datasets available, are they sacrificing estimation accuracy to estimate a variable that is only indirectly related to the microwave emission? In section 3.2 above, the average global density of snow depth recording stations was reported as 1 site per 160 000 km^2. Most of these WMO sites do not record SWE. The historical and regional FSU dataset reported SWE measurements routinely, but the USA COOP data consist of only snow depth. During validation of large-scale global retrieval algorithms, error metrics tend to be calculated for differences between areal microwave snow depth estimates and point snow depth estimates rather than differences in SWE since snow depth measurements have been the only potentially reliable, widespread data source available at this scale. Scientists who have chosen to estimate SWE have done so often with access to supporting SWE measurements at their scales of interest. These SWE estimation algorithms either have been developed for regional studies (e.g. Canadian prairies) or they have been developed for limited life experimental projects (e.g. BOREAS). For routine regional and global studies, these datasets are not available and so the estimation schemes tend to focus on estimating snow depth.

For satellite estimates, historically, most SWE or snow depth retrieval schemes have been based on some empirical formulation similar to the approach developed by Chang *et al.* (1987). In this approach single, homogeneous layered snow depth was expressed as a function of the horizontally polarized brightness temperature calculated from radiative transfer theory. The result of this experiment was a simple linear expression such that:

$$SD = a(\Delta TB) + b \text{ [cm]} \tag{3}$$

where SD is snow depth, b is generally regarded as zero and $a = 1.59$ cm K^{-1} and the assumption is made that the snow grain radius is 0.3 mm and snow density is 300 kg m^{-3}. If SWE is required in the retrieval, the units become mm and the a term is 4.8 mm K^{-1} for the same form. ΔTB represents the difference in brightness temperature between 19 GHz and 37 GHz channels at horizontal polarization or Tb19H and Tb37H, respectively. This model works well under simple snow conditions (single uniform layer) and where the terrain is flat and unforested. In locations where the snow physical characteristics conform to these parameter values, reasonable results are obtained, namely locations where the snow is characterized by a grain radius of 0.3 mm or bulk density of 300 kg m^{-3}. At the regional to global scale, the model parameters can be considered an average global seasonal grain size and density so at the average global seasonal scale, the results ought to be 'reasonable' (accepting that snow maps tend to underestimate). However, where these parameters are not locally representative at a given time, retrieval errors of between 10 to 40 cm snow depth or more can be expected. If hydrologists are interested in instantaneous local estimates, refinements to this approach are essential. Chang *et al.* (1996) included a forest cover compensation factor in their updated algorithm and Foster *et al.* (1997) made progress to spatially vary the coefficient a in equation (3) at broad continental scales. However, refinements are still required to reduce the errors further and improve the potentially utility of the data.

3.4.1 Spatial Representivity of SSM/I Snow Depth Estimates: An Example

To compare the spatial variability of snow depth estimates from an SSM/I algorithm with measured station snow depth, two datasets were generated consisting of global snow depth and local snow depth estimates from SSM/I data. Using the Foster *et al.* (1997) algorithm, snow depth was estimated for global snow cover on 2 February, 2001 and for the Red River catchment, located in the mid-western United States of America, for 28 January, 1994. The Red River estimates do not coincide exactly with the COOP station data for 12 February, 1994 but they are within approximately 2 weeks and it is known, from station records, that conditions did not change dramatically over this period. Assessment of the similarities in variogram structure between station data and remote sensing data should determine whether any conclusions could be drawn about the relative spatial representativity of each dataset.

Figure 3.4a and b show the variograms for the 2 February, 2001 global data and the 1994 Red River datasets respectively. In general, it is noticeable that in both cases the experimental variogram data are much smoother in character than the data for the discrete points for the same dates shown in Figure 3.2a and d. For the global data in Figure 3.4a, there is evidence of nested variation in the data, that is, variation of snow depth that occurs over different spatial scales (see Oliver, 2001, for details about nested variation). In essence, the nested variation is modelled by the variogram and produces breaks of slope at 108 km, 448 km and 1952 km. Figure 3.4a, therefore, could be decomposed into three separate variograms each having ranges of 108 km, 448 km and 1952 km relating to different scales of snow depth variation. Each variogram would represent the variation at these different spatial scales. The two smaller scales of variation (108 km and 448 km) represent local scale spatial variations of snow depth, probably caused by vegetation or topography controls. The larger-scale variation, represented by the sill with a range of 1952 km, is probably

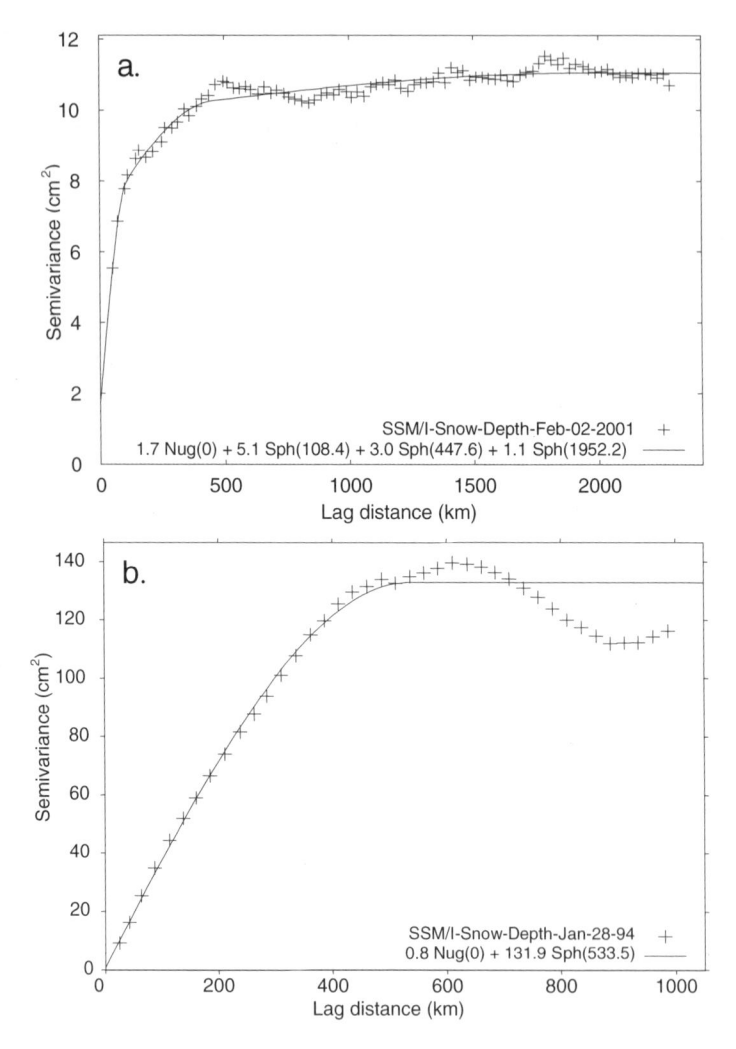

48   *Spatial Modelling of the Terrestrial Environment*

Figure 3.4 Variograms of 'static' SSM/I snow depth estimates for (a) global snow depth on 2 February, 2001 and (b) Red River area snow depth for 28 January, 1994

caused climate controls. Thus, the variogram structure implies that an SSM/I map of snow depth for this date would contain information at the local to regional scales of variation. In addition, comparison of Figure 3.4a with Figure 3.2a shows that the spatial detail in the SSM/I estimates are finer than those found in the station data; the sampling density of the station data are not capable of resolving different scales of variability identified by McKay and Gray (1981). Furthermore, the variogram structures are quite different in character and the snow depth semivariances in the discrete data are at least an order of magnitude greater than those in the SSM/I data. This implies that the spatial variation is greater for the station data than for the remote sensing estimates, a feature that is not unexpected given the spatial smoothing implicit in the SSM/I estimates. The discrete data variogram in Figure 3.2a has sill at a range of 3703 km but only a weak variogram structure can be discerned as noted in section 3.2. In this case study, therefore, the SSM/I snow depth estimates appear to represent more effectively and perhaps more accurately (although this cannot be stated conclusively) the spatial variability of snow depth at the global scale.

Figure 3.4b shows the variogram of snow depth for the Red River basin area for 1994. Since the COOP data are more spatially intensive than the WMO data, similarities between remote sensing and station data should be evident. In Figure 3.4b the variogram range is located at 533 km while in Figure 3.2d it is estimated at 496 km. Given the homogenous nature of the terrain in this region (Josberger and Mognard, 2002), it is suggested that little or no nested variation should be present in the variogram, which is indeed the case. For the same reason as before, a major difference between variograms is that the semivariance in the station data is four to five times greater than the semivariance in the SSM/I estimated snow depth data. At this scale of variation, the SSM/I estimates appear to be reasonable and comparable with the station data which are capable of representing snow depth variations at the given sampling spatial density.

From these two case studies, it appears that the SSM/I is capable of representing snow depth at spatial variations ranging from local scale to regional and global scales. The station data, however, are only good at representing snow depth at local scales. This is an important conclusion since it impacts directly on how global snow depth or SWE algorithms might be tested or even validated (see below). Whether or not the passive microwave estimates are 'good enough' is an issue that cannot be addressed here. All we can say is that the scale of variability appears to be reasonable based on our understanding of how snow cover varies in space. Given the local scale comparisons of SWE or snow depth based on recently developed snow depth or SWE algorithms, it is very probable that the passive microwave estimates need improvement before they are adopted by hydrologists. To begin improvement, therefore, the algorithms need to be made more dynamic and sensitive to those changing snow pack properties that affect the microwave emission signal in both in space and time.

3.5 Improving Estimates of Snow Depth/SWE at All Scales: Combining Models and Observations

It is apparent that different approaches to SWE or snow depth estimation work best at different spatial scales. While the physically based energy balance approaches are accurate at the micro-scale of variation, their performance at local to regional scales is uncertain. Remote sensing approaches, however, while not capable of producing estimates at the micro-scale, are able to estimate SWE successfully at local scales. At regional scales there is some uncertainty about their performance in general which is related to the static nature to date of the developed algorithms. To improve the SWE or snow depth estimates at regional scales, there are three possible approaches that could be taken. First, the application of snow pack energy balance models at large scales with climate models forcing as input can be developed, and second, the application of microwave remote sensing algorithms can be improved with dynamic parameterization schemes added. Third, and perhaps optimally, a combination of these two approaches can be developed and implemented. In fact, several research groups have developed retrieval schemes based on this hybrid approach with some success.

Section 3.3 explained in general terms how physically based models operate and concluded by suggesting that they are accurate at estimating snow pack properties at microscales of variation. Hence, the largest improvement to algorithms that estimate SWE or snow depth at local to regional scales, is through the transformation of static methodologies into more spatially and dynamic algorithms. In other words, the algorithms need to be flexible enough to estimate snow depth or SWE from snow packs that are continually changing. Algorithm parameters require constant adjusting through the winter season to reflect these changes. One approach that can achieve this dynamism is through the use of microwave emission models that are parameterized by independently derived snow pack properties or by optimizing the match of modelled and observed brightness temperatures by adjusting the microwave emission model physical snow pack parameters.

Early theoretical studies of the microwave emission from snow used radiative transfer models with some success (see Ulaby *et al.*, 1981). Through improved understanding of the physical snow pack evolution processes, these models have developed to new levels of sophistication and accuracy in the simulation of the microwave response from snow (e.g. Tsang *et al.*, 2000; Wiesman and Mätzler, 1999). Emission models require parameter inputs describing the physical properties of a snow pack (e.g. average grain size radius, volumetric fraction of snow, vertical temperature profile) and this information can be derived from the output of an energy or mass balance model of the snow. Hence, by using the emission models in conjunction with energy or mass balance models of snow, several studies have shown demonstrable improvements in SWE or snow depth estimation accuracy. While it is not the purpose of this chapter to give detailed descriptions of these models, Figure 3.5a shows a generalized implementation approach of a coupled snow model and microwave radiative transfer model.

Weisman *et al.* (2000) coupled the microwave emission model of layered snowpacks (MEMLS) with SNTHERM and Crocus (Brun *et al.*, 1989) to estimate snow depth at an alpine site in Switzerland. Half-hourly meteorological data were collected at the site and input to SNTHERM or Crocus. These models predict a variety of snow pack properties including number of layers, thickness, temperature, density, liquid water content, and grain size and shape of each layer. These estimates are then passed to the MEMLS model which calculates brightness temperatures in the range between 5 GHz to 100 GHz at a given linear polarization and at a prescribed incidence angle. These radiative transfer calculations are based on empirically-derived scattering coefficients and physically based absorption coefficients. Results showed that the estimated brightness temperatures matched well with measurements made by ground-based radiometers nearby. Thus, the results demonstrate that our understanding of the microwave interaction processes can be translated into physical models. Chen *et al.* (2001) adopted a similar approach but this time used only the SNTHERM model and a different form of radiative transfer model, namely the dense media radiative transfer (DMRT) model based on the quasicrystalline approximation with a sticky particle model. Tsang *et al.* (2000) give a full description of this two-layer model. Compared with the MEMLS model, the DMRT has been applied to a larger area in western USA (local scale) where SNOTEL data are input into the SNTHERM model. Also, estimated emissions from the DMRT model are compared with actual SSM/I brightness temperature observations and an inversion procedure is applied to produce a final snow depth estimate. The results obtained from the approach are an improvement over static, regression-based approaches. This is especially noteworthy since the region of

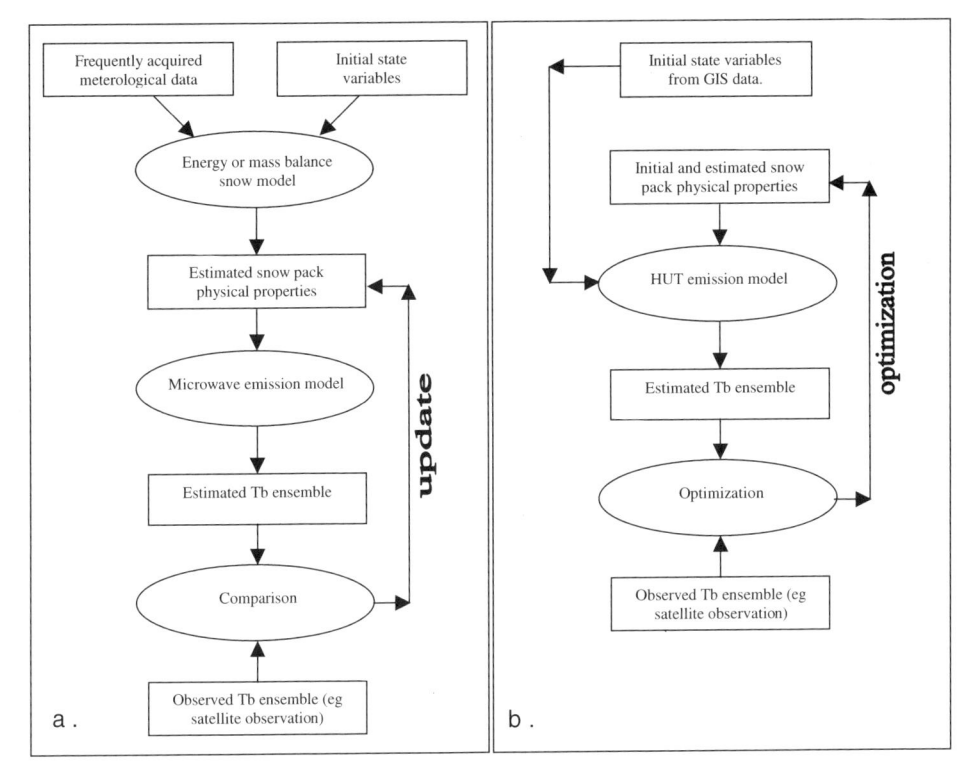

Figure 3.5 Methodological approaches to SWE or snow depth estimation using hybrid snow energy and mass balance models and microwave emission models. (a) represents the general approach adopted by Chen et al. (2001) (Reproduced by permission of IEEE), and (b) represents the method adopted by Pulliainen and Hallikainen (2001) (Reproduced permission of Elsevier Science). Reprinted from Remote Sensing of Environment 75, J. Pulliainen and M. Hallikainen, Retrieval of regional snow water equivalent from space-borne passive microwave observations, 76–85, © (2001), with permission from Elsevier

implementation is mountainous terrain where standard 'static' algorithms have difficulty in successful snow depth estimation. A constraint to the approach, however, is that it produces the best results for medium snow grain sizes; further refinement to the model is required if larger faceted grains develop.

The Helsinki University of Technology (HUT) snow emission model is different from the two previous models in that it is semi-empirical in approach and it does not require frequent externally derived snow state variables to estimate SWE. The model is based on radiative transfer emission from snow plus semi-empirical parameterizations of forest, atmosphere and soil (Pulliainen *et al.*, 1999). It is parameterized by geographical information (perhaps stored in a GIS) on forest stem volume, soil roughness and atmospheric optical thickness. Estimates of brightness temperature ensembles are computed at each pixel and compared with observed microwave data. An optimization routine then minimizes the error between modelled and observed brightness temperature ensembles by iteratively adjusting the snow physical properties in the model. Once the differences have been minimized, the iteration ceases and that SWE is the final estimate. Figure 3.5b shows a generalized diagram of the methodology. SWE retrieval errors have been found to be less than static approaches (Pulliainen and Hallikainen, 2001) with the smallest errors calculated for estimates in Finland to be 20 mm SWE.

Each of the approaches outlined above has shown improvements over static SWE or snow depth passive microwave algorithms. However, further improvement is still required as the errors are still substantial. Also, the models described so far have only been applied to local or small regional areas of interest where frequent and or high spatial resolution ancillary data are available (meteorological inputs, terrain and vegetation cover data). The challenge is to apply these methods to regions where the initialization data required for the emission model or the snow energy or mass balance models is largely non-existent or must be derived from outdated historical maps, large regional scale re-analysis or climate model data. In the case of MEMLS and the DMRT implementations, this poses a significant challenge since these models rely on frequent meteorological inputs from nearby stations for forcing SNTHERM or Crocus. For the HUT model, detailed information about stem volume is required along with atmospheric factors that are often unavailable at the scale required. Therefore, even though these hybrid approaches are improvements on the previous static algorithms, further research activity is required to enhance these models so that they can be implemented globally. Kelly et al. (2002) have demonstrated that a microwave emission model can be parameterized by a simple empirical model of grain growth for which grain radius is estimated from evolution through the season and the history of estimated temperature differences through the pack. It is based on the DMRT approach of Tsang et al. (2000) but rather than use data from meteorological stations, surface temperatures are estimated from an empirical relationship between SSM/I brightness temperatures and surface temperature. The implementation is a simplified DMRT model applied deterministically to estimate snow depth (inversion is not required). Despite its simplicity in its current form, the results so far are an improvement on those from 'static' approaches; for the 2000–2001 winter in the northern hemisphere, the simplified DMRT approach produced global seasonal estimates that were 3 cm of snow depth better than the Foster et al. (1997) algorithm. It is expected therefore that for this global algorithm, and the HUT and DMRT implementations, estimates of snow depth or SWE will improve as parameterization and initialization datasets improve. Additionally, with radiometrically enhanced passive microwave instrument technology such as AMSR-E, the errors will be reduced even further.

3.6 Conclusion

In this chapter, we have examined the way in which (spatial) models are used in the estimation of snow cover and especially snow depth or SWE. The variogram analysis suggested that for point measurements of snow depth at meteorological stations, only local-scale networks, where the station network is sufficiently dense (less than 1 per 40 km × 40 km area in the case of the COOP network), can quantify the spatial variability of snow depth. In regions where the cover is sparse, the representativeness of point data is uncertain and should not be relied on to characterize the spatial variability of snow depth. In the case of spatial estimates of snow depth or SWE, micro-scale variations are successfully represented by physically based models, which are dynamic in operation and can estimate suites of snow hydrological parameters. At the local scale of variation, these physically based models can also be used to estimate the changing SWE or snow depth of a pack provided sufficient meteorological data inputs are available. As the area scales to the regional domain, it is uncertain whether these models can be used successfully to represent snow depth or SWE variability. Remote sensing retrievals, on the other hand, are well suited to local to regional scale applications as demonstrated by the structure in the variograms of snow depth estimates in Figure 3.4. However, the errors produced are still large and need reduction before these estimates are adopted more generally. It seems that a promising area of research in the form of hybrid physically based models and microwave emission models should improve our estimates of SWE and snow depth at regional and global scales. This is an interesting prospect since traditionally, snow mass or energy balance models have been subordinated by remote sensing application scientists or vice versa. Davis *et al.* contend that "Ironically, we may soon see the use of high resolution, process-detailed snow models to aid in interpolating ground measurements for validating remote sensing algorithms to recover SWE" (2001, p. 283). Given the nature of snow depth or SWE spatial variability, this seems entirely reasonable. If snow hydrology models are capable of estimating SWE at finer local scales, then it makes sense for passive microwave estimates to be tested in regions where the hydrological models are accurate. Ultimately, this will lend more weight to the validation of the passive microwave algorithms. Comprehensively validated algorithms should then be able to estimate SWE more accurately in areas of the world where the hydrological models are incapable of providing reliable estimates on account of sparse meteorological station networks. An even stronger reason for using snow hydrology models is through their combination with microwave emission models. Since the microwave emission signal from a snow pack is related in some way to the SWE, a snow energy or mass balance hydrology model can be used to 'unravel' the microwave signal and provide a better, more dynamic estimate of SWE or snow depth.

It appears, therefore, that good potential exists for estimating SWE or snow depth and characterizing their spatial variability using remote sensing and hydrological models of snow energy or mass balance dynamics. However, while these models are in development, there is a conspicuous lack of attention to validation frameworks of SWE or snow depth estimates. This statement is an important one and perhaps liberal in scope since energy balance models of snow have a strong tradition of rigorous validation. However, at the scale at which they have been applied, the validation is relatively straightforward. At local or regional scales, validation is less straightforward and has often been implemented in not the most robust of ways. Typically, most validation exercises to test passive microwave estimates have assumed that point measurements of SWE are 'representative' at the passive microwave support scale (grid sizes of 25 km × 25 km). In some regions of the world this may indeed be the case. However, passive microwave sub-grid scale heterogeneity of snow depth or SWE is an important issue that requires some formal attention; how can hydrologists use existing point measurements at the footprint scale? Simple comparisons between point measurements and areal estimates are often completely inappropriate, perhaps when the site location is in a region where land cover is highly variable or terrain steep and dissected and snow depth is highly variable; one point estimate of snow depth is unlikely to represent accurately the spatial footprint average. Furthermore, if we persist with the general approach of comparing point measurements of SWE with areal estimates from passive microwave models and algorithms, it is necessary to determine how many points are needed to adequately perform the validation. More specifically, what is the mechanism that determines the required density of snow measurement sites for a given accuracy requisite of passive microwave estimate and how might this density vary with changing SWE conditions or under different vegetation cover or terrain? This issue is an important one that requires attention. One potential solution to this problem is to examine the possibility of developing a more analytical approach to error propagation for the remote sensing of SWE and snow depth. There is a large literature on the characterization of error in spatial environemtal modelling (e.g. Heuvelink, 1998). It would probably be of great interest to the remote sensing and snow hydrology community for a framework to be developed that could be universally adopted for validation studies in this field. A key challenge with such an approach is that all component variables in the algorithm, coupled model and ancillary data require associated error estimates. In the case of the coupled SNTHERM and DMRT approach, this would mean that for every variable used in the method, an error would need to be derived and combined in an analytical method (such as the Taylor expansion) to produce an overall estimation error statistic. This is quite some challenge since it is often very difficult to provide robust error quantities for many of the terms in SWE estimation algorithms or hybrid models described previously. However, the advantage of such an approach would be to make the estimates of SWE or snow depth highly attractive to many data users since for each estimate of SWE an error term would also be generated. Specifically, snow product users such as land surface modellers or climate modelling scientists who often require error estimates along with the actual variable estimate would find this error approach of great value.

In conclusion, then, there are various reasons why spatial models are used in the representation of snow depth or SWE at various spatial scales. Some of the outstanding issues concerning the estimation of SWE and snow depth will be addressed directly and indirectly in the near future thanks to several new technological and scientific developments in planning or under way. First, with the launch of Aqua and the availability of AMSR-E data in the near future, the spatial scale of observation of snow from space will be the most detailed ever. The potential to explore some of the spatial variability issues further will be greatly enhanced. Furthermore, with science field experiments such as the NASA Cold Lands Processes Experiment (CLPX), many of the questions regarding spatial variability and spatial representativity of snow properties from micro- to local scales of variation will be addressed (Cline et al., 2002). This is an important experiment involving many teams of scientists undertaking laborious field experiments at the micro- and local scale. However, with these prospects, it might be possible to begin developing a framework for error estimation. Ultimately, the errors associated with snow depth and SWE estimation from spacecraft instruments and terrestrial models should be reduced thereby enhancing our ability to quantify the role of snow in the global hydrological cycle.

Acknowledgements

This research was supported under NASA's Office of Earth Sciences program and the EOS AMSR-E algorithm development project. The authors would like to thank Richard Armstrong for useful critical comments.

References

Armstrong, R.L. and Brodzik, M.J., 1995, An earth-gridded SSM/I data set for cryospheric studies and global change monitoring, *Advances in Space Research*, **10**, 155–163.

Armstrong, R.L. and Brodzik, M.J., 2001, Recent Northern Hemisphere snow extent: a comparison of data derived from visible and microwave satellite sensors, *Geophysical Research Letters*, **28**, 3673–3676.

Atkinson, P.M., 2001, Geostatistical regularization in remote sensing, in N.J. Tate and P.M. Atkinson (eds), *Modelling Scale in Geographical Information Science* (Chichester: John Wiley & Sons), 238–260.

Bales. R.C. and Harrington, R.F., 1995, Recent progress in snow hydrology, *Reviews of Geophysics*, **33**, 1011–1021.

Brasnett, B., 1999, A global analysis of snow depth for numerical weather prediction, *Journal of Applied Meteorology*, **38**, 726–740.

Brun, E., Martin, E., Simon, V., Gendre, C. and Coléou, C., 1989, An energy and mass model of snow cover suitable for operational avalanche forecasting, *Journal of Glaciology*, **35**, 333–342.

Carroll, S.S. and Cressie, N., 1997, Spatial modeling of snow water equivalent using covariances estimated from spatial geomorphic attributes, *Journal of Hydrology*, **190**, 42–59.

Chang, A.T.C. and Rango, A., 2000, Algorithm Theoretical Basis Document (ATBD) for the AMSR-E Snow Water Equivalent Algorithm, NASA/GSFC, Nov. 2000.

Chang, A.T.C., Foster, J.L. and Hall, D.K., 1987, Nimbus-7 derived global snow cover parameters, *Annals of Glaciology*, **9**, 39–44.

Chang, A.T.C., Foster, J.L. and Hall, D.K., 1996, Effects of forest on the snow parameters derived from microwave measurements during the BOREAS winter field experiment, *Hydrological Processes*, **10**, 1565–1574.

Chen, C., Nijssen, B., Guo, J., Tsang, L., Wood, A.W., Hwang J. and Lettenmaier, D.P., 2001, Passive microwave remote sensing of snow constrained by hydrological simulations, *IEEE Transactions on Geoscience and Remote Sensing*, **39**, 1744–1756.

Cline, D., Armstrong, R., Davis, R., Elder, K. and Liston, G., 2002., *Cold Land Processes Field Experiment. In situ* data edited by M. Parsons and M.J. Brodzik (Boulder, CO: National Snow and Ice Data Center), Digital Media.

Cohen, J., 1994, Snow cover and climate, *Weather*, **49**, 150–155.

Davis, R.E., Jordan, R., Daly, S. and Koenig, G., 2001, Validation of snow models, in M.G. Anderson and P.D. Bates (eds), *Model Validation: Perspectives in Hydrological Science* (Chichester: John Wiley & Sons), 261–292.

Derksen, C. and LeDrew, E., 2000, Variability and change in terrestrial snow cover: data acquisition and links to the atmosphere, *Progress in Physical Geography*, **24**, 469–498.

Dewey, K.F. and Heim Jr., R., 1981, Satellite observations of variations in Northern Hemisphere seasonal snow cover, NOAA Technical Report NESS 87.

Dewey, K.F. and Heim Jr., R., 1983, Satellite observations of variations in Southern Hemisphere snow cover, NOAA Technical Report NESDIS 1.

Dozier, J., 1987, Recent research in snow hydrology, *Reviews of Geophysics*, **25**, 153–161.

Elder, K., Rosenthal, W. and Davis, R.E., 1998, Estimating the spatial distribution of snow water equivalence in a montane watershed, *Hydrological Processes*, **12**, 1793–1808.

Foster, J.L, Chang, A.T.C. and Hall, D.K., 1997, Comparison of snow mass estimates from a prototype passive microwave snow algorithm, a revised algorithm and snow depth climatology, *Remote Sensing of Environment*, **62**, 132–142.

Foster, J.L., Liston, G., Koster, R., Essery, R., Behr, H., Dumenil, L., Verseghy, D., Thompson, S., Pollard, D. and Cohen, J., 1996, Snow cover and snow mass intercomparisons of general circulation models and remotely sensed data sets, *Journal of Climate*, **9**, 409–426.

Frei, A. and Robinson, D.A., 1999, Northern hemisphere snow extent: regional variability 1972–1994, *International Journal of Climatology*, **19**, 1535–1560.

Gloersen, P. and Barath, F.T., 1977, A scanning multichannel microwave radiometer for Nimbus-G and Seasat-A, *IEEE Journal of Oceanic Engineering*, **2**, 172–178.

Goodison, B.E. and Walker, A.E., 1994, Canadian development and use of snow cover information from passive microwave satellite data, in B. Choudhury, Y. Kerr, E. Njoku and P. Pampaloni (eds), *Passive Microwave Remote Sensing of Land-Atmosphere Interactions* (Utrecht:) VSP BV 245–62.

Grody, N.C. and Basist, A.N., 1996, Global identification of snowcover using SSM/I measurements, *IEEE Transactions on Geoscience and Remote Sensing*, **34**, 237–249.

Hall, D.K., Kelly, R.E.J., Riggs, G.A., Chang, A.T.C. and Foster, J.L., 2002a, Assessment of the relative accuracy of hemispheric-scale snow-cover maps, *Annals of Glaciology*, **34**, 24–30.

Hall, D.K., Riggs, G.A., Salomonson, V.V., DiGiromamo, N. and Bayr, K.J., 2002b, MODIS Snow Cover Products, *Remote Sensing of Environment*, **83**, 181–194.

Hallikainen, M.T. and Jolma, P.A., 1992, Comparison of algorithms for retrieval of snow water equivalent from Nimbus-7 SMMR data in Finland, *IEEE Transactions on Geoscience and Remote Sensing*, **30**, 124–131.

Heuvelink, G.B.M., 1998, *Error Propagation in Environmental Modelling with GIS, Research Monographs in Geographical Information Systems* (London: Taylor and Francis).

Hollinger, J.P., Pierce, J.L. and Poe, G.A., 1990, SSM/I instrument evaluation, *IEEE Transactions on Geoscience and Remote Sensing*, **28**, 781–790.

Isaaks, E.H. and Srivastava, R.M., 1989, *Applied Geostatistics* (Oxford: Oxford University Press).

Jordan, R., 1991, One-dimensional temperature model for a snow cover: technical documentation for SNTHERM 89. *CRREL Special Report*, SR91-16 *US Army Cold Regions Research and Engineering Laboratory*.

Josberger, E.G. and Mognard, N.M., 2002, A passive microwave snow depth algorithm with a proxy for snow metamorphism, *Hydrological Processes*, **16**, 1557–1568.

Judson, A. and Doeksen, N., 2000, Density of freshly fallen snow in the Central Rocky Mountains, *Bulletin of the American Meteorological Society*, **81**, 1577–1587.

Kelly, R.E.J. and Atkinson, P.M., 1999, Modelling and efficient mapping of snow cover in the UK for remote sensing validation. in P.M. Atkinson and N.J. Tate (eds), *Advances in Remote Sensing and GIS Analysis* (Chichester: John Wiley & Sons), 75–96.

Kelly, R.E.J., Chang, A.T.C., Tsang, L. and Chen, C.T., 2002, Parameterization of snowpack grain size for global satellite microwave estimates of snow depth, *Proceedings of IGARSS 2002 and the 24th Canadian Symposium on Remote Sensing*, Toronto, Canada, 24–28 June, 686–688.

König, M., Winther, J-G. and Isaksson, E., 2001, Measuring snow and glacier ice properties from satellite, *Reviews of Geophysics*, **39**, 1–27.

Luce, C.H., Tarbotton, D.G. and Cooley, K.R., 1998, The influence of the spatial distribution of snow on basin-averaged snowmelt, *Hydrological Processes*, **12**, 1671–1685.

Marks, D., Cooley, K.R., Robertson, D.C. and Winstral, A., 2001, Long-term snow data base, Reynolds Creek Experimental Watershed, Idaho, United States, *Water Resources Research*, **37**, 2835–2838.

McKay, G.A. and Gray, D.M., 1981, The distribution of snowcover, in D.M. Gray and D.H. Male (eds), *Handbook of Snow* (Toronto: Pergamon Press), 153–190.

Oliver, M.A., 2001, Determining the spatial scale of variation in environmental properties using the variogram, in N.J. Tate and P.M. Atkinson (eds), *Modelling Scale in Geographical Information Science* (Chichester: John Wiley & Sons). 193–219.

Pulliainen, J. and Hallikainen, M., 2001, Retrieval of regional snow water equivalent from space-borne passive microwave observations, *Remote Sensing of Environment*, **75**, 76–85.

Pulliainen, J., Grandell, J. and Hallikainen, M.T., 1999, HUT snow emission model and its applicability to snow water equivalent retrieval, *IEEE Transactions on Geoscience and Remote Sensing*, **37**, 1378–1390.

Ramsay, B., 1998, The interactive multisensor snow and ice mapping system, *Hydrological Processes*, **12**, 1537–1546.

Robinson, D.A., 1999, Northern hemisphere snow cover during the satellite era, *Proc. 5th Conf. Polar Met. and Ocean*, Dallas, TX, American Meteorological Society, Boston, MA, 255–260.

Shi, J. and Dozier, J., 2000, Estimation of snow water equivalent using SIR-C/X-SAR, Part I: inferring snow density and subsurface properties, *IEEE Transactions on Geoscience and Remote Sensing*, **38**, 2465–2474.

Slater, A.G., Schlosser, C.A., Desborough, C.E., Pitman, A.J., Henderson-Sellers, A., Robock, A., Vinnikov, K. Ya., Speranskaya, N.A., Mitchell, K., Boone, A., Braden, H., Chen, F., Cox, P., de Rosnay, P., Dickinson, R.E., Dai, Y-J., Duan, Q., Entin, J., Etchevers, P., Gedney, N., Gusev, Ye. M., Habets, F., Kim, J., Koren, V., Kowalczyk, E., Nasonova, O.N., Noilhan, J., Shaake, J., Shmakin, A.B., Smirnova, T., Verseghy, D., Wetzel, P., Xue, Y., Yang, Z-L. and Zeng, Q., 2001, The representation of snow in land surface schemes: results from PILPS 2(d), *Journal of Hydrometeorology*, **2**, 7–25.

Tait, A., 1997, Estimation of snow water equivalent using passive microwave radiometer data, *Remote Sensing of Environment*, **64**, 286–291.

Tsang, L., Chen, C., Chang, A.T.C., Guo, J. and Ding, K., 2000, Dense media radiative transfer theory based on quasicrystalline approximation with applications to passive microwave remote sensing of snow, *Radio Science*, **35**, 731–749.

Ulaby, F.T. and Stiles, W.H., 1980, The active and passive microwave response to snow parameters, 2, water equivalent of dry snow, *Journal of Geophysical Research*, **85**, 1045–1049.

Ulaby, F.T., Moore, R.K. and Fung, A., 1981, *Microwave Remote Sensing* (Reading, MA: Addison-Wesley).

Walsh, J.E., 1984, Snow cover and atmospheric variability, *American Scientist*, **72**, 50–57.

Wiesman, A. and Mätzler, C., 1999, Microwave emission model of layered snowpacks, *Remote Sensing of Environment*, **70**, 307–316.

Wiesman, A., Fierz, C. and Mätzler, C., 2000, Simulation of microwave emission from physically modeled snowpacks, *Annals of Glaciology*, **31**, 397–405.

4

Using Coupled Land Surface and Microwave Emission Models to Address Issues in Satellite-Based Estimates of Soil Moisture

Eleanor J. Burke, R. Chawn Harlow and W. James Shuttleworth

4.1 Introduction

Atmospheric models would benefit from improved initialization and description of the evolution of soil-moisture status, the main impact being through the local and regional availability of the soil moisture that can reach the atmosphere by evaporation from the soil or by transpiration from plants (Beljaars *et al.*, 1996; Betts *et al.*, 1996; Liu and Avissar, 1999; Koster *et al.*, 2000). At present, the most practical method of improving the accuracy of soil moisture in models is via the use of land data assimilation systems (LDAS). LDAS are two-dimensional arrays of the land-surface scheme used in the relevant weather- or climate-forecast model forced, to the maximum extent possible, by observations. The resulting modelled soil moisture status is less biased by poor simulation of the near-surface atmospheric forcing, especially precipitation. An example of an LDAS running in near real-time can be found at (http://ldas.gsfc.nasa.gov): it consists of several physically based land-surface models running on a common 0.125°-resolution grid covering the contiguous United States. It is driven by common surface forcing fields, which include observed hourly gauge-radar precipitation, and observed GOES-based satellite-derived surface solar insolation (Mitchell *et al.*, 2000). In coming years, such LDAS may well become the routine mechanism by which many predictive weather and climate models will be initiated. If this is so, it will be via assimilation into LDAS that other data relevant to the current status of the

---

Spatial Modelling of the Terrestrial Environment. Edited by R. Kelly, N. Drake, S. Barr.
© 2004 John Wiley & Sons, Ltd. ISBN: 0-470-84348-9.

land surface, such as remotely sensed estimates of soil moisture, will find value. Coupling a description of microwave emission to the land surface models used in an LDAS should allow direct assimilation of measured microwave brightness temperatures and improve estimates of the soil moisture fields calculated by the LDAS (Houser *et al.*, 1998; Reichle *et al.*, 2001; Galantowicz *et al.*, 1999; Burke *et al.*, 2003).

Remotely sensed observations of global surface soil moisture using an L-band (1.4 GHz) passive microwave radiometer are likely to become available within the next decade. Hence, there is considerable interest in developing methods to use this information effectively. However, such measurements have limitations, specifically:

1. The relationship between the measured microwave brightness temperature and soil moisture is strongly influenced by the presence of vegetation. A simple one-parameter (optical depth) model is usually used to specify the effect of the vegetation, with the optical depth taken to be proportional to the vegetation water content. However, the constant of proportionality (the opacity coefficient) is an uncertain function of vegetation type and radiometer characteristics.
2. L-band microwave brightness temperatures are only directly related to soil moisture in the top few centimetres of the soil and information about the soil moisture deeper in the soil profile has to be inferred indirectly through the use of land surface models.
3. In the near future, any potential L-band mission will likely have a resolution ~30–60 km for technical reasons. At this resolution, the vegetation cover, soil type, and soil water content within each pixel will necessarily be heterogeneous for all pixels and this could introduce errors. For some regional and local-scale applications, such as modelling runoff from a catchment, for example, or for initializing a mesoscale model to predict convective precipitation, higher resolution fields of soil moisture are important.

This chapter describes progress towards potential solutions to these issues through the use of coupled land surface and microwave emission models, where possible in the context of potential upcoming L-band satellite missions (SMOS – Kerr *et al.*, 2001 and HYDROS – HYDROS homepage).

4.2 Models and Methods

4.2.1 Coupled Land Surface and Microwave Emission Models

Coupled land-surface and microwave emission models can be used as a tool to explore the potential of L-band microwave radiometry and to evaluate the relationship between microwave brightness temperatures and soil moisture. The land surface model is forced by meteorological data (incoming solar radiation, incoming longwave radiation, air temperature, relative humidity, precipitation and wind speed). It uses these data to provide calculations of the evolving (profiles of) soil water content and soil temperature and vegetation temperature, which are input to the microwave emission model. The microwave emission model then calculates the microwave brightness temperature of the soil–vegetation–atmosphere interface. Such forward modelling of microwave brightness temperature incorporates model physics neglected in simple retrievals of near surface soil moisture, including representation of the effect of non-uniform near-surface profiles of soil moisture, and ensures consistency between the parameterizations used within the modelling system. Two coupled models are commonly discussed in the literature, MICRO-SWEAT (Burke et al., 1997; 1998; 2003) and LSP/R (Judge et al., 1999, 2001; Liou et al., 1998; Liou and England, 1998). Both of these models produce excellent agreement with measurements: MICRO-SWEAT has been successfully compared with 1.4 GHz (L-band) truck-based data and LSP/R with 19 GHz tower-based data. To date, most research has focused on running these models at point scale, but Burke et al. (2003) used data collected during the Southern Great Plains 1997 experiment (SGP97 homepage) to investigate the effectiveness of using a 2-D array of MICRO-SWEAT models to predict distributed L-band microwave brightness temperatures measured by an aircraft. They found discrepancies between the spatial distributions of modelled and measured brightness temperature which likely result from imperfect knowledge of the spatial distributions of soil properties, precipitation, and the representation of vegetation.

In theory, any land-surface model [e.g. BATS (Dickinson et al., 1986); CLM (Zeng et al., 2002); LSM (Bonan, 1996)] could be coupled with a microwave emission model. However, in practice, because the microwave brightness temperature is mainly sensitive to near-surface soil moisture (in top 2–5 cm) and decreases with depth in the soil, the most accurate predictions of microwave brightness temperature result when using a land surface model that represents very fine near-surface soil layers. For example, in their studies, Burke et al. (1997; 1998) used SWEAT (Daamen and Simmonds, 1996), a land surface model that has 8 layers in the top 5 cm, with layer thickness increasing with depth from a 1 mm thick top layer.

In general, the microwave emission from the soil can be described using coherent (Wilheit, 1978; Njoku and Kong, 1977; England, 1976) or non-coherent models (Burke et al., 1979; England, 1975). Schmugge and Choudhury (1981) and Ulaby et al. (1986) compared these two types of models, as follows. Non-coherent models estimate emissivity using the dielectric contrast at the air/soil interface and are accurate only when the (variable) sampling depth within the soil is well known. Most retrieval algorithms are based on non-coherent models, with the near-surface soil water content assumed to be uniform to a specified depth. In the case of L-band, this is usually 5 cm (e.g. Jackson et al., 1999). In coherent models, the value of the emissivity at the surface is coupled to the dielectric properties of the soil below the surface; consequently, they provide better estimates of the emissivity when the soil water content profile is not uniform. One significant disadvantage of coherent models is that interference effects can occur, but these have only rarely been witnessed in nature (Schmugge et al., 1998).

The emission component of MICRO-SWEAT is based on the Wilheit (1978) coherent model of electromagnetic propagation through a plain stratified medium. The Wilheit (1978) model relates the brightness temperature emergent at the soil surface to the dielectric properties and temperature of the underlying soil layers. For radiation with polarization, p, the microwave brightness temperature, T_{Bp} (K), is given (Wilheit, 1978) by:

$$T_{Bp}(\theta) = \sum_{i=1}^{N} f_{ip}(\theta) T_i \qquad (1)$$

where T_i is the temperature of the ith layer (K), f_{ip} is the fraction of energy absorbed from an incident microwave with polarization p by the ith layer (a function of the dielectric

62   Spatial Modelling of the Terrestrial Environment

constant of the soil), and N is the number of the layers in the semi-infinite medium. The Wilheit model assumes scattering is negligible, consequently the model is only applicable at longer wavelengths. The Dobson *et al.* (1985) model can be used to calculate the profile of soil dielectric from knowledge of the profile of soil water content and soil particle size distribution. The Dobson *et al.* (1985) model is a simple, non-linear dielectric-mixing model that takes into account the proportions of free water (dielectric ~80), bound water (dielectric ~4), soil solids (dielectric ~4), and air (dielectric ~1) in the soil medium. The soil particle size distribution and the water content determine the relative proportions of free and bound water.

A rough soil surface will increase the microwave emission because the emitting surface area is greater. Its effect can be parameterized using the two-parameter Mo and Schmugge (1987) model, which includes the mixing of vertical and horizontal polarizations. At L-band, for a relatively smooth soil, the effect of soil surface roughness is small and, for simplicity in the analyses discussed here, the soil surface roughness is assumed to be zero.

Figures 4.1a and b show the time series of modelled and measured microwave brightness temperature for a smooth bare sandy loam soil (Burke *et al.*, 1998) compared to measurements made using a truck-mounted radiometer during two separate 10-day drying periods. The model calculations are made using detailed profiles of soil water content and

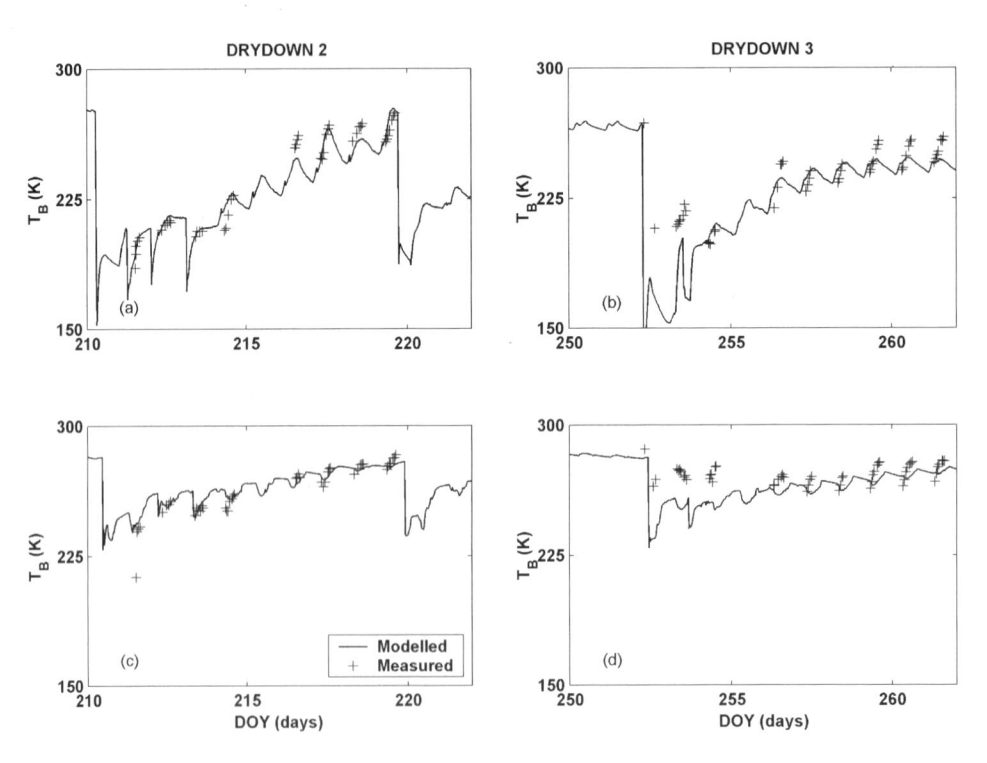

Figure 4.1 Time series of modelled and measured brightness temperatures for a bare soil (a and b) and the same soil covered with a soybean canopy (c and d) for two different drying periods (redrawn from Lee et al., 2002a). The moral right of the authors to be identified as the authors of such figure is asserted.

temperature calculated by a land surface model and the Wilheit (1978) microwave emission model (Lee et al., 2002a). The sharp decrease in brightness temperatures on DOY 210 and 252 is caused by irrigation while the decrease on DOY 253 results from rainfall. The near-surface water content ranges from 40% to 10% for both drying periods. Diurnal variation in the near-surface soil water content and temperature are also reflected in the modelled microwave brightness temperatures.

4.2.2 Effect of Vegetation on Microwave Emission from the Soil

The presence of vegetation has a significant impact on the relationship between near-surface soil moisture and microwave brightness temperature. For a bare soil, the dynamic range in brightness temperature is approximately 180 K, whereas for a crop with canopy water content of only 3.3 kg m^{-2}, the dynamic range is significantly reduced, to approximately 60 K. In general, the effects of the vegetation on the microwave emission from the soil can be described using radiative transfer equations. These can be solved either numerically or after applying simplifying assumptions. The following sections describe radiative transfer theory, and its application to describing the effect of vegetation on microwave emission from the soil.

Radiative transfer theory. A vegetation canopy will scatter and absorb microwave emission from the soil. It will also contribute with its own emission, which will be scattered and absorbed by the canopy through which it passes. Within an infinitesimal volume of the canopy, the energy balance for upward radiative transfer with horizontal polarization is:

$$\mu \frac{dT_{Bh}(\theta, z)}{dz} = K_a T_{can}(z) - K_e(\theta) T_{Bh}(\theta, z)$$

$$+ \int_{-1}^{1} ((h, h') T_{Bh}(\theta', z) + (h, v') T_{Bv}(\theta', z)) d\mu'. \quad (2)$$

The energy balance for downward radiative transfer with horizontal polarization is:

$$-\mu \frac{dT_{Bh}(\pi - \theta, z)}{dz} = K_a T_{can}(z) - K_e(\pi - \theta) T_{Bh}(\pi - \theta, z)$$

$$+ \int_{-1}^{1} ((h(\pi - \theta), h') T_{Bh}(\theta', z) + (h(\pi - \theta), v') T_{Bv}(\theta', z)) d\mu' \quad (3)$$

where T_{Bh} is the horizontally polarized microwave brightness temperature (K), T_{can} is the physical temperature of the canopy (K), z is the depth within the canopy (m), K_a is the absorption coefficient, K_s is the scattering coefficient, $K_e (= K_a + K_s)$ is the extinction coefficient, θ and θ' are incidence angles, with $\mu = \cos\theta$ and $\mu' = \cos\theta'$, and (h,h') and (h,v') are scattering phase functions, where (p, q') represents the scattering probability of p polarized radiation being scattered into q polarized radiation. (Note: p and q represent horizontally (h) and vertically (v) polarized radiation interchangeably.) The first term on the right-hand side in equations (2) and (3) describes the thermal emission from the canopy, the second term represents the energy absorbed and scattered by the canopy, and the final term represents the redistribution of scattered energy among different look angles.

There are no analytical solutions to equations (1) and (2). However, they can be solved numerically as they are, for example, in the complex 'discrete' model of Wigneron et al.

(1993), or they can be simplified by assuming that the scattering effects (and hence the phase functions) are negligible, as they are in the 'simple' model of Ulaby et al. (1986) and the 'extended Wilheit' (1978) model (Lee et al., 2002a). Both the simple model and the extended Wilheit (1978) model are only valid at longer wavelengths where the dimensions of portions of the canopy are of the same order of magnitude as the wavelength of the radiation detected. These three approaches are discussed further below.

Discrete Model. Discrete models estimate the scattering phase functions and the absorption and scattering coefficients from measurements of the vegetation characteristics. Discs are used to represent leaves and finite cylinders the stems of vegetation. Numerical solutions of equations (2) and (3) can then be calculated and the microwave brightness temperature above the canopy estimated. Wigneron et al. (1993) and Ferrazzoli et al. (2000) discussed two different versions of a discrete model in detail. An example of the application of the discrete model is given by Burke et al. (1999) who used the Wigneron et al. (1993) model to successfully predict the time series of brightness temperatures measured over a soybean canopy using a truck-mounted L-band radiometer.

Simple Model. The simple model assumes that the scattering phase functions and, thus, the final term in equations 2 and 3 above are negligible. The brightness temperature above the vegetation canopy is then given by the sum of the emission from the soil (the first term in equation (4) below), the upward emission from the canopy (the second term in equation (4)), and the downward emission from the canopy that is reflected by the soil (the third term in equation (4)), the microwave emission being then attenuated by any vegetation it passes through. The simple model therefore gives:

$$T_{Bp} = \Gamma T_{Bpsoil} + (1 - \Gamma)(1 - \Omega)T_{can} + (1 - \Gamma)(1 - \Omega)\Gamma T_{can} r_{sp} \qquad (4)$$

with:

$$\Gamma = e^{-\tau \sec \theta} \qquad (5)$$

$$\tau = K_e d \qquad (6)$$

$$\Omega = \frac{K_s}{K_e} \qquad (7)$$

where T_{Bpsoil} (K) is the brightness temperature of the soil for polarization p, r_s is the reflectivity of the soil surface, Γ is the transmissivity of the canopy, τ is the optical depth, d is the canopy height (m), Ω is the single scattering albedo, K_e is the extinction coefficient of the canopy, and K_s is the scattering coefficient of the canopy. The single scattering albedo (Ω) determines the relative importance of absorption and scattering by the canopy. For L-band radiation, K_s and, therefore, Ω are close to zero (equation (7) – Jackson and Schmugge, 1991). However, Ω has rarely been estimated (Kerr and Wigneron, 1994) and its dependence on vegetation characteristics is unknown. The optical depth, τ, represents the amount of absorption and emission by the canopy and is often approximated by:

$$\tau = \beta \theta_{veg} \qquad (8)$$

where θ_{veg} is the vegetation water content and β is the opacity coefficient. One output from the Wigneron *et al.* (1993) discrete model is the extinction coefficient (K_e). Burke *et al.* (1999) showed that the optical depth calculated using the extinction coefficient (equation (7)) could be used in the simple model to predict accurate time series of microwave brightness temperatures.

Extended Wilheit (1978) Model. Unlike the previous two models, which assume the canopy consists of one uniform layer; the extended Wilheit (1978) model assumes a multi-layered vegetation canopy. The energy balance of each layer is determined using simplified radiative transfer theory. This model requires as input profiles of both the dielectric and temperature within the vegetation. Currently, there is only limited information on the dielectric permittivity of either the vegetation matter (El-Rayes and Ulaby, 1987; Ulaby and Jedlicka, 1984; Chuah *et al.*, 1997; Colpitts and Coleman, 1997; Franchois *et al.*, 1998; Ulaby *et al.*, 1986) or the canopy as a whole (Ulaby *et al.*, 1986; Ulaby and Jedlicka, 1984; Brunfeldt and Ulaby, 1984; Schmugge and Jackson, 1992). In order to model the dielectric properties of a vegetation canopy, two separate mixing effects must be recognized. The first is the mixing between constituents of the vegetation, which determines the dielectric of the vegetation matter itself. One plausible approach to modelling the dielectric of the vegetation matter is analogous to a linear version of the Dobson *et al.* (1985) mixing model for soils, i.e., to assume:

$$\varepsilon_v = \varepsilon_{dry} V_{dry} + \varepsilon_{fw} V_{fw} + \varepsilon_{bw} V_{bw} \tag{9}$$

where ε_v is the dielectric permittivity for leaf material as a whole, ε_{dry}, ε_{fw}, and ε_{bw} are the dielectric permittivities, and V_{dry}, V_{fw}, and V_{bw} are the volume fractions of dry matter, free water and bound water, respectively. It is assumed that ε_{dry}, ε_{fw}, and ε_{bw} are independent of the vegetation water content. The second mixing model determines the dielectric permittivity of the mixture of vegetation matter and air that make up the vegetation canopy (ε_{can}). If non-linear mixing is assumed:

$$\varepsilon_{can}^{\alpha} = \varepsilon_v^{\alpha} V_V + \varepsilon_{air}^{\alpha}(1 - V_V)$$

where ε_{air} is the dielectric permittivity of air; V_v is the fractional volume of vegetation elements per unit volume canopy; and α is a so-called "shape factor". [Note: in the case of the Dobson *et al.* (1985) mixing model for soils, $\alpha = 0.65$.] Schmugge and Jackson (1992) suggested that the refractive model [$\alpha = 0.5$] provides a better representation of the dielectric properties of the canopy than a linear model [$\alpha = 1$]. Lee *et al.* (2002a) optimized time series of modelled brightness temperatures against measurements from the field experiment evaluated by Burke *et al.* (1998) and retrieved values ranging between 1.1 and 2.2.

Equations (9) and (10) together calculate the dielectric constant for the canopy as a whole and, in the extended Wilheit (1978) model, this amount of dielectric is then distributed vertically among the plane parallel layers above the soil. The heights of the top and bottom of the canopy are specified, but gradual changes in dielectric permittivity are simulated around these levels by introducing broadening that follows a Gaussian distribution with specified standard deviations. This broadening reflects the natural variability between the individual plants that make up the canopy, and its presence is also critical to the reliable operation of this coherent emission model because avoiding sharp transitions in dielectric at canopy boundaries suppresses internal reflections in the canopy and associated interference patterns in the microwave emission (Lee et al., 2002a).

Figures 4.1c and d show the time series of modelled and measured brightness temperatures for the same time periods as in Figure 4.1a and b but in the presence of a soy bean canopy. During drying period 2 (Figure 4.1c), the fresh weight of the canopy was 3.45 kg m^{-2} and the height was 0.89 m, while during drying period 3 (Figure 4.1d), the fresh weight of the canopy was 4.41 kg m^{-2} and the height was 1.15 m. The effect of vegetation on the microwave emission from the soil is included using the extended Wilheit (1978) model (Lee et al., 2002a). As in the bare soil examples, the soil near-surface water contents range from 40% to 10% for drying period 2 (Figure 4.1c), and span a similar range for drying period 3 (Figure 4.1d). However, this is not as readily apparent in the presence of vegetation because the vegetation perturbs the emission from the soil.

4.2.3 Potential Near-Future L-Band Missions

The European Space Agency (ESA) has approved the Soil Moisture Ocean Salinity (SMOS) mission (SMOS homepage) with a proposed launch date between 2005 and 2007. The SMOS mission will be based on a dual polarization, L-band radiometer with an innovative aperture synthesis concept (a two-dimensional interferometer) that can achieve an on-the-ground resolution of around 50 km and provide global measurements of microwave brightness temperature at a range of different angles. The proposed SMOS retrieval algorithm is based on a non-coherent model of emission from soil and vegetation using the simple model discussed in section 4.2.2 and will simultaneously retrieve the soil moisture and the vegetation optical depth by exploiting the range of available look-angles (Kerr et al., 2001; Wigneron et al., 2000). Wigneron et al. (2000) used simulated SMOS observations (created by adding random and systematic errors to the model proposed for the SMOS retrieval algorithm) and showed that the retrieval lost accuracy as the number of available independent measures of microwave brightness temperatures at different angles decreased.

NASA has selected the HYDROspheric States mission (HYDROS) as a reserve mission. It will consist of an L-band passive microwave radiometer with a resolution of around 40 km and an active microwave radar with a resolution of up to 3 km. Both instruments will measure at multi-polarizations but constant look-angle.

4.3 Discussion

The use of coupled land surface and microwave emission models to explore potential solutions to three of the limitations of L-band passive microwave remote sensing is discussed.

4.3.1 Accounting for Effects of Vegetation in Retrieval Algorithms

Retrieval algorithms typically use the simple optical depth model to account for the effects of vegetation. If the only brightness temperature measurements available are at a single polarization and look-angle, such as those measured during the Southern Great Plains 1997 (SGP97) field experiment, the algorithms require ancillary information to estimate the optical depth. One commonly used method is to define the vegetation water content and opacity coefficient for each class of vegetation within a land cover classification, and

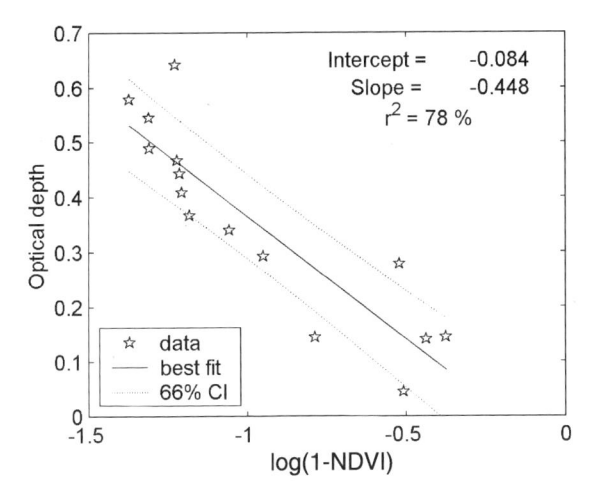

Figure 4.2 Relationship between derived optical depth and Landsat-TM measured NDVI for 15 of the sites intensively monitored during SGP97 (redrawn from Burke et al., 2001a). The moral right of the authors to be identified as the authors of such figure is asserted.

to estimate the optical depth using equation (8), see, for example, Jackson et al. (1999). An alternative method is to use vegetation indices derived from visible wavelength satellite data. These two methods for deriving the optical depth can both potentially introduce significant errors into the retrieved soil moisture. For example, Burke et al. (2001a) used the Normalized Difference Vegetation Index (NDVI) to derive optical depths for selected sites intensively monitored during SGP97. They assumed that, at specific measurement sites, the only unknown in the relationship between the measured soil moisture and measured microwave brightness temperature was the optical depth. They then related the optical depth derived in this way to the local NDVI estimated by Landsat-TM (Figure 4.2). A linear function was fitted to this relationship, with 78% of the variation in the derived optical depth explained by the variation in NDVI. Figure 4.2 also shows the 66% confidence intervals of the fitted relationship which correspond to one standard deviation of the optical depth, or approximately ± 0.08. This empirical relationship was found to be dependent on the resolution of the microwave sensor used to measure brightness temperature (Burke et al., 2001a).

The SMOS mission provides an opportunity to retrieve estimates of the optical depth as part of the retrieval algorithm because it will measure brightness temperatures at multiple look-angles and both polarizations over similar areas at similar times. Therefore, there is enough information for the proposed retrieval algorithm to retrieve both the near-surface soil water content and the optical depth. The proposed retrieval algorithm assumes that the opacity coefficient of the vegetation (equation 8) does not depend on either the polarization of the radiation or the look-angle of the sensor. There is, however, some evidence that it depends on both of these radiometer characteristics (van de Griend and Owe, 1996; Lee et al., 2002b). Figure 4.3 shows the look-angle and polarization dependence of the mean opacity coefficient retrieved (using the method described by Lee et al. (2002b)) from the time series of brightness temperatures given in Figures 4.1c and d, this time series having been calculated using the arguably more realistic extended Wilheit (1978) model. Values

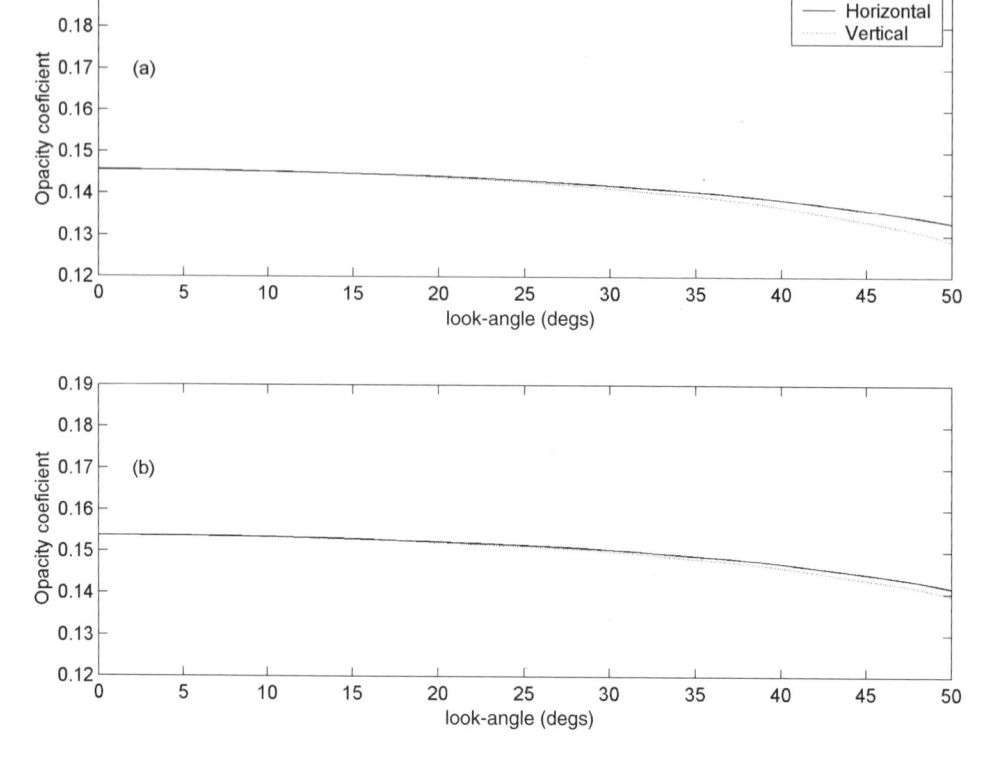

Figure 4.3 *Opacity coefficient derived for the soyabean canopies in (a) Figure 4.1c and (b) Figure 4.1d as a function of look angle and polarization*

found at look-angle 10° (0.36 for drying period 2, and 0.43 for drying period 3) fall within the range of those found by Burke *et al.* (1999) using the simple model for the same dataset, i.e., 0.31–0.39 for drying period 2, and 0.44–0.52 for drying period 3. They are also similar to those found by Burke *et al.* (1999) using the discrete model of Wigneron *et al.* (1993), i.e., 0.36 and 0.49. However, Figure 4.3 also shows that the opacity coefficient is a distinct function of both look angle and polarization. Moreover, the opacity coefficient is greater for drying period 3 when the canopy is denser, implying that the opacity coefficient might also be a function of vegetation water content, as suggested by Le Vine and Karam (1996) and Wigneron *et al.* (1996; 2000). Wigneron (2002, personal communication) also found a dependence of the optical depth on look angle using the discrete model of Ferrazzoli *et al.* (2000).

To quantify the errors introduced by assumptions about the vegetation made in soil moisture retrieval algorithms, field experiments are required that focus on the dependence of the opacity coefficient on look-angle and polarization, canopy size and type and ambient temperature. Quantifying these dependences will provide useful information relevant to a potential soil moisture product, in particular, on its reliability when used in data assimilation systems. Burke *et al.* (2003) quantified the errors in the distributed optical depth estimated using a distributed measure of NDVI and the relationship shown in Figure 4.2, translated these into errors in the retrieved soil moisture, and used this information to calculate an 'Assimilation Value Index' which defines the worth of assimilating the soil moisture information retrieved from microwave brightness temperature data. Burke *et al.* (2003) showed that, for their example, errors in the estimated optical depth resulted in the microwave brightness temperature data having little use over about 50% of the area modelled.

4.3.2 Extending Estimates of Soil Moisture Deeper Within the Profile

Only soil moisture in the top (~5 cm) of the soil are amenable to L-band remote sensing, but studies suggest (e.g. Houser *et al.*, 1998; Walker *et al.*, 2001) that merely modifying surface soil moisture during data assimilation into a land surface model does not provide a sufficiently strong updating of the deeper soil moisture profile. A method to propagate the surface layer information deeper into the profile is therefore needed.

The simplest approach is to use statistical methods (Kostov and Jackson, 1993). One example is that proposed by Burke *et al.* (2001b), who suggest that, in practice, there always exists a modelled relationship between near-surface soil water content and deeper soil water content for each different patch of vegetation and for each grid cell in a particular LDAS (2-D array of land surface models). This relationship can be estimated from historic simulations but is clearly model-dependent. The main concern when using this method to estimate the deeper soil water content is that the LDAS modelled soil profile, and hence the derived relationship, may not be realistic, but this issue is not new as it is already present whenever the LDAS approach is used. Figure 4.4a shows the relationship between surface

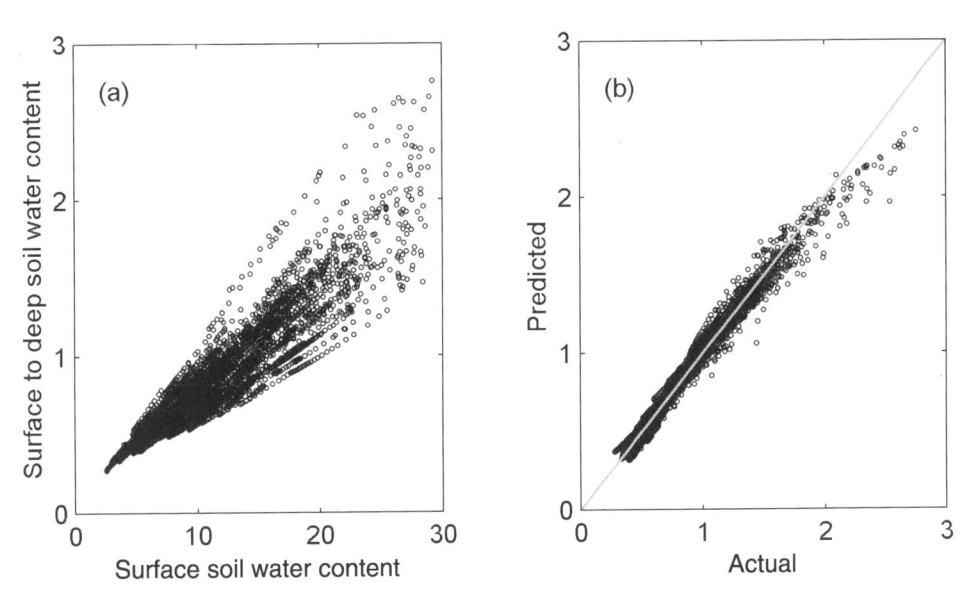

Figure 4.4 (a) Relationship between near-surface water content and the ratio of near-surface to deep water content; (b) agreement between predicted and actual ratio of near surface to deep water content (redrawn from Burke et al., 2002a). © 2003 IEEE, reproduced with permission

soil moisture (0–5 cm) and deeper soil moisture (5–120 cm) predicted using MICRO-SWEAT derived over one year for a typical crop growing in sandy soil using hourly driving data from the SGP97 study area. There is a significant, but noisy, relationship that is structured around several separate linear correlations, which are themselves related to the water content in the soil profile immediately after the most recent precipitation event. (Note: at the time of precipitation the modelled relationship between surface and deep soil moisture depends strongly on the amount of precipitation and modelled runoff, but once precipitation has ceased, the modelled relationship between the surface and deep soil moisture changes mainly as a result of the surface soil drying.) Figure 4.4b shows the good agreement between the MICRO-SWEAT modelled values of the ratio of near-surface to deep-soil moisture and the values calculated using a simple derived relationship that acknowledges the effect of recent precipitation.

Other methods of extending estimates of surface soil moisture deep in the soil profile generally involve assimilation techniques, such as direct insertion of the surface soil moisture (Walker *et al.*, 2001) or use of the extended Kalman filter (Walker *et al.*, 2001; Walker and Houser, 2001; Hoeben and Troch, 2000) or variational assimilation (Reichle *et al.*, 2001). Walker *et al.* (2001) compared the direct insertion and Kalman filter assimilation methods using synthetic data and demonstrated that using the Kalman filtering is superior, with correction of the soil moisture profile being achieved in 12 hours as compared to 8 days or more with direct insertion. Variational data assimilation is computationally more effective than using the Kalman filter, but variational assimilation requires an adjoint model that is numerically well behaved and there are currently no adjoint models available for the commonly used land surface models.

Assimilation procedures either introduce retrieved soil moisture into the land surface model (Walker *et al.*, 2001; Montaldo *et al.*, 2001; Hoeben and Troch, 2000; Wigneron *et al.*, 1999; Li and Islam; 2002, Calvet and Noilhan, 2000), or the measured microwave brightness temperatures is itself assimilated through the use of a coupled land surface and microwave emission model (Crosson *et al.*, 2002; Galantowicz *et al.*, 1999). They generally build on information already present in a land surface model and, in general, result in improved profile soil moisture estimates by the land surface model, regardless of the assimilation methods used. It has been demonstrated (Walker and Houser, 2001) that through the assimilation of near-surface soil moisture observations, errors in forecast soil moisture profiles that result from poor initialization may be removed, and the resulting predictions of runoff and evapotranspiration by a hydrological model improved.

4.3.3 Subpixel Heterogeneity

In the near future, any passive microwave satellite mission is likely to measure with resolution between approximately 30 and 60 km. At this resolution, the land surface is strongly heterogeneous and the impact of this heterogeneity on the accuracy of the retrieved soil moisture is a significant issue.

In the case of a heterogeneous bare soil, both Burke and Simmonds (2002) and Galantowicz *et al.* (2000) demonstrated that errors in the retrieved soil moisture are negligible. For a pixel with heterogeneous vegetation, Drusch *et al.* (1999), Liou *et al.* (1998), Crow *et al.* (2001), Crow and Wood (2002), and Burke and Simmonds (2002) suggested that errors in the retrieved soil moisture are generally less than 0.03 cm^3 cm^{-3}. This is a comparatively small error, but potentially significant, particularly when there are already other sources of errors in the retrieval algorithm.

Crow and Wood (2002) demonstrated that area-average soil moisture, when used in a land surface model, can lead to potentially large errors in surface energy fluxes of up to 50 W m^{-2}. One method of incorporating subgrid scale heterogeneity into land surface models is to represent such heterogeneity using a probability density function (PDF) (e.g. Famiglietti and Wood, 1995). However, if the model grid scale corresponds to the scale of the satellite footprint for which soil moisture information is available, estimating subgrid statistics is not a trivial task. Crow and Wood (2002) discuss a possible soil moisture downscaling procedure based on an assumption of spatial scaling (i.e., a power-law relationship between statistical moments and scale), and demonstrate that the downscaled soil moisture derived from coarse-resolution soil moisture imagery can improve prediction of grid-scale surface energy fluxes. Recently, Kim and Barros (2002) developed a modified fractal interpolation technique to downscale soil moisture estimates, based on an analysis of the effects of topography, soil properties, and vegetation on the measured distribution of soil moisture. Reichle *et al.* (2001) used a four-dimensional assimilation algorithm to show that soil moisture can be satisfactorily estimated at scales finer than the resolution of the microwave brightness temperature image. Their downscaling experiment suggests that brightness temperature images with a resolution of tens of kilometres can yield soil moisture profile estimates on a scale of a few kilometres, provided that micrometeorological, soil texture, and land cover inputs are available at the finer scale.

The high-resolution active and low-resolution passive combination of the HYDROS mission may provide an opportunity to develop a novel method of downscaling. Bindlish and Barros (2002) have demonstrated the potential of this using an L-band satellite-based radar and an L-band aircraft-based radiometer. They successfully demonstrated downscaling of soil moisture estimates from 200 m to 40 m.

The SMOS mission may also provide enough information to allow downscaled estimates of soil moisture. Burke *et al.* (2002a) explored this potential by using MICRO-SWEAT in a year-long simulation to define the patch-specific soil moisture, optical depth, and the synthetic, pixel-average microwave brightness temperatures for the range of angles that will be provided by SMOS. The microwave emission component of MICRO-SWEAT was used as the basis of an exploratory SMOS retrieval algorithm in which the RMSE between the synthetic and modelled pixel-average microwave brightness temperatures was minimized by optimizing the soil moisture and optical depth in different patches of vegetation. The optimization was made using the Shuffled Complex Evolution optimization procedure (Duan *et al.*, 1993). An example five-patch pixel comprising equal areas of water, bare soil, short grass, crop, and shrub was studied, assuming a uniform sandy loam soil (75% sand, 5% clay). Simulation of the growth of the vegetation through the year was also included. It is assumed that the portion of the pixel occupied by each land-cover type was known and that each vegetation type was homogeneous. Figure 4.5 shows the prescribed and retrieved near-surface soil moisture and vegetation optical depth for the four land patches (it was assumed that the contribution of water to the area-average brightness temperature is known). The last plot in Figure 4.5 is the area weighted mean soil moisture (excluding the water patch) for the pixel. The SCE-UA algorithm, which was run ten times, gave each time, ten possible values of soil moisture, optical depth and ten possible values of the errors in brightness temperature. The ten possible solutions are plotted as individual

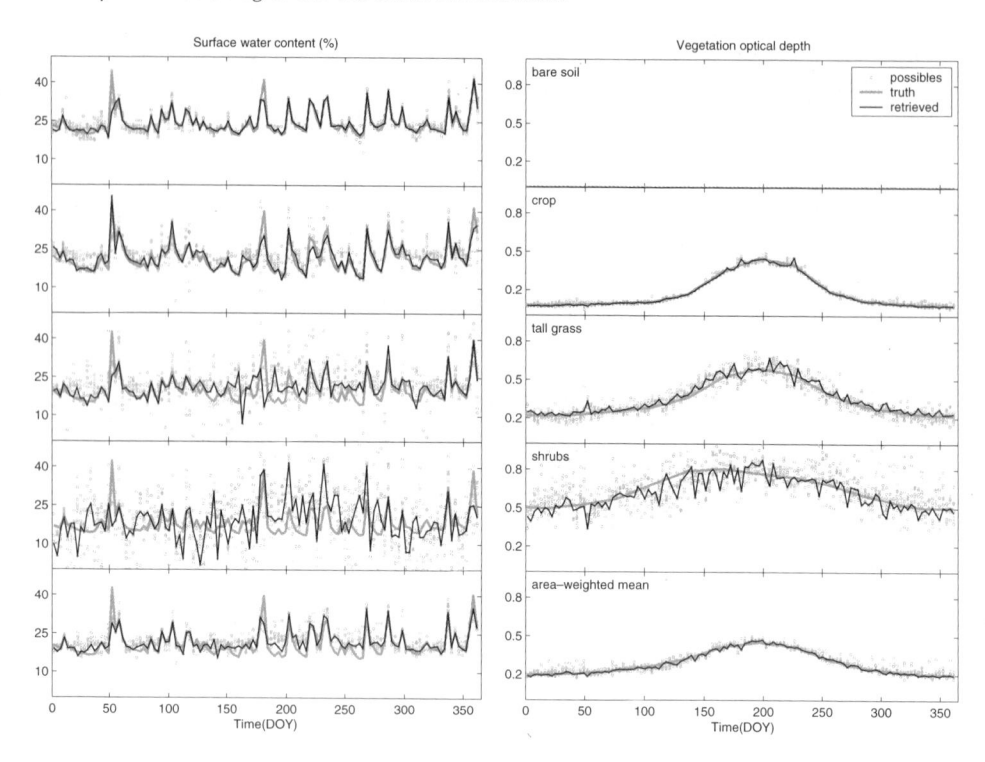

Figure 4.5 Patch-specific soil water content and vegetation optical depth for a synthetic 5 patch pixel with 20% bare soil; 20% water; 20% grass; 20% crop and 20% shrub. Also shown are the weighted averages. © 2002 IEEE, reproduced with permission

symbols, with the black line linking those with the smallest error in brightness temperature. A wide range of possible solutions indicates there is little information in the microwave brightness temperatures about the value of the soil moisture/optical depth at a particular time, whereas a narrow range indicates there is much more information in the brightness temperature observations. The results suggest that, at higher optical depths, the microwave brightness temperature is less sensitive to the soil moisture than at lower optical depths as expected. When the preferred values are compared with the true values, i.e., the values for which the original area-average brightness temperatures were calculated, it is gratifying to see a significant level of agreement between the two. If these results were used for data assimilation, potential errors could be evaluated and included in the assimilation process. This study assumes that the optical depth is independent of both look-angle and polarization. However, if the relationship between optical depth and look-angle/polarization could be parameterized, a similar, but more realistic, study could be made.

4.4 Conclusion

Passive microwave remote sensing is an important technique for the retrieval of soil moisture with application at a wide range of scales and in a variety of disciplines. However, at present, there are no satellite-based sensors optimally configured for the retrieval of soil moisture and there are still issues associated with the technology and retrieval methods to be resolved, some of which are discussed in this chapter. In summary, the primary conclusions of this chapter are as follows:

1. The impact of vegetation on the microwave emission from soil is significant and small errors in its estimation can lead to significant errors in the retrieved soil moisture. Three models accounting for the effect of vegetation were discussed. The most commonly used method is a simple, one-parameter, optical-depth model, and several methods are suggested for obtaining the required value of optical depth (a) using visible remote sensing; (b) by comparison with more complex models; and (c) simultaneous retrievals of both soil moisture and vegetation optical depth using multi-angle dual polarization brightness temperatures. A further issue with the optical depth approach emerges, namely that there is now evidence that the optical depth of vegetation is a function of, amongst other things, the look-angle and polarization of the radiometer.
2. Although passive microwave remote sensing can provide estimates of near-surface soil moisture, by itself, this is not enough to significantly improve the performance of land surface models. Additional information about the whole soil profile is needed. Simple statistically based methods and more complex assimilation methods have been proposed to address this issue.
3. Land surface heterogeneity will impact the estimate of area-average soil moisture. Recently, downscaling methods have been developed using either statistical techniques or land surface modelling techniques with higher resolution information on vegetation, soils and topography. Both the HYDROS and SMOS missions will provide simultaneous observations that should allow some exploration of downscaling techniques.

Acknowledgements

Primary support for Dr. Eleanor Burke, while preparing this chapter, came from NASA grant No. NAG5-8214.

References

Beljaars, A.C.M., Viterbo, P., Miller, M.J. and Betts, A.K., 1996, Anomalous rainfall over the U.S. during July 1993: sensitivity to land surface parameterization, *Monthly Weather Review*, **124**, 364–383.

Betts, A.K., Ball, J.H., Beljaars, A.C.M., Miller, M.J. and Viterbo, P., 1996, The land–surface–atmosphere interaction: a review based on observational and global modeling perspectives, *Journal of Geophysical Research*, **101**, 7209–7225.

Bindlish, R. and Barros, A.P., 2002, Subpixel variability of remotely sensed soil moisture: an intercomparison study of SAR and ESTAR, *IEEE Transactions on Geoscience and Remote Sensing*, **40**, 326–337.

Bonan, G.B., 1996, A land surface model [LSM version 1.0] for ecological, hydrological, and atmospheric studies: technical description and user's guide, NCAR Technical Note, NCAR/TN-417+STR, Boulder, CO.

Brunfeldt, D.R. and Ulaby, F.T., 1984, Measured microwave emission and scattering in vegetation canopies, *IEEE Transactions on Geoscience and Remote Sensing*, **22**, 520–524.

Burke, E.J. and Simmonds, L.P., 2003, Effects of sub-pixel heterogeneity on the retrieval of soil moisture from passive microwave radiometry, *International Journal of Remote Sensing*, **24**, 2085–2104.

Burke, E.J., Bastidas, L.A. and Shuttleworth, W.J., 2002a, Multi-patch retrieval for the SMOS mission, *IEEE Transactions on Geoscience and Remote Sensing*, **40**, 1114–1120.

Burke, W.J., Schmugge, T.J. and Paris, J.F., 1979, Comparison of 2.8 and 21 cm microwave radiometer observations over soils with emission model calculations, *Journal of Geophysical Research*, **84**, 287–294.

Burke, E.J., Shuttleworth, W.J. and French, A.N. 2001a, Vegetation indices for soil moisture retrievals from passive microwave radiometry, *Hydrology and Earth Systems Sciences*, **5**, 671–677.

Burke, E.J., Shuttleworth, W.J. and Harlow, R.C., 2003, Modeling microwave brightness temperatures measured during SGP97 using MICRO-SWEAT, *Journal of Hydrometeorology*, **2**, 460–472.

Burke, E.J., Wigneron, J.-P. and Gurney, R.J., 1999, The comparison of two models that determine the effects of a vegetation canopy on passive microwave emission, *Hydrology and Earth System Sciences*, **3**, 439–444.

Burke, E.J., Gurney, R.J., Simmonds, L.P. and Jackson, T.J., 1997, Calibrating a soil water and energy budget model with remotely sensed data to obtain quantitative information about the soil, *Water Resources Research*, **33**, 1689–1697.

Burke, E.J., Gurney, R.J., Simmonds, L.P. and O'Neill, P.E., 1998, Using a modeling approach to predict soil hydraulic properties from passive microwave measurements, *IEEE Transactions on Geoscience and Remote Sensing*, **36**, 454–462.

Burke, E.J., Shuttleworth, W.J., Lee, K. and Bastidas, L.A., 2001b, Using area-average remotely sensed surface soil moisture in multi-patch Land Data Assimilation Systems, *IEEE Transactions on Geoscience and Remote Sensing*, **39**, 2091–2100.

Calvet, J.-C. and Noilhan, J., 2000, From near-surface to root-zone soil moisture using year-round data, *Journal of Hydrometeorology*, **1**, 393–411.

Chuah, H.T., Kam, S.W. and Chye, Y.H., 1997, Microwave dielectric properties of rubber and oil palm leaf samples: measurement and modelling, *International Journal of Remote Sensing*, **18**, 2623–2639.

Colpitts, B.G., and Coleman, W.K., 1997, Complex permittivity of the potato leaf during imposed drought stress, *IEEE Transactions on Geoscience and Remote Sensing*, **35**, 1059–1064.

Crosson, W.L., Laymon, C.A., Inguva, R. and Schamschula, M.P., 2002, Assimilating remote sensing data in a surface flux-soil moisture model, *Hydrological Processes*, **16**, 1645–1662.

Crow, W.T., and Wood, E.F., 2002, The value of coarse-scale soil moisture observations for regional surface energy balance modeling, *Journal of Hydrometeorology*, **3**, 467–482.

Crow, W.T., Drusch, M. and Wood, E.F., 2001, An observation system simulation experiment for the impact of land surface heterogeneity on AMSR-E soil moisture retrieval, *IEEE Transactions on Geoscience and Remote Sensing*, **39**, 1622–1631.

Daamen, C.C. and Simmonds, L.P., 1996, Measurement of evaporation from bare soil and its estimation using surface-resistance, *Water Resources Research*, **32**, 1393–1402.

Dickinson, R.E., Henderson-Sellers, A., Kennedy, P.J. and Wilson, M.F., 1986, Biosphere-Atmosphere Transfer Scheme [BATS] for the NCAR Community Climate Model, Technical Note, NCAR/TN-275+STR, National Center for Atmospheric Research, Boulder, CO.

Dobson, M.C., Ulaby, F.T., Hallikainen, M.T. and El-Rayes, M.A., 1985, Microwave behavior of wet soil, 2: dielectric mixing models, *IEEE Transactions in Geoscience and Remote Sensing*, **23**, 35–46.

Drusch, M., Wood, E.F. and Simmer, C., 1999, Up-scaling effects in passive microwave remote sensing: ESTAR 1.4 GHz measurements during SGP '97, *Geophysical Research Letters*, **26**, 879–882.

Duan, Q.Y., Gupta, V.K. and Sorooshian, S., 1993, Shuffled complex evolution approach for effective and efficient global minimization, *Journal of Optimization Theory Applications*, **76**, 501–521.

El-Rayes, M.A. and Ulaby, F.T., 1987, Microwave dielectric spectrum of vegetation, 1: experimental observations, *IEEE Transactions on Geoscience and Remote Sensing*, **25**, 541–549.

England, A.W., 1975, Thermal microwave emission from a scattering layer, *Journal of Geophysical Research*, **32**, 4484–4496.
England, A.W., 1976, Relative influence upon microwave emissivity of fine-scale stratigraphy, *Pure and Applied Geophysics*, **114**, 287–299.
Famiglietti, J.F. and Wood, E.F., 1995, Effects of spatial variability and scale on areally averaged evapotranspiration, *Water Resources Research*, **31**, 699–712.
Ferrazzoli, P., Wigneron, J.-P., Guerriero, L. and Chanzy, A., 2000, Multifrequency emission of wheat: modeling and applications, *IEEE Transactions on Geoscience and Remote Sensing*, **38**, 2598–2607.
Franchois, A., Pineiro, Y. and Lang, R.H., 1998, Microwave permittivity measurements of two conifers, *IEEE Transactions on Geoscience and Remote Sensing*, **36**, 1384–1395.
Galantowicz, J.F., Entekhabi, D. and Njoku, E.G., 1999, Tests of sequential data assimilation for retrieving profile soil moisture and temperature from observed L-band radiobrightness, *IEEE Transactions in Geoscience and Remote Sensing*, **37**, 1860–1870.
Galantowicz, J.F., Entekhabi, D. and Njoku, E.G., 2000, Estimation of soil-type heterogeneity effects in the retrieval of soil moisture from radiobrightness, *IEEE Transactions in Geoscience and Remote Sensing*, **38**, 312–316.
Hoeben, R. and Troch, P.A., 2000, Assimilation of active microwave observation data for soil moisture profile estimation, *Water Resources Research*, **36**, 2805–2819.
Houser, P.R., Shuttleworth, W.J., Famiglietti, J.S., Gupta, H.V., Syed, K.H. and Goodrich, D.C., 1998, Integration of soil moisture remote sensing and hydrological modeling using data assimilation, *Water Resources Research*, **34**, 3405–3420.
HYDROS homepage, http://hydros.gsfc.nasa.gov
Jackson, T.J. and Schmugge, T.J., 1991, Vegetation effects on the microwave emission of soils, *Remote Sensing of the Environment*, **36**, 203–212.
Jackson, T.J., Le Vine, D.M., Hsu, A.Y., Oldak, A., Starks, P.J., Swift, C.T., Isham, J.D. and Haken, M., 1999, Soil moisture mapping at regional scales using microwave radiometry: the Southern Great Plains hydrology experiment, *IEEE Transactions on Geoscience and Remote Sensing*, **37**, 2136–2151.
Judge, J., England, A.W., Crosson, W.L., Laymon, C.A., Hornbuckle, B.K., Boprie, D.L., Kim, E.J. and Liou, Y.A., 1999, A growing season Land Surface Process/Radiobrightness model for wheat stubble in the Southern Great Plains, *IEEE Transactions on Geoscience and Remote Sensing*, **37**, 2152–2158.
Judge, J., Galantowicz, J.F. and England, A.W., 2001, A comparison of ground-based and satellite-borne microwave radiometric observations in the Great Plains, *IEEE Transactions in Geoscience and Remote Sensing*, **9**, 1686–1696.
Kerr, Y.H. and Wigneron, J.-P., 1994, Vegetation models and observations: a review, in B.J.Choudhury, Y.H. Kerr, E.G. Njoku and P. Pampaloni (eds), *ESA/NASA International Workshop on Passive Microwave Remote Sensing Research Related to Land-Atmosphere Interactions*, 317–344.
Kerr, Y.H., Waldteufel, P., Wigneron, J.-P., Martinuzzi, J.-M., Font, J. and Berger, M., 2001, Soil moisture retrieval from space: the Soil Moisture and Ocean Salinity (SMOS), *IEEE Transactions in Geoscience and Remote Sensing*, **39**, 1729–1735.
Kim, G. and Barros, A.P., 2002, Space-time characterization of soil moisture from passive microwave remotely sensed imagery and ancillary data, *Remote Sensing of the Environment*, **81**, 393–403.
Koster, R.D., Suarez, M.J. and Heiser, M., 2000, Variance and predictability of precipitation at seasonal-to-interannual timescales, *Journal of Hydrometeorology*, **1**, 26–46.
Kostov, K.G. and Jackson, T.J., 1993, Estimating profile soil moisture from surface layer measurements – a review, in H.N. Nasr (ed.), *Ground Sensing*, SPIE Proceedings, **1941**, 125–136.
LDAS Homepage, http://ldas.gsfc.nasa.gov
Le Vine, D.M. and Karam, M.A., 1996, Dependence of attenuation in a vegetation canopy on frequency and plant water content, *IEEE Transactions in Geoscience and Remote Sensing*, **34**, 1090–1096.
Lee, K.H., Burke, E.J., Shuttleworth, W.J. and Harlow, R.C., 2002b, Influence of vegetation on SMOS mission retrievals, *Hydrology and Earth Systems Sciences*, **6**, 153–166.

Lee, K.H., Harlow, R.C., Burke, E.J. and Shuttleworth, W.J. 2002a. A plane stratified emission model for use in the prediction of vegetation effects on passive microwave radiometry, *Hydrology and Earth Systems Sciences*, **6**, 139–151.

Li, J.K. and Isham, S., 2002, Estimation of root zone soil moisture and surface fluxes partitioning using near surface soil moisture measurements, *Journal of Hydrology*, **259**, 1–14.

Liou, Y.A. and England, A.W., 1998, A land surface process radiobrightness model with coupled heat and moisture transport in soils, *IEEE Transactions in Geoscience and Remote Sensing*, **36**, 273–286.

Liou, Y.A., Kim, E.J. and England, A.W., 1998, Radiobrightness of prairie soil and grassland during dry-down simulations, *Radio Science*, **33**, 259–265.

Liu, Y.Q. and Avissar, R., 1999, A study of persistence in the land-atmosphere system with a fourth-order analytical model, *Journal of Climate*, **12**, 2154–2168.

Mitchell, K., Marshall, C., Lohmann, D., Ek, M., Lin, Y., Grunmann, P., Houser, P., Wood, E., Schaake, J., Lettenmaier, D., Tarpley, D., Higgins, W., Pinker, R., Robock, A., Cosgrove, B., Entin, J. and Duan, Q., 2000, The collaborative GCIP land data assimilation (LDAS) project and supportive NCEP uncoupled land-surface modeling initiatives, *Proceedings of the 15th American Meteorological Society Conference on Hydrology*.

Mo, T. and Schmugge, T.J., 1987, A parameterization of the effect of surface roughness on microwave emission, *IEEE Transactions on Geoscience and Remote Sensing*, **25**, 481–486.

Montaldo, N., Albertson, J.D., Mancini, M. and Kiely, G., 2001, Robust simulation of root zone soil moisture with assimilation of surface soil moisture data, *Water Resources Research*, **37**, 2889–2900.

Njoku, E.G. and Kong, J.A., 1977, Theory for passive microwave sensing of near surface soil moisture, *Journal of Geophysical Research*, **82**, 3108–3118.

Reichle, R., Entekhabi, D. and McLaughlin, D.B., 2001, Downscaling of radiobrightness measurements for soil moisture estimation: a four-dimensional variational data assimilation approach, *Water Resources Research*, **37**, 2353–2364.

Schmugge, T.J., and Choudhury, B.J., 1981, A comparison of radiative transfer models for predicting the microwave emission from soils, *Radio Science*, **16**, 927–938.

Schmugge, T.J. and Jackson, T.J., 1992, A dielectric model of the vegetation effects on the microwave emission from soils, *IEEE Transactions on Geoscience and Remote Sensing*, **30**, 757–760.

Schmugge, T.J., Jackson, T.J., O'Neill, P.E. and Parlange, M.B., 1998, Observations of coherent emissions from soils, *Radio Science*, **33**, 267–272.

SGP97 Homepage, http://daac.gsfc.nasa.gov/CAMPAIGN_DOCS/SGP97/sgp97.html

SMOS Homepage, http://www-sv.cict.fr/cesbio/smos

Ulaby, F.T., and Jedlicka, R.P., 1984, Microwave dielectric-properties of plant materials, *IEEE Transactions on Geoscience and Remote Sensing*, **22**, 406–415.

Ulaby, F.T., Moore, R.K. and Fung, A.K., 1986, *Microwave Remote Sensing: Active and Passive* (Reading, MA: Addison-Wesley).

van de Griend, A.A. and Owe, M., 1996, Measurement and behavior of dual-polarization vegetation optical depth and single scattering albedo at 1.4 and 5 GHz microwave frequencies, *IEEE Transactions on Geoscience and Remote Sensing*, **34**, 957–965.

Walker, J.P. and Houser, P.R., 2001, A methodology for initializing soil moisture in a global climate model: assimilation of near-surface soil moisture observations, *Journal of Geophysical Research*, **106**, 11761–11774.

Walker, J.P., Willgoose, G.R. and Kalma, J.D., 2001, One-dimensional soil moisture profile retrieval by assimilation of near-surface observations: a comparison of retrieval algorithms, *Advances in Water Resources*, **24**, 631–650.

Wigneron, J.-P., Calvet, J.-C. and Kerr, Y.H., 1996, Monitoring water interception by crop fields from passive microwave observations, *Agricultural Forest Meteorology*, **80**, 177–194.

Wigneron, J.-P., Olioso, A., Calvet, J.-C. and Bertuzzi, P., 1999, Estimating root zone soil moisture from surface soil moisture data and soil-vegetation-atmosphere transfer modeling, *Water Resources Research*, **35**, 3735–3745.

Wigneron, J.-P., Waldteufel, P., Chanzy, A.C. and Calvet, J.-C., 2000., Two-dimensional microwave interferometer retrieval capabilities over land surfaces (SMOS mission), *Remote Sensing of the Environment*, **73**, 270–282.

Wigneron, J.-P., Calvert, J.-C., Kerr, Y.H., Chanzy A. and Lopes, A., 1993, Microwave emission of vegetation, sensitivity to leaf characteristics, *IEEE Transactions in Geoscience and Remote Sensing*, **31**, 716–726.

Wilheit, T.T., 1978, Radiative transfer in a plane stratified dielectric, *IEEE Transactions on Geoscience and Electronics*, **16**, 138–143.

Zeng, X.B., Shaikh, M., Dai, Y.J., Dickinson, R.E. and Myneni, R., 2002, Coupling of the common land model to the NCAR community climate model, *Journal of Climate*, **15**, 1832–1854.

5
Flood Inundation Modelling Using LiDAR and SAR Data

Paul D. Bates, M.S. Horritt, D. Cobby and D. Mason

5.1 Introduction

Flood inundation is a major hazard in the UK and worldwide. Its prediction and prevention require considerable investment, even aside from the socio-economic consequences of severe flooding episodes. Better flood extent prediction is relevant to a significant percentage of the global population, as well as raising fundamental scientific issues and challenges relating to remote sensing, distributed environmental modelling, risk analysis and uncertainty. In recent years, remote sensing of floodplain environments has increasingly become an operational tool that may begin to resolve some fundamental problems in flood conveyance estimation. In a number of areas, in particular topography and validation data, we are moving rapidly from a data-poor to a data-rich and spatially complex modelling environment with attendant possibilities for model testing and development. The availability of rich and relatively accurate spatial data sources in digital format may lead to significant changes in scientific and engineering practice in flood conveyance estimation over the next 5–10 years and it is the purpose of this chapter to identify potential areas in which integration of hydraulic models and remotely sensed spatial data may be of benefit.

Traditional methods of hydraulic investigation for real river reaches have focussed largely on model validation against bulk flow measurements such as stage and discharge measured at only a single point in the system, typically the reach outflow (e.g. Bates *et al.*, 1998a). The main calibration parameter for these applications is boundary friction, and it is relatively easy to match the available validation data with a spatially lumped parameter set that has been calibrated to maximize some objective measure of fit between the observed and predicted outflow hydrographs. Whilst hydraulic resistance does have a sound

---

Spatial Modelling of the Terrestrial Environment. Edited by R. Kelly, N. Drake, S. Barr.
© 2004 John Wiley & Sons, Ltd. ISBN: 0-470-84348-9.

physical basis, it is extremely difficult to measure in the field, is highly variable in space and time and is rather sensitive as a model parameter. In addition, how frictional effects vary with scale in terms of 'effective' grid scale parameters (Beven, 1989) is also poorly understood. There are usually, therefore, a wide range of parameter values for any given problem that could be considered 'physically acceptable' and the calibration space for the model is thus poorly constrained. Typical applications have tended to use one friction value for the channel, which should control the point at which bankful discharge is exceeded and water moves onto adjacent floodplain sections, and one for the floodplain, which should control the floodplain flow velocity and depth. In practice, however, even this simple spatial disaggregation is probably not warranted by the available data and one could likely obtain an equally good fit to an external bulk flow hydrograph with a single calibrated value for boundary friction. In terms of discriminatory power, such data tests the ability of a calibrated hydraulic model to replicate flow routing behaviour and very little else. Moreover, as Beven and Binley (1992) point out, it is likely that there will be many available parameter sets that provide an equally good match to the available data at a single location, yet the performance of these parameter sets will differ elsewhere in the distributed model domain and during further events, particularly if these are substantially different to the calibration event. The predictive uncertainty resulting from such equifinal behaviour is a significant problem for hydraulic modellers and it is clear that validation data from available gauging stations cannot discriminate well between different models and different parameterizations. Neither will non-distributed validation data at the model external boundary require complex (or even accurate) topographic data to complete the model specification. In such cases, topographic errors are easily subsumed within the calibration process and it is impossible to discriminate between topographic, boundary condition, parameterization or model conceptual errors. Using such validation we are thus unable to identify sources of error and hence cannot design new schemes or even rigorously inter-compare those hydraulic models which are available. It is not clear whether the instrumentation of reaches with multiple gauges is capable of solving this problem, as it will still be relatively easy to calibrate friction in the sub-reaches between gauges and we merely reduce the scale of the calibration problem rather than change its essential basis. However, this possibility is yet to be adequately tested and is deserving of further research.

The above discussion raises interesting questions concerning the assimilation of spatial data into distributed models, the relative merits of lumped versus distributed parameterization, calibration, validation and uncertainty analysis and how errors propagate through complex non-linear models. However, the approach taken to hydraulic model calibration and validation also has significant practical consequences for estimation of flood envelopes and maps of flood risk. These are now statutory requirements in a number of countries, including the UK, and are based on delimiting the flood extent (ie. a single line on a map) resulting from a particular discharge, typically the 100-year flow. For most reaches such high magnitude flows do not occur within the hydrometric record, and even if they do, the resulting inundation has rarely been mapped in a consistent manner. Flood envelope construction therefore relies on hydraulic modelling calibrated in the manner outlined above. Inundation extent is predicted by a single model run at the design discharge, with a single set of parameter values derived by calibration against a limited number of gauging station records for a small number of events for which hydrometric data are available. These will most likely be of lower magnitude than the design flood and we must assume that the calibration is stationary with respect to event magnitude. The calibrated parameter values are typically independently tested for a limited number of validation events. Whilst such split sample calibration and validation is good practice, the approach ignores calibration equifinality and hence masks predictive uncertainty which may be considerable (Romanowicz et al., 1996; Romanowicz and Beven, 1997; Aronica et al., 1998; Aronica et al., 2002). Use of the term 'validation' to describe this process imparts a level of confidence that is probably not warranted and, as Konikow and Bredehoeft (1992) note, may give a false impression of security to the end-users of such information. A flood envelope is thus more correctly conceived as a fuzzy map where areas of the floodplain are assigned a probability of being inundated which accounts for all possible errors in the modelling process.

These then are the fundamental problems that increasingly remote sensing is being called upon to solve. Of overwhelming importance is the need for model validation data that goes beyond traditional bulk flow measurements and here synoptic maps of inundation extent derived from satellite and airborne sources are an obvious first solution. With such data in place, two other developments can follow. First, we require methods to better constrain the parameterization of hydraulic resistance in both time and space. Second, to model detailed inundation patterns it is clear that we need topographic data of a commensurate accuracy and scale to the hydraulic flow features we wish to simulate. We also need to examine the scaling behaviour of both topography and boundary friction in a variety of hydraulic models and frame our conclusions in a context which acknowledges the uncertainty that will undoubtedly still be present in our simulations. These developments have the potential, in parallel with more traditional methods of hydraulic investigation, to generate new thinking and new approaches in hydraulic modelling.

We commence the chapter with an evaluation of the information content relevant to flood conveyance that can, or will in the near future, be available from remote sensing (section 5.2). We then explore how such data might be integrated with hydraulic models and discuss preliminary studies that have attempted such work (section 5.3). Based on this review we suggest some immediate and longer-term research needs in this area (section 5.4).

5.2 Development of Spatial Data Fields for Flood Inundation Modelling

5.2.1 Model Validation Data

As we argue above, hydraulic models of reach scale flood inundation are currently only validated using either water levels or discharge from maintained national gauging stations (Gee et al., 1990; Bates et al., 1992; Feldhaus et al., 1992; Bates et al., 1998a). More rarely, individual flood levels are recorded during major events, although the reliability of such records can be questioned and such data have been relatively little used in the validation process. With the notable exception of Robert Sellin's data from the River Blackwater (see Bates et al., 1998a), the spacing of national gauging stations is defined by their flood warning role and these are typically between 10 and 40 kms apart. The network was never designed with hydraulic model validation in mind and the low density means that very few data points are available internal to a reach scale model domain. Whilst such data, particularly stage, are accurate, they are only one-dimensional in time and zero-dimensional in space and do not directly test a model's ability to predict distributed hydraulics. Remote sensing has the potential to supplement these existing point data sources with two other, broad-area, types of data: vector data of inundation extent and flow velocity fields.

Inundation Extent. In the late 1990s inundation data were seen as a potential solution to the problem of hydraulic model validation as, whilst these were 0D in time, their 2D spatial format could provide 'whole reach' data for both distributed calibration and validation of distributed predictions. Inundation extent was also seen as a sensitive test of a hydraulic model, as small errors in predicted water surface elevation would lead to large errors in shoreline position over flat floodplain topography. The sensors available for this task are reviewed by Pearson *et al.* (2001) so only a brief synopsis is provided here. The available sensors are:

- optical imagers on airborne and satellite platforms;
- synthetic aperture radar (SAR) imagers on airborne and satellite platforms;
- digital video cameras mounted on surveillance aircraft.

These are compared in Table 5.1.

Table 5.1 demonstrates that the use of satellite optical platforms for inundation extent monitoring is problematic, apart from for large rivers and non-storm conditions. For example, Bates *et al.* (1997) used a Landsat Thematic Mapper (TM) image of the Missouri River to validate a two-dimensional finite element model of a 60-km reach under normal flow conditions by comparing the extent of islands and mid-channel bars in both the model predictions and satellite data. In parallel work, Smith *et al.* (1997) used suspended sediment concentration to map flow patterns where a turbid tributary entered the relatively clear water of the Missouri main stem and hence were able to qualitatively validate the velocity patterns predicted by the model. Such studies would not have been possible during storm conditions and could not be considered a rigorous test.

Rather better data can be acquired from optical airborne platforms as occasionally such systems can be flown below the cloud base during flood conditions, particularly if the reach in question is some way down the drainage basin and away from major flood generation, and hence rainfall, areas. A limited amount of such data has thus been collected on an *ad hoc* basis. For example, Biggin and Blyth (1996) acquired air photo data co-incident with an overpass of the ERS-1 SAR satellite for a 1-in-5-year flood on the upper Thames, UK. Other regulatory agencies, such as the Environment Agency in the UK, also possess a limited number of air photo datasets of flooding. However, conversion of oblique air photos to synoptic maps of inundation extent is by no means straightforward, and the quality of such data is highly dependent on illumination conditions. As a consequence of these errors, Horritt (2000a) estimated that the flood shoreline that could be derived from the Biggin and Blyth (1996) data was only accurate to ±12.5 m, equivalent to the pixel size of ERS SAR. Nor can acquisition of visual imagery under the storm conditions prevalent during flooding episodes be guaranteed.

Table 5.1 Characteristics of remote sensing systems capable of determining flood inundation extent

Type	Spatial resolution	Frequency	Accuracy	All weather day/night capability?	Survey rate/cost
Satellite optical	~30 m (TM)	16 days (Landsat)	Not yet determined	No	0.2–2 k per image
Airborne optical	~3 m	Responsive mode, mounted in light aircraft	Not yet determined	No	50 km² per hour, £200 per km² (£10 k minimum cost)
Satellite SAR (ERS, RADARSAT)	12.5	1 overpass per 35 days with ERS-1 RADARSAT has steerable antennae so potentially more responsive	±25 m (2 pixel) resolution plus misclassification errors as it may be difficult to discriminate emergent vegetation or wet soil from flooded areas Likely to correctly map 85–90% of true flooded area	Yes	~0.3–3 k per image
Airborne SAR	~0.5–1.0 m	Responsive mode, mounted in large aircraft	Not yet determined, but potentially ±1 m resolution with elimination of misclassification errors	Yes	~150 km² per hour, ~£80 per km² (£35 k minimum cost)
Airborne digital video	~3 m	Responsive mode, mounted in light aircraft	Not yet determined	No	50 km² per hour, £100 per km² (£10 k minimum cost)

The most promising sensors are thus Synthetic Aperture Radars due to their high resolution and all weather, day/night capability and their potential to generate consistent datasets. However, current satellite-borne radars such as ERS-1 and RADARSAT have same track repeat cycles measured in weeks, and despite RADARSAT's steerable antenna capability, the generation of multiple imagery of a flood event is unfeasible with today's satellite missions. Such techniques cannot, therefore, be used to capture dynamic flooding processes. Current satellite SAR systems can give higher frequency coverage (about 7 days) by using data from both ascending and descending tracks. However, multi-temporal data from the same track, either ascending or descending, is best for extent mapping (Veitch, 2001). Further, whilst statistical active contour methods or 'snakes' have been developed to extract inundation from satellite SAR data (Horritt, 1999; Horritt et al., 2001), these are currently limited to an accuracy of 85–90% of the true inundated area due to the resolution of the imagery and misclassification errors. The latter results from a lack of fundamental understanding of the interaction of the radar signal with wet soil, emergent and submerged vegetation and the single frequency and polarization modes and fixed or limited range of incidence angles available on current satellite systems. Small incidence angles will reduce the path length between the point at which microwaves enter the canopy and the water surface. This reduces scattering from the canopy and makes a specular reflection away from the sensor more likely, thus aiding flood detection (Töyrä et al., 2001). Frequency and polarization are also important factors, with L-band radar more successful at penetrating the canopy than C-band (Ulaby et al., 1996) and the canopy backscatter greater for VV polarization than for HH (Engman and Chauhan, 1995). A double bounce mechanism (Alsdorf et al., 2000) may also enhance backscatter, with inundated forests being identified with high returns in the L-band (Richards et al., 1987). This effect is reduced at C-band due to the increased volume scattering in the canopy, and Ormsby et al. (1985) found that the X-band ($\sim$3 cm) gave bright returns for flooded marshland but no backscatter enhancement in forests. The use of polarimetric or multi-frequency SARs may therefore assist in discriminating between open water, emergent vegetation and flooded areas beneath forests and may prove essential to accurate flood mapping. Further research is therefore required to understand the microwave scattering properties of the floodplain environment, particularly because in the next few years, data from a number of new satellite SAR systems will become available, e.g. ENVISAT, RADARSAT-2, ALOS and TerraSAR, which potentially will provide the required frequency of observation and polarimetric or high-resolution capabilities. Similarly, airborne SAR systems have the spatial and temporal resolution to more accurately map dynamic flooding processes and may allow the evaluation of different instrument configurations which may better discriminate flooded and non-flooded areas.

Currently available radar data consists of single and double images of flooding on particular river reaches from the ERS and RADARSAT satellite platforms and the UK Defence Establishment Research Agency (DERA) airborne SAR. To the authors' knowledge, good satellite images exist for a 1-in-5-year event on the River Thames (co-incident with the Biggin and Blyth air photo data), the 1995 Meuse and Oder floods, the October 1998 River Severn floods and the November 2000 Severn floods. Unusually, for the latter series of events three satellite images are available; ERS-2 and RADARSAT data on 9 November 2000 and a further ERS-2 image on 16 November 2000. Of these sensors, RADARSAT is acknowledged as the better system for monitoring floods (Veitch, 2001), and preliminary analysis of the 16/11/00 ERS-2 image at Bristol has shown this to be the case here. Only limited information may thus be available from these ERS-2 images. Airborne SAR data from the UK Defence Establishment Research Agency (DERA) instrument were also collected at the behest of the UK Environment Agency for the rivers Thames, Severn and Ouse during the December 2000 flooding. Overflights of all three rivers were conducted on the 8 November 2000, and the Severn between Stourport and Upton-on-Severn was flown again on the 14 December 2000. To summarize, for most river reaches the data currently available consist of a single satellite image, and, of the available systems, RADARSAT seems to have advantages for flood extent delineation. We have only been able to identify one river, the Severn, where more than one SAR image of flooding exists. Here four good images, two RADARSAT and two airborne SAR, are potentially available and these are of considerable value as they potentially allow split sample validation using various combinations of extent and hydrometric data. However, the images are still only single 'snapshots' of inundation extent from separate events and thus still do not properly capture dynamic flooding processes over the course of a flood. It should also be noted that, with the exception of the Thames event, all other floods for which data have so far been identified are of high recurrence interval. This is actually a problem when using such data for model validation, as the flood tends to completely fill the valley floor. The flood boundary therefore lies on relatively high slopes and large changes in water level are required to effect a detectable (1 pixel) change in flood boundary position. The lower the image resolution, the more significant this problem becomes.

No class of instrument used for flood inundation monitoring is thus problem-free, and data availability is the major constraint on validation studies, regardless of the system employed. We do not yet possess a benchmark dataset capable of convincing validation/falsification of the dynamic behaviour of current flood inundation models. Perhaps more fundamentally, we have not yet even observed the development of flood for a real river reach in a systematic manner, and thus are unsure about the basic physical mechanisms and energy losses that we need to represent. Nevertheless, in the last few years, available inundation data have begun to be used in the model validation process and important preliminary conclusions can be drawn from these studies. These are explored in section 5.3.

Flow Velocity and Free Surface Elevation. Present methods of determining water velocities rely on making *in-situ* point measurements, which are necessarily limited in spatial extent and can be dangerous to undertake. A resolution to this problem is potentially provided by Microwave Doppler radar which offers a remote means of measuring water-surface velocities based on the Bragg scattering from short waves produced by turbulence. When the transmitted radar signal is scattered from a rough water surface, a Doppler frequency shift is produced by the centimetre-length surface waves that backscatter. The magnitude of the shift is a function of the stream current. The principle has been used to map surface velocities of coastal currents for a number of years (e.g. Hwang *et al.*, 2000), and has recently been applied to the measurement of the spatial distribution of river currents using a radar mounted on a van parked next to a river (Costa *et al.*, 2000). By aiming the radar across the river in two directions about 30 degrees apart, one looking upstream and the other downstream, Costa *et al.* found that the difference in the Doppler shifts of the two return signals gave the downstream velocity component.

Experiments are also now being conducted on the use of along-track airborne radar interferometry for measuring water-surface velocities (Srokosz, 1997). These employ aircraft having microwave coherent real-aperture radar with two antennae aimed a few degrees apart in the along-track direction. If the two radars are arranged to look across- rather than along-track, it is possible to perform across-track interferometry instead, allowing a map of water-surface elevations to be constructed remotely. Such studies are still in the research stage and a useable technique is still some distance away. Nevertheless, the model validation potential of such surface current measurements is considerable and deserves further exploration.

5.2.2 Topography

Traditionally, hydraulic models have been parameterized using ground survey of cross-sections perpendicular to the channel at spacings of between 100 and 1000 m. Such data are accurate to within a few millimetres in both horizontal and vertical levels and integrate well with one-dimensional hydraulic models such as HEC-RAS and ISIS. These schemes calculate flows between such cross-sections and remain the industry-standard for flood routing and flood extent modelling studies. However, ground survey data are expensive and time-consuming to collect and of relatively low spatial resolution. Considerable potential therefore exists for more automated, broad-area mapping of topography from satellite and, more importantly, airborne platforms. Two such techniques are airborne laser altimetry (LiDAR) and airborne stereo-photogrammetry. These are now regarded as reasonably standard methods for determining topography at the river reach scale. Of these, LiDAR has the advantage of higher survey rates and lower cost, but with comparable accuracy to airborne stereo-photogrammetry (Gomes Pereira and Wicherson, 1999). Accuracy is, as one would expect, much lower than for ground survey, with rms errors for LiDAR data quoted as being of the order of 15 cm in the vertical and 40 cm in the horizontal. The resulting datasets are dense (typically 1 point per 4 m^2 with a 5 KHz instrument flown at 800 m) and can be collected at sampling rates of up to 50 km^2 per hour. Initial problems with the operational use of such systems (ellipsoid conversion, vegetation removal, etc.) have been for the most part resolved and, in the UK at least, a national data collection programme is now producing large amounts of high quality data. Processing techniques to extract topography from LiDAR returns are discussed further in section 5.2.3 and the integration of LiDAR data with hydraulic models in section 5.3.

5.2.3 Friction

Friction is usually the only unconstrained parameter in a hydraulic model. Two- and three-dimensional codes which use a zero equation turbulence closure may additionally require specification of an 'eddy viscosity' parameter which describes the transport of momentum within the flow by turbulent dispersion, however, this prerequisite disappears for most higher-order turbulence models of practical interest. Hydraulic resistance is a lumped term that represents the sum of a number of effects: skin friction, form drag and the impact of acceleration and deceleration of the flow. These combine to give an overall drag force C_d, that in hydraulics is usually expressed in terms of resistance coefficients such as Manning's n and Chezy's C and which are derived from uniform flow theory. The precise effects represented by the friction coefficient for a particular model depend on the model's dimensionality and resolution. Thus, the extent to which form drag is represented depends on how the channel cross-sectional shape, meanders and long profile are incorporated into the model discretization. For example, a one-dimensional code will not include frictional losses due to channel meandering in the same way as a two-dimensional code. In the one-dimensional code these frictional losses need to be incorporated into the hydraulic resistance term. Similarly, a high-resolution discretization will explicitly represent a greater proportion of the form drag component than a low-resolution discretization using the same model. Hence, complex questions of scaling and dimensionality arise which may be somewhat difficult to disentangle. Certain components of the hydraulic resistance are, however, more tractable. In particular, skin friction for in-channel flows is a strong function of bed material grain size, and a number of relationships exist which express the resistance coefficient in terms of the bed material median grain size, D_{50} (e.g. Hey, 1979). Equally, on floodplain sections, form drag due to vegetation is likely to dominate the resistance term (Kouwen, 2000). Determining the drag coefficient of vegetation is, however, rather complex, as the frictional losses result from an interaction between plant biophysical properties and the flow. For example, at high flows the vegetation momentum absorbing area will reduce due to plant bending and flattening. To account for such effects Kouwen and Li (1980) and Kouwen (1988) calculated the Darcy-Weisbach friction factor f for short vegetation, such as floodplain grasses and crops, by treating such vegetation as flexible, and assuming that it may be submerged or non-submerged. f is dependent on water depth and velocity, vegetation height and a product MEI, where M is plant flexural rigidity in bending, E is the stem modulus of elasticity and I is the stem area's second moment of inertia. Whilst MEI often cannot be measured directly, it has been shown to correlate well with vegetation height (Temple, 1987).

Whilst bed material size is not amenable to analysis using remote sensing, a number of technologies are capable of determining vegetation properties at a variety of scales. As this is the principal component of frictional resistance on floodplains, the extent to which the data derived from such techniques can be converted into hydraulic friction coefficients is a key question. Optical remote sensing techniques are a standard methodology for vegetation classification and a large body of research has built up in this area. Additionally, recent research developments allow extraction of specific biophysical parameters from remotely sensed data. Both may allow spatially variable hydraulic resistance to be automatically mapped and are thus discussed below.

Vegetation Classification. A large number of satellite (AVHRR, Landsat MSS and TM, SPOT) and airborne (CASI) sensors measure light reflected from a target in various bands within the visible part of the electromagnetic spectrum. These sensors vary in terms of their spatial resolution (~1 km for AVHRR, ~3 m for CASI) and the number and width of the spectral bands resolved. A common property, however, is that such data can be used to discriminate vegetation of different type and hence produce a classified map of vegetation at various scales. The technique relies on the fact that different earth surface features (e.g. bare ground, vegetation, etc.) differentially absorb and reflect certain wavelengths of light and can thus be discriminated. This is now a relatively well-established technique and for the UK and Europe a number of operational products exist. In the UK, the Institute of Terrestrial Ecology (ITE, now Centre for Ecology and Hydrology, Monks Wood) has produced a national land cover map at 25 m resolution from Landsat TM data (http://www.ceh.ac.uk/data/lcm/). This consists of 25 vegetation classes whose spatial distribution was determined based on winter and summer images. An updated version of the dataset was released in October 2001. This dataset formed the UK contribution to the Europe-wide CORINNE land cover database (http://www.ceh.ac.uk/subsites/corine/backgr.html; http://etc.satellus.se/) which amalgamated pixels in the ITE data to a slightly coarser resolution (1:100 000) and disaggregated them into slightly more classes (44) in conjunction with further contextual datasets. Whilst there are acknowledged mis-classification problems (Fuller *et al.*, 1990, 1994a, 1994b) with such data, they are consistent, high resolution and have complete national coverage that is likely to be repeated on an eight-year cycle. As yet no research has been conducted to associate a hydraulic resistance, or range of resistance, with each vegetation class. However, such a development would be relatively straightforward and rapid to accomplish and could provide a first-order estimate of floodplain frictional resistance for the whole of the UK at 25 m resolution.

Vegetation Biophysical Attributes. Whilst the association of vegetation classes with typical hydraulic resistance values could be seen as an extension and standardization of the Chow (1959) typography of river channels, other remote sensing techniques offer the potential to directly measure vegetation biophysical parameters such as vegetation height, density and stiffness. These measures may correlate with frictional resistance (Temple, 1987) and hence potentially provide a physically based method of determining spatial variable resistance coefficients that overcomes the subjective nature of typographic approaches. However, such approaches require further work over the medium term before they can be used operationally.

To date, most work in this area has concentrated on the determination of vegetation height as part of the standard processing chain for LiDAR data. The problem in processing LiDAR data is how to separate ground hits from surface object hits on vegetation or buildings. Ground hits can be used to construct a digital elevation model (DEM) of the underlying ground surface, while surface object hits taken in conjunction with nearby ground hits allow object heights to be determined. Papers by Mason *et al.* (1999), Cobby *et al.* (2000) and Cobby *et al.* (2001) describe the development of a LiDAR range image segmentation system to perform this separation (see Figure 5.1), and this is reviewed briefly here.

The system converts the input height image into two output raster images of surface topography and vegetation height at each point. The river channel and the model domain extent are also determined, as an aid to constructing a model discretization. The segmenter is semi-automatic and requires minimal user intervention. Flood modelling does not require a perfect segmentation as it involves large-area studies over many regions, so a straightforward segmentation technique able to cope with large ($\sim$Gb) datasets and designed to segment rural scenes has been implemented.

The action of the segmenter can be illustrated using a 6 × 6 km sub-image from a large 140 km^2 LiDAR image of the Severn basin. The data were acquired in June 1999 by the UK Environment Agency using an Optech ALTM1020 LiDAR system measuring time of last return only. At this time of year leaf canopies were dense, so that relatively few LiDAR pulses penetrated to the ground. The aircraft was flown at approximately 65 m s^{-1} and at 800 m height whilst scanning to a maximum of 19° off-nadir at a rate of 13 Hz. The laser was pulsed at 5 kHz, which resulted in a mean cross-track point spacing of 3 m. Because current one- and two-dimensional flood models are capable of predicting inundation extent for river reaches up to 20 km long, with floodplains spanning associated large areas, the segmenter has been designed to use LiDAR data with lower resolution than that used for other applications (Gomes Pereira and Wicherson, 1999). Figure 5.2a shows (on its left-hand side) the raw 3-m gridded sub-image used as input to the segmenter. The sub-image contains a 17-km reach of the Severn near Bicton west of Shrewsbury. The bright

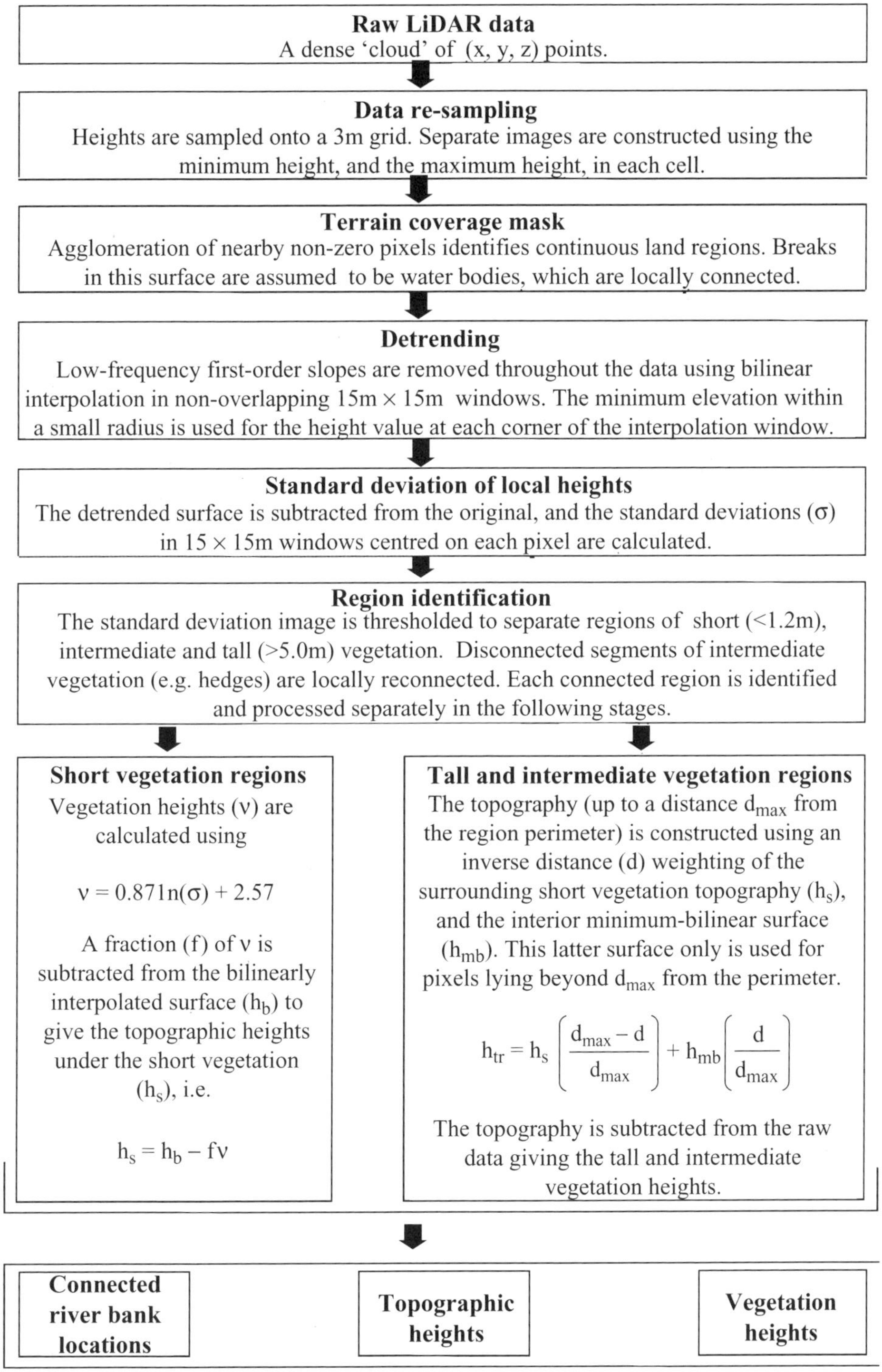

Raw LiDAR data
A dense 'cloud' of (x, y, z) points.

⬇

Data re-sampling
Heights are sampled onto a 3m grid. Separate images are constructed using the minimum height, and the maximum height, in each cell.

⬇

Terrain coverage mask
Agglomeration of nearby non-zero pixels identifies continuous land regions. Breaks in this surface are assumed to be water bodies, which are locally connected.

⬇

Detrending
Low-frequency first-order slopes are removed throughout the data using bilinear interpolation in non-overlapping 15m × 15m windows. The minimum elevation within a small radius is used for the height value at each corner of the interpolation window.

⬇

Standard deviation of local heights
The detrended surface is subtracted from the original, and the standard deviations (σ) in 15 × 15m windows centred on each pixel are calculated.

⬇

Region identification
The standard deviation image is thresholded to separate regions of short (<1.2m), intermediate and tall (>5.0m) vegetation. Disconnected segments of intermediate vegetation (e.g. hedges) are locally reconnected. Each connected region is identified and processed separately in the following stages.

⬇

Short vegetation regions
Vegetation heights (v) are calculated using

$$v = 0.87 \ln(\sigma) + 2.57$$

A fraction (f) of v is subtracted from the bilinearly interpolated surface (h_b) to give the topographic heights under the short vegetation (h_s), i.e.

$$h_s = h_b - fv$$

Tall and intermediate vegetation regions
The topography (up to a distance d_{max} from the region perimeter) is constructed using an inverse distance (d) weighting of the surrounding short vegetation topography (h_s), and the interior minimum-bilinear surface (h_{mb}). This latter surface only is used for pixels lying beyond d_{max} from the perimeter.

$$h_{tr} = h_s \left(\frac{d_{max} - d}{d_{max}} \right) + h_{mb} \left(\frac{d}{d_{max}} \right)$$

The topography is subtracted from the raw data giving the tall and intermediate vegetation heights.

⬇

| Connected river bank locations | Topographic heights | Vegetation heights |

Figure 5.1 Flow chart showing the main stages of the LiDAR segmentation system (from Mason et al., 2003). © John Wiley & Sons, Ltd.

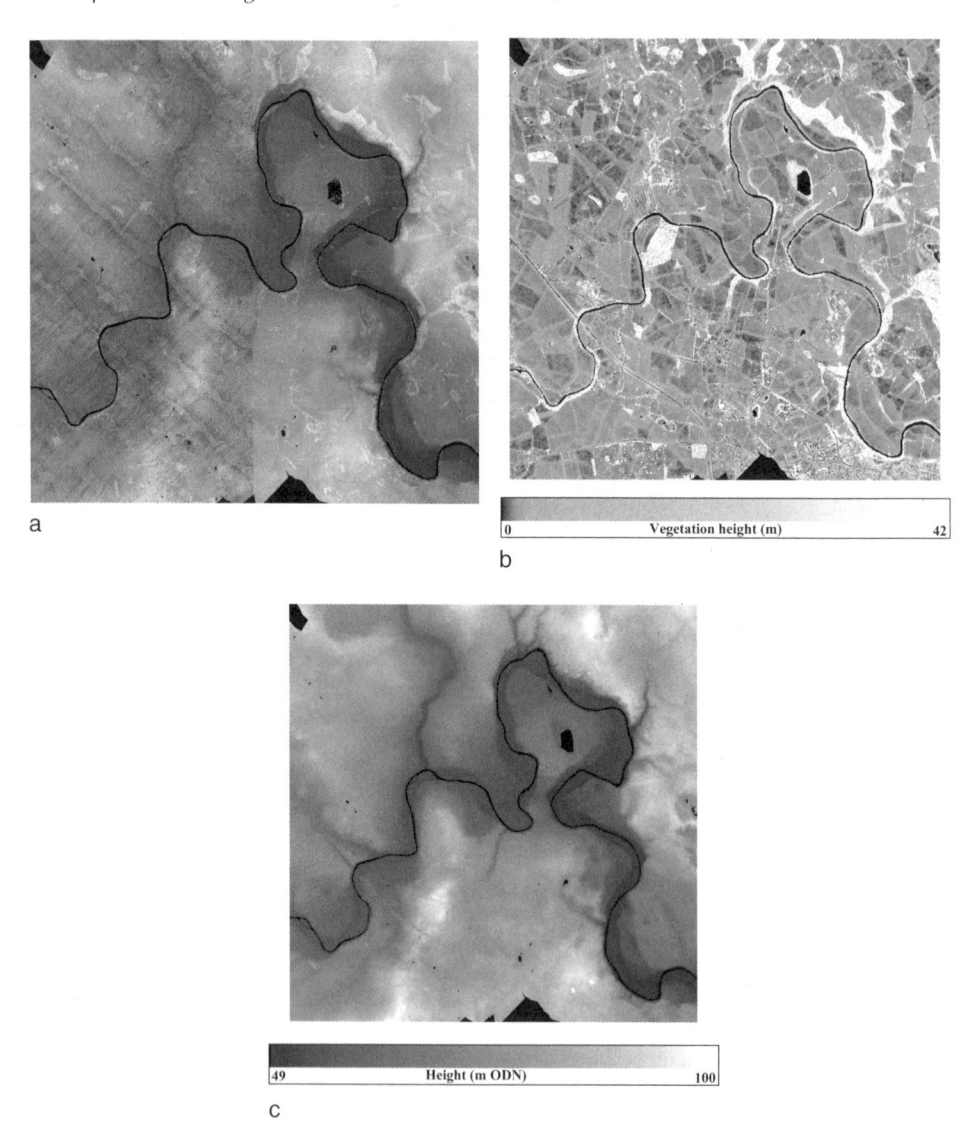

90 *Spatial Modelling of the Terrestrial Environment*

Figure 5.2 (a) Raw 3-m gridded LiDAR data of a 6 ×6 km area in the Severn Basin, UK (left-hand side). The right-hand side shows an interpolated region of the raw data. (b) Vegetation height map. (c) Topographic height map (from Mason et al., 2003). See also Plate 1. © John Wiley & Sons, Ltd.

areas are the overlaps between adjacent LiDAR swaths, which are typically about 550 m wide. The image only appears sensible to the eye when interpolated, as shown on the right-hand side. However, to avoid loss of resolution in interpolation the raw data are used in the segmentation.

The segmenter first identifies water bodies by their zero return (due to specular reflection of the laser pulse away from the sensor), and agglomerates them into connected regions. To remove low frequency trends, the image is subjected to a detrending step that produces an initial rough estimate of ground heights over the whole image. The detrended height image is then segmented based on its local texture. The texture measure used is the standard deviation of the pixel heights in a small window of side 15 m centred on a pixel. Regions of short vegetation should have low standard deviations, and regions of tall vegetation larger values. The image is thresholded into regions of short, intermediate and tall vegetation, and connected regions of each class are found (Figure 5.2(b), see Plate 1. Short vegetation less than 1.2 m high includes most agricultural crops and grasses and is a predominant feature on floodplains. Intermediate vegetation includes hedges and shrubs, which may be expected to be significant obstacles to flow over floodplains. Tall vegetation greater than 5 m high includes trees and buildings. No attempt is made to distinguish buildings from trees, as there are generally few buildings in rural floodplains such as the one used in this study.

An advantage of segmentation is that it allows different topographic and vegetation height extraction algorithms to be used in regions of different cover type. Figure 5.2(b) shows the vegetation height map derived from the segmentation. Short vegetation heights are calculated using an empirically-derived relationship between the LiDAR standard deviation and measured crop height, which predicts vegetation heights up to 1.2 m with an r.m.s. accuracy of 14 cm. In contrast to most LiDAR vegetation height algorithms, this relationship does not rely on a certain fraction of LiDAR pulses being reflected from the ground. The topographic height map (Figure 5.2(c) is constructed in regions of short vegetation by subtracting an empirically determined fraction of the vegetation height from the initial rough estimate of ground heights. A comparison with ground control points (GCPs) gave an r.m.s. height accuracy of 17 cm for the topographic surface in regions of short vegetation. This is close to the 10 cm rms accuracy aimed for in DEMs for model bathymetry (Bates *et al.*, 1997). The cutoff of 10 cm provides a realistic lower limit for DEM accuracy, as beyond this the sensor signal becomes indistinguishable from background 'noise' generated by microtopography features such as furrows in a ploughed field.

In regions of tall and intermediate vegetation, the topographic height map is constructed by interpolation between local minima (assumed to be ground hits) and topographic heights in nearby short vegetation regions. However, ground height accuracy falls off in wooded regions due to poor penetration of the 5 Hz LiDAR pulse through the canopy. The UK Environment Agency has recently acquired a 33 Hz LiDAR that should substantially overcome this problem. Measured over a range of tree canopy densities, the LiDAR-generated topography lay on average 64 cm above the corresponding GCPs, with an r.m.s. error of 195 cm. The larger error under trees can be seen in Figure 5.2(c), though fortunately the example floodplain contains few large areas of woodland. Accurate calculation of tree height is well documented, and assumes that pulses reflect from both the tree canopy and the ground beneath (Naesset, 1997; Magnussen and Boudewyn, 1998). Tall and intermediate vegetation heights are derived by subtracting the ground heights from the canopy returns, and are accurate to about 10%.

The vegetation height map (Figure 5.2(b) is partly corrupted by bands of increased height in areas of overlap between adjacent LiDAR swaths. These are caused by systematic height errors between adjacent swaths increasing height texture in overlap regions. The problem can be substantially reduced by the removal of heights measured at large scan angles (Cobby *et al.*, 2001). Unfortunately, the correction algorithm requires swaths supplied as individual data files that were unavailable for this test area. It is hoped that future surveys will provide data in a format suitable for the correction to be applied. Additional research may also generate the ability to retrieve further biophysical parameters from remotely sensed data. For example, a combination of LiDAR-derived vegetation height estimates and vegetation classification from optical systems such as CASI has the potential to further reduce uncertainty in estimates of vegetation properties, but this has yet to be investigated.

5.3 Integration of Spatial Data with Hydraulic Models

The above data sources typically require further manipulation before they can be directly incorporated into hydraulic models. New data sources are also a stimulus for additional technical developments of both parameterizations and models. These integration studies and developments are explored in the following section.

5.3.1 High Resolution Topographic Data

LiDAR can be regarded as a largely operational technique that can be integrated with a variety of standard hydraulic models. Indeed, its use is now relatively standard in the UK in the creation of indicative floodplain maps. The UK Environment Agency is obliged by Section 105 of the 1991 Water Resources Act to determine the extent of flooding resulting from the 1-in-100-year recurrence interval flood. Such studies are typically accomplished with a one-dimensional code as outlined above, with the model calibrated and validated against bulk hydrometric data as outlined in section 5.1. Marks and Bates (2000) have also explored the integration of LiDAR data with two-dimensional finite-element models. Whilst LiDAR topographic accuracy has been validated against ground truth data, no studies have been conducted to examine the impact of LiDAR or other topographic data sources (and their error structures) on model predictions. Hence, although we know that LiDAR is subject to r.m.s. errors in the region of 15 cm as opposed to <1 cm for ground survey, we do not know whether such errors lead to differences in inundation extent predictions that can be conclusively discriminated on the basis of available validation data. One would expect that the increase in topographic error caused by model parameterization using LiDAR data rather than a commensurate resolution ground survey would be subsumed within the calibration process for models validated against bulk hydrometric data, although this is yet to be tested. The same may also be true for models validated against synoptic inundation extent data and it is at present unclear what level of topographic error will cause problems for particular hydraulic model applications, given the uncertainty in the calibration process. Put another way, we do not yet know the level of data provision necessary to predict inundation extent with a particular class of model to a given accuracy.

A number of recent studies have begun addressing this problem in terms of topography, although it should be noted that very little research has been conducted in other areas, such as boundary friction parameterization, where similar questions apply. Thus, Marks (2001) has conducted variogram analysis of LiDAR data and shown that for horizontal length scales smaller than ~10 m, floodplain topographic variability is indistinguishable from noise. Topographic features that are likely to be hydraulically significant will thus typically be larger than this. This begins to indicate appropriate grid resolutions for use in hydraulic modelling. In this respect, there is an analogy here between topography parameterization and large eddy simulation of turbulence. In both, grid resolution distinguishes between those flow features at the grid scale or above that are captured explicitly, and those at sub-grid scales, which are more homogeneous and can merely be treated statistically in terms of their impact on the mean flow field (see discussion on parameterization of sub-grid wetting and drying models below) or as part of the resistance term. Hence, differently scaled elements of the total frictional loss can be represented as combinations of these three, very different, approaches. Nor are these different representations discrete, but rather they inter-grade and interact.

In terms of the necessary spatial resolution for topographic data, Horritt and Bates (2001b) have validated various resolution implementations of a raster-based flood inundation model against hydrometric and satellite SAR inundation data for a 40-km reach of the River Severn between the gauging stations at Montford Bridge and Buildwas. The topographic data used to drive these simulations were derived from the 1999 LiDAR survey conducted by the UK Environment Agency processed using the algorithm described in section 5.2.3 to yield a 10 m raster digital elevation model of the whole reach. Horritt and Bates tested 10, 25, 50, 100, 250, 500 and 1000 m resolution models generated by spatial averaging of the DEM and found that the optimum calibration was stable with respect to changes in scale when the model was calibrated against the observed inundated area. Observed and simulated inundated areas were compared using the measure of fit:

$$F^{(2)} = \frac{A_{\text{obs}} \cap A_{\text{mod}}}{A_{\text{obs}} \cup A_{\text{mod}}} \tag{1}$$

Here $F^{(2)}$ is thus a zero-dimensional global performance measure, where A_{obs} and A_{mod} represent the sets of pixels observed to be inundated and predicted as inundated respectively. $F^{(2)}$ therefore varies between 0 for a model with no overlap between predicted and observed inundated areas and 100 for a model where these coincide perfectly.

In the case of the River Severn simulations $F^{(2)}$ reached a maximum of ~72% at a resolution of 100 m, after which no improvement was seen with increasing resolution. Projecting predicted water levels onto a high resolution DEM improved performance further, and a model resolution of 500 m proved adequate for predicting water levels. Predicted floodwave travel times were, however, strongly dependent on model resolution, and water storage in low lying floodplain areas near the channel was identified as an important mechanism affecting wave propagation velocity as this is strongly influenced by the near channel micro-topography and hence model resolution. The impact of changing topographic representation therefore differs depending on what the modeller wishes to predict.

High-resolution topographic data has also stimulated two other developments. First, the digital raster format of LiDAR has led to the creation of new flood inundation models which directly integrate with this data format in a GIS-type framework (e.g. FLOODSIM – Bechteler *et al.*, 1994; GISPLANA – Estrela and Quintas, 1994; and LISFLOOD-FP - Bates and De Roo, 2000; Horritt and Bates, 2001a and 2001b). These models use channel routing combined with storage cell or 2D diffusive wave routing on floodplains to simulate dynamic flooding in a computationally efficient manner. Such codes could also be readily integrated with socio-economic datasets to provide complete management systems. Second, whilst a number of algorithms have been developed which correct for dynamic wetting and drying processes in fixed grid models (e.g. King and Roig, 1988; Defina et al., 1994; Bates and Hervouet, 1999; Defina, 2000) using proportionality coefficients based on the sub-grid topography, lack of data has usually meant that these coefficients are assumed rather than parameterized from actual measurements. Unlike ground survey data, which are frequently unavailable at even the grid resolution, LiDAR data are typically available at a much higher resolution than most model grids. It is therefore possible to use the redundant, sub-grid scale data to directly parameterize these sub-grid scale algorithms and better correct for wetting and drying effects in two-dimensional Shallow Water models (Bates and Hervouet, 1999; Bates, 2000). Both areas of model development are still in their infancy and further research is required.

5.3.2 Spatially Distributed Friction Data

The traditional method of parameterizing bottom friction in hydraulic models involves specifying friction factors for floodplain and channel. These are treated as free parameters to be adjusted to provide the best fit between model predictions and observations. As noted above, use of sparse hydrometric data for model calibration means that the assigned parameters are typically spatially and temporally lumped so that one value is used to represent the channel and one to represent the floodplain. A more physically-based friction parameterization can now be obtained using the vegetation height data generated from airborne LiDAR as described in section 5.2.3, which can be used (see Figure 5.3) to specify an 'effective' individual friction factor at each node of a computational model (Mason et al., 2003). A different equation set can be used to model flow resistance at a node depending on the vegetation within the node neighbourhood. Thus, for in channel areas flow resistance is calculated as a function of bed material size using the relationship of Hey (1979). For short vegetation, such as floodplain grasses and crops, the Darcy-Weisbach friction factor f is modelled using the approach of Kouwen (1988) described above with the term MEI determined using an empirical correlation with vegetation height determined by Temple (1987). Finally, tall vegetation is treated using the method of Fathi-Moghadam and Kouwen (Fathi-Moghadam and Kouwen, 1997; Kouwen and Fathi-Moghadam, 2000). This calculates the hydraulic resistance of such vegetation as a linear function of velocity. The surface properties in the vicinity of each node are then integrated to give a value for f which then varies in time as a function of the local hydraulics. Specifically, Figure 5.3 shows the computational mesh of a two-dimensional finite element hydraulic model of the 17 km reach of the River Severn upstream of Shrewsbury, UK. This uses data from the LiDAR survey described in section 5.2.3 and shown in Figures 5.1 and 5.2, along with data from the flow gauging station at Montford Bridge to assign an upstream boundary condition to the model. This was then used to simulate a dynamic 1-in 50-year flood event, which occurred in October 1998 for which a RADARSAT SAR image was available for model validation. Use of the Mason et al. parameterization generated a reach-average 16 cm difference in water levels over a control simulation with a spatially lumped and calibrated friction surface. This is a significant difference, however, both parameterizations fitted the available satellite SAR validation data equally well given its resolution and error. This was because the floodplain slope at the waterline (~ 0.03 m m^{-1}) for this 50-year event means that a 16-cm elevation change does not generate a 1 pixel change in inundation extent and is thus undetectable. An essential requirement is, therefore, the acquisition of a benchmark model validation dataset

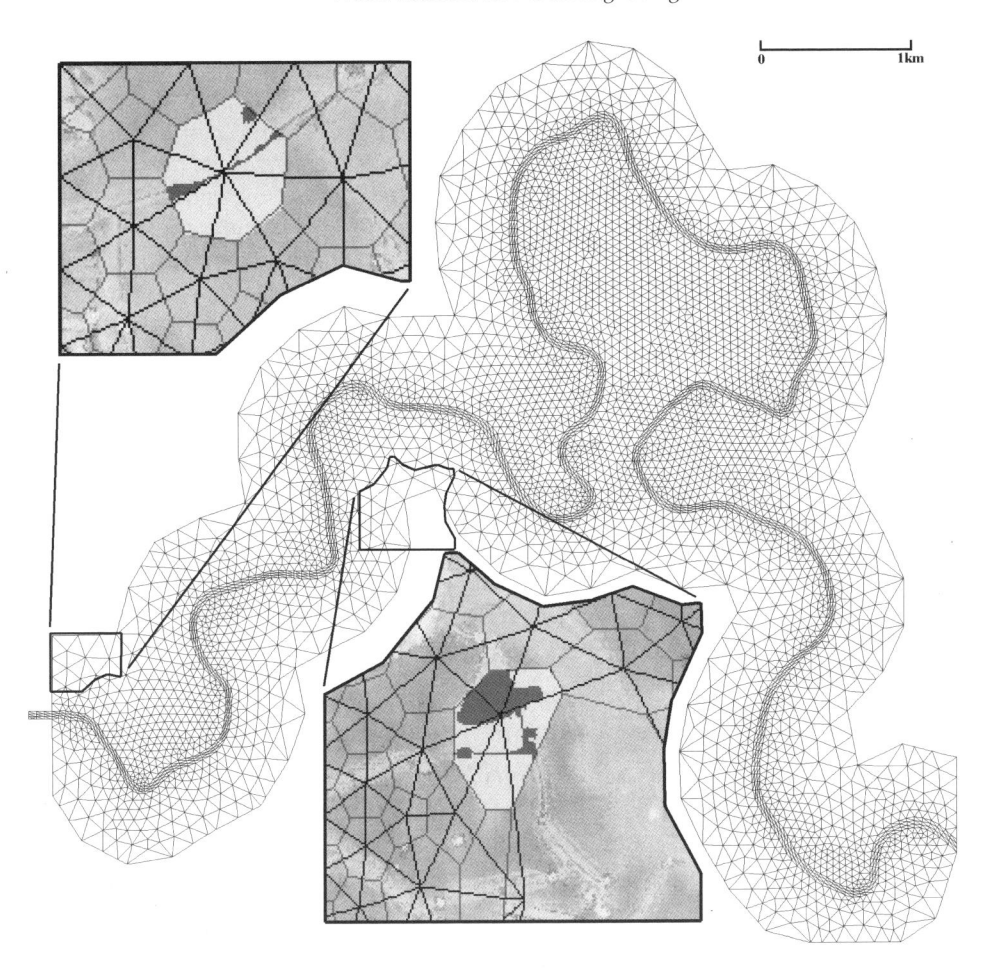

Figure 5.3 Assignment of an 'effective' friction value for each element of a finite-element mesh using the vegetation height data shown in Figure 5.2c and the friction mapping algorithm of Mason et al. (2003). For each mesh node, an instantaneous friction factor is calculated at each timestep, given the frictional material in the neighbourhood of the node and the current water depth and flow velocity there. A node's neighbourhood is defined as a polygon whose vertices are the centroids of the elements surrounding the node. Sub-regions of connected regions in the vegetation height map are formed by intersecting the node polygon map with the vegetation height map. Each sub-region in a polygon may contain either sediment or one of the three vegetation height classes. If a sub-region contains vegetation, its region's average vegetation height is attributed to the sub-region (from Mason et al.,) 2003. © John Wiley & Sons, Ltd.

in order to rigorously test and refine such schemes and discriminate between competing approaches. Whilst we should not expect the method proposed by Mason *et al.* (2003) to fully overcome the problem of specifying and calibrating frictional resistance, it may be able to substantially reduce the predictive uncertainty associated with friction estimates for floodplain areas and hence allow the remaining sources of uncertainty to be more rigorously analyed.

5.3.3 Automatic Mesh Generation

Two-dimensional structured and unstructured grids form the basis of many flood models. The automatic mesh generator of Horritt (1998) requires the meandering position of the river channel banks and some (arbitrary) contour to describe the extent of the model domain. Rather than digitizing this information from maps, the LiDAR segmentation system of Cobby *et al.* (2001) described above can be used to automatically identify any river channel and provide the chainage description of its banks in a format that can be used by the mesh generator. Furthermore, the map of LiDAR topographic slope is used to define the domain extent contour, so that the domain is narrower where steep slopes bound the channel than where the floodplain is flat and wide. The number of elements in the domain is further reduced by using larger elements away from regions of high process gradient, for example, at the highly curved sections of a river (Horritt, 2000b). Figure 5.3 gives an example of a two-dimensional finite-element mesh generated in this manner. Such methodologies could also be developed for one-dimensional models and one can envisage a situation where model discretizations can be generated at least semi-automatically from LiDAR data. In the case of finite-element techniques, the mesh generation process could even incorporate the vegetation pattern (hedge lines, woodland) information data from the LiDAR and CASI surveys into the mesh. Thus, elements would follow the boundaries of such features and enable spatially correct and well discretized friction fields to be specified.

5.3.4 Model Calibration and Validation Studies

Very little direct validation of inundation models against inundation extent data has been conducted (or at least formally reported in peer-reviewed journals), and such schemes are typically validated only indirectly against hydrometric data. As noted above, given the relatively sparse hydrometric data available, this may be an insufficient test of an inundation model. Recently, the first comprehensive trials of hydraulic (2D raster storage cell and 2D FE) models against consistent inundation datasets have been conducted and published (Bates and De Roo, 2000; Horritt and Bates, 2001a and b; Horritt and Bates, 2002). These used satellite SAR images of flooding on the Rivers Meuse (Netherlands), Thames (UK) and Severn (UK) and showed that it was possible to replicate these single, rather inaccurate inundation images with relatively simple 2D raster storage cell models such as LISFLOOD-FP (see Figure 5.4). The maximum predictive ability for all models calculated using the $F^{(2)}$ performance measure given in equation (1) was in the range 70%–85% of inundated area predicted correctly and therefore approached the 85%–90% accuracy of inundation extent derived from currently available satellite SAR sources. More recent research using the LISFLOOD-FP to model a single RADARSAT image of the October 1998 flood on the River Severn, UK, has shown that, because of these errors, current satellite-derived inundation images are no better than hydrometric data at constraining the predictions of hydraulic models over some reaches.

As we note in section 5.2.1, for particular reaches two flooding images do exist and an essential next step is to attempt model validation using these, rather than single image data. This will allow split sample calibration and validation using inundation extent information and allow calibration portability to be assessed. Hence, we should be able to ascertain the

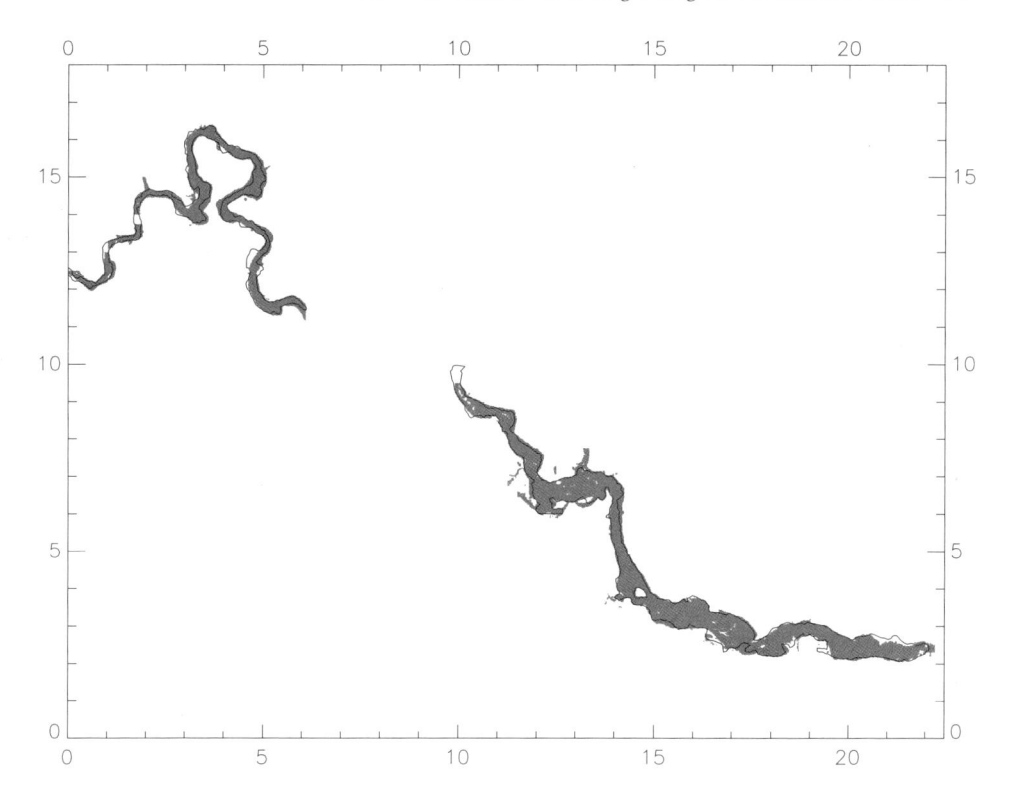

Figure 5.4 Comparison of flood extent predicted by the LISFLOOD-FP inundation model with a SAR-derived shoreline acquired in October 1998 during flooding on a 40-km reach of the River Severn in Shropshire, west-central England downstream of the Montford Bridge gauging station. Peak discharge at Montford Bridge was measured at 435 m^3 s^{-1} and the recurrence interval of the event was estimated at 1 in 50 years. Over much of the domain the model provides an excellent match to the SAR shoreline (solid black line) (from Horritt and Bates, 2002). © Elsevier Science.

errors involved in predicting inundation for a design event from a model conditioned on an historic inundation image. Moreover, in the next few years a number of new satellite SAR systems will be available, e.g. ENVISAT, RADARSAT-2, ALOS and TerraSAR. These may potentially provide a higher frequency of observation and their polarimetric and high-resolution capabilities may offer improved prospects for detection and delineation of flooded areas. Whether the frequency, resolution and accuracy of observations provided by these systems will be sufficient to capture multiple images through a flood, and particularly on the short duration rising limb, is still unclear.

For this reason, airborne SAR data is perhaps of even greater potential. Such data are of higher resolution and potentially of much greater accuracy than that from satellite sources and may thus be able to discriminate better between competing models. Model validation studies thus need to be conducted using the existing airborne SAR data, which again at best consist of two 'snapshot' images taken six days apart and do not properly capture dynamic flooding processes. However, the responsive mode operation of such systems would also allow high temporal and spatial resolution sampling through individual flood events to be conducted during a dedicated airborne campaign. This may be the only way to generate multiple synoptic images of flood extent that show the dynamic extension and retreat of the flow field. We currently lack such data, yet they are essential if we wish to rigorously validate the dynamic behaviour of flood inundation models and develop them beyond a relatively modest accuracy.

Finally, only a relatively limited number of model classes have been tested in this way and the process needs to be extended to incorporate other standard model types such as ISIS and HEC-RAS and the results inter-compared. This needs to be accomplished within a formal uncertainty analysis framework and the result is likely to be an appraisal of limitations with existing models. This will then form an objective basis for a model development programme aimed at producing better inundation models

5.3.5 Uncertainty Estimation Using Spatial Data and Distributed Risk Mapping

Prior to rigorous validation of the friction parameterization methodology of Mason *et al.* (in press), poorly constrained calibration will remain a necessary step in any modelling study. Even with this in place, we may still be left with a residual, if better constrained, calibration problem, which will also lead to uncertainty in model predictions. Calibration is, therefore, always likely to be present in any model application to real-world data, and as observed data are typically sparse and contain errors, potentially many different calibrations and model structures will fit available data equally well. Beven and Binley (1992) term this the 'equifinality problem'. These multiple acceptable model structures will, however, lead to different predictions. Hence, a deterministic approach to conveyance estimation, whereby single sets of calibrated coefficients are used to make single flood extent predictions, may be problematic. This is particularly true if a set of calibrated coefficients from one event is used to predict flood inundation for a further, larger event for which observed data are unavailable. There will be multiple acceptable flood envelopes that fit the available calibration data and hence, design flood extent is better conceived as a fuzzy map. This approach has been advocated by a number of authors who have deployed Monte Carlo analysis to derive inundation probability maps that take account of model uncertainties (Romanowicz *et al.*, 1996; Romanowicz and Beven, 1997; Aronica *et al.*, 2002).

For example, Aronica *et al.* (2002) explored the parameter space of the two-dimensional storage cell model, LISFLOOD-FP, developed by Bates and De Roo (2000). The model was applied in two basins to separate floods, one of which (the Imera River, Sicily) represented a basin-filling problem controlled by embankments, whilst the other (the River Thames, UK) represented a more typical out-of-bank flow in a compound channel. The friction parameter space was treated as two-dimensional and comprised single values of Manning's *n* for the channel and floodplain. A Monte Carlo ensemble of 500 realizations of the model was conducted to explore this parameter space using a random sampling scheme. The simulations were then validated against a single observed inundation extent available in each basin using the $F^{(2)}$ performance measure given in equation (1).

Plots of this objective function were then mapped over the parameter space and are shown in Figures 5.5 and 5.6. These show the calibration response of the LISFLOOD-FP model

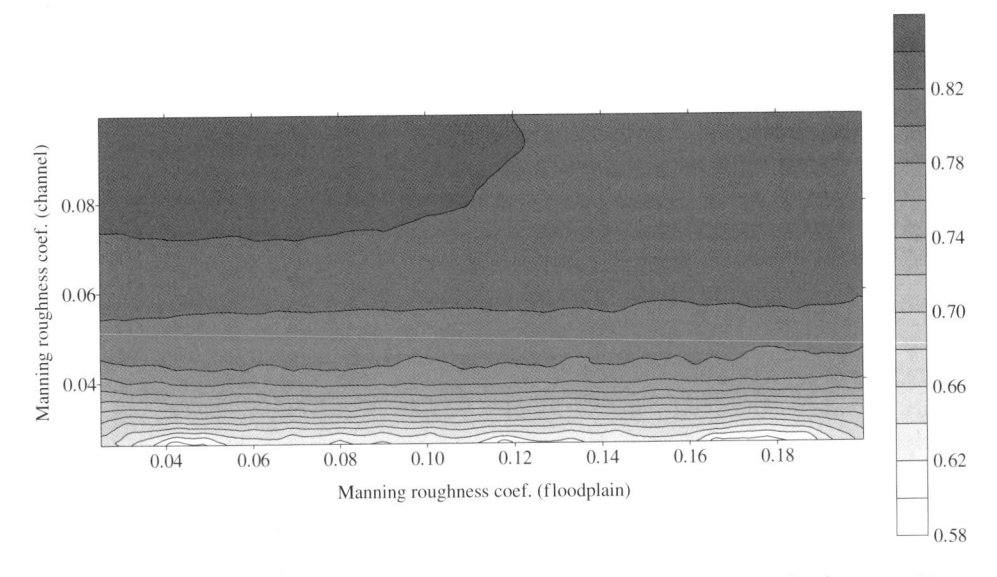

Figure 5.5 Plot of the $F^{(2)}$ performance measure over the parameter space for the Imera River model (from Aronica et al., 2002). © John Wiley & Sons, Ltd.

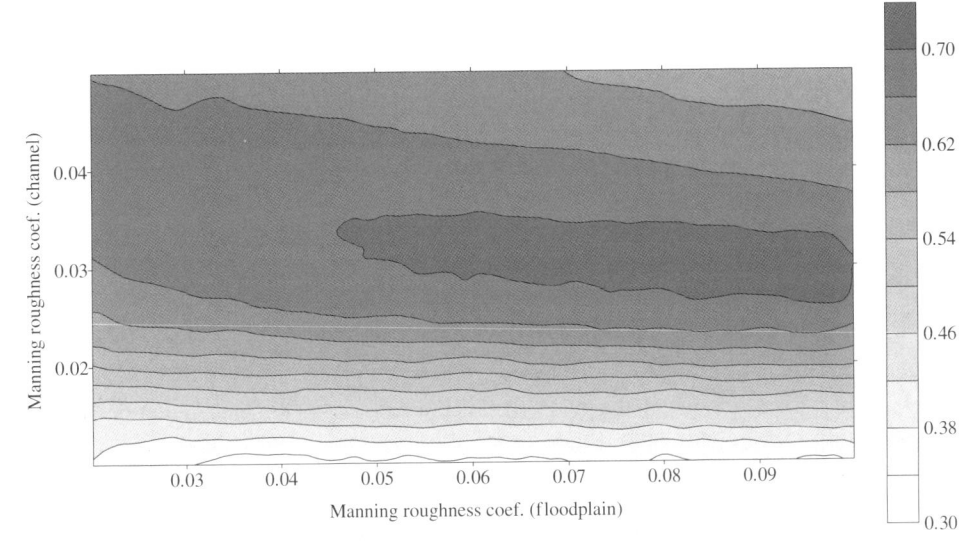

Figure 5.6 Plot of the $F^{(2)}$ performance measure over the parameter space for the River Thames model (from Aronica et al., 2002). © John Wiley & Sons, Ltd.

in each case to be a broad-peaked ridge, with the model displaying greater sensitivity to channel rather than floodplain friction. This response seems typical of storage cell models (see for example Romanowicz *et al.*, 1996). Figures 5.5 and 5.6 also show the equifinal nature of inundation modelling, even when using distributed validation data. Thus, instead of a single optimum set of parameters, which produces a maximum value of $F^{(2)}$, we identify a broad region of the parameter space where values of $F^{(2)}$ are greater than 70%. The complexity of the parameter space also appears to be a function of model non-linearity. Thus, Horritt and Bates (2001a) showed that the parameter space for the weakly non-linear LISFLOOD-FP model typically contains a single optimum region, whilst the same parameter space for the more highly non-linear TELEMAC-2D model contained multiple optima.

Such parameter space mappings do not, however, give information on the distributed risk of inundation across the floodplain given model uncertainty. The problem here is to design a measure that both captures distributed uncertainty and, at the same time, contains information on global simulation likelihood that can be used to condition predictions of further events for which observed data are not available. Thus, to unearth the spatial uncertainty in model predictions for the particular flood being modelled, we take the flood state as predicted by the model for each pixel for each realization, and weight it according to the measure-of-fit $F^{(2)}$ to give a flood risk for each pixel i, P_i^{flood}:

$$P_i^{flood} = \frac{\sum_j f_{ij} F_j^{(2)}}{\sum_j F_j^{(2)}} \qquad (2)$$

where f_{ij} takes a value of 1 for a flooded pixel and is zero otherwise and $F_j^{(2)}$ is the global performance measure for simulation j. Here P_i^{flood} will assume a value of 1 for pixels that are predicted as flooded in all simulations and 0 for pixels always predicted as dry. Model uncertainty (here defined by the interaction of the global performance measure and spatially distributed probabilities of the event being modelled) will manifest itself as a region of pixels with intermediate values, maximum uncertainty being indicated by pixels with $P_i^{flood} \approx 0.5$. When events different from the conditioning event are to be modelled, the same values of $F_j^{(2)}$ can be used (these are associated with each parameterization), but the predicted risk map f_{ij} will differ. Whilst the technique has been developed with the problem of flood inundation modelling in mind, it should equally well apply to any distributed model conditioned against binary pattern data. Plots of P_i^{flood} over the real space of the Imera and Thames models are shown in Figures 5.7 and 5.8, respectively. These maps indicate the risk that a given pixel is predicted as wet in the simulations whilst retaining information on the global performance of the model determined by $F^{(2)}$. When compared to the observed shoreline (solid line), this gives an indication of the degree of model under- or over-prediction. Figures 5.7 and 5.8 give much more information on the spatial structure in simulation uncertainty than can be obtained from a global measure and significant discrepancies between model predictions and the observed shoreline are indicated at the limits of the parameter space.

Finally, P_i^{flood} may be updated as more information becomes available and one is thus able to determine the extent to which additional data act to reduce predictive uncertainty. Such techniques are thus capable of telling us something about the value of data, and here further studies are thus required that examine combinations of data, both hydrometric and inundation extent, and calibration strategies that act to reduce uncertainty in model predictions.

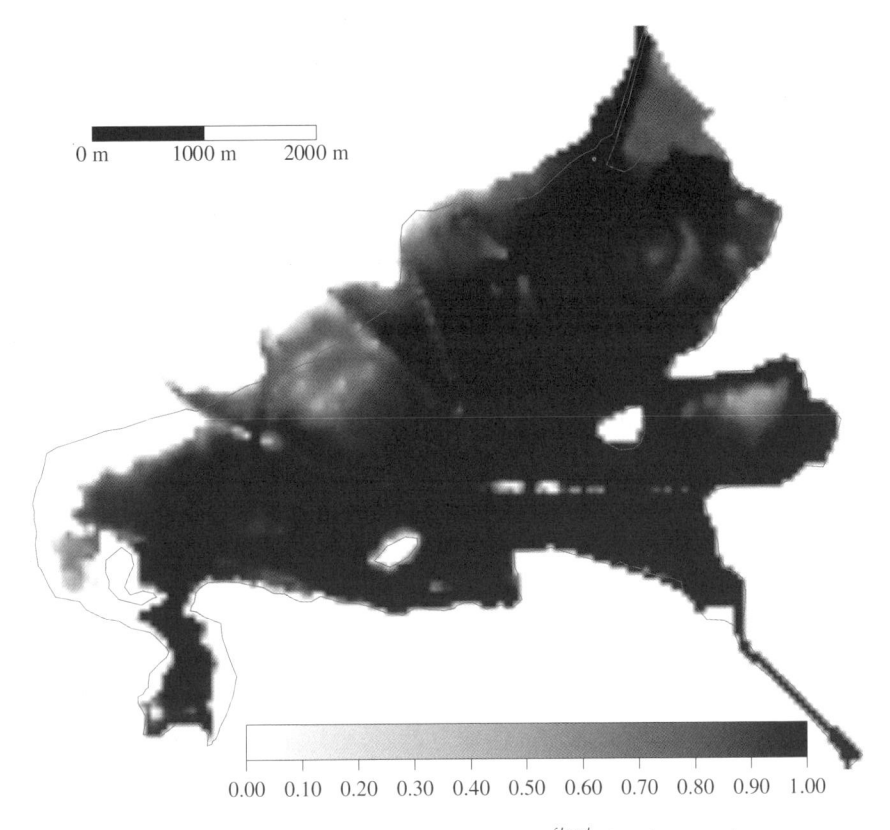

Figure 5.7 Probability map of predicted inundation, P_i^{flood}, for the October 1991 event for the Imera River (from Aronica et al., 2002). © John Wiley & Sons, Ltd.

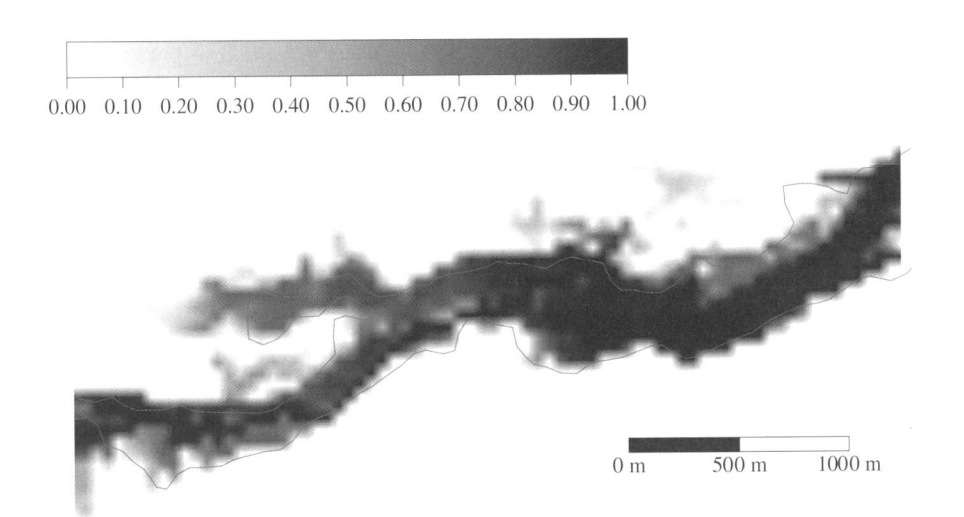

Figure 5.8 Probability map of predicted inundation, P_i^{flood}, for the December 1992 event for the River Thames (from Aronica et al., 2002). © John Wiley & Sons, Ltd.

5.4 Research Needs in Flood Inundation Modelling

It is clear from the above discussion that full incorporation of spatial data into flood inundation models is constrained in a number of ways. Underpinning all strategic developments in the area of inundation modelling is the need for benchmark model validation data to discriminate between competing models and parameterizations. In terms of hydrometric data, this should consist of a relatively dense network of stage recorders. Flood extent data needs to be of high resolution (<5 m pixel size), should avoid misclassification errors and be multi-temporal in order to depict dynamic flooding processes. Such studies are required for a number of reaches with different hydraulic conditions and for multiple events on the same reach to enable calibration portability to be assessed. With such data in place, a number of other developments can then follow. In particular, remote sensing algorithm development is needed in two specific areas: SAR image processing and friction determination from remotely sensed plant biophysical properties. In the case of SAR data, algorithms require further development to eliminate misclassification errors and this research should consist of backscatter modelling combined with laboratory and field experiments using ground-based SAR systems. Second, and given a benchmark validation dataset, further research should be conducted into the development of methods to extract plant biophysical parameters from remotely sensed data and relate these to hydraulic friction. Studies should then examine the impact of different parameterization and calibration strategies on model results in a formal uncertainty analysis framework.

With adequate parameterization and validation data, it will then be possible to conduct studies of the inundation modelling process. Here, numerical experiments will enable questions of data redundancy and scaling to be formally addressed to determine minimum data and model configurations capable of providing reliable predictions. Such numerical experiments should be conducted for a variety of models in terms of both topographic and friction data provision. Following from this is a need to undertake comparative validation of hydraulic models in order to provide an objective basis on which to plan and undertake future model developments. The richness of particular remotely sensed datasets also stimulates innovative developments as well as extensions to established codes. Both such avenues need to be encouraged and integrated with the development objectives emergent from more traditional modes of hydraulic investigation.

5.5 The Value of Spatial Data

Flood inundation modelling provides an excellent example of the incorporation of spatial data into environmental models and the benefits and research challenges that this generates. This is because such models usually have only a single significant parameter, are relatively parsimonious in terms of data requirements (topography and flow boundary conditions) and the surface flows involved are easy to observe using remote sensing techniques. They thus provide a test bed in which to explore the integration of rich and spatially complex datasets with non-linear models without the need to account for multiple interacting parameters. It is clear from the above discussion that many long-established concepts and techniques in environmental modelling need wholesale revision if we are to fully benefit from the potential offered by such techniques. This is particularly true in the case of topography where until very recently lack of data has been the central problem rather than the reverse. Nor can standard sensitivity, calibration and uncertainty analysis techniques be smoothly transferred to distributed problems. In many cases these are designed to operate with limited amounts of point data rather than distributed fields. It is likely that as we increasingly move to distributed calibration and validation, then new methodologies will be required to achieve, for example, distributed calibration against distributed data, use of distributed classification and shape data in uncertainty analysis and to assess spatially distributed, and often highly non-linear (Bates *et al.*, 1998b), sensitivities in distributed models. None of these methodologies has yet been adequately researched.

In moving to an increasingly complex analysis of model spatial behaviour we should also not lose sight of the fact that end-users will require better guidance in order to evaluate correctly the information being imparted to them. Scientific uncertainty and risk have proved difficult concepts to communicate, yet to continue with a view of models as single deterministic predictors seems flawed and, one suspects, susceptible to legal challenge. The value of spatial data is that it forces modellers to uncover and analyse uncertainties previously suppressed in the calibration process and our challenge is to both incorporate it in our models and provide a more realistic view of the science that we are able to achieve.

Acknowledgements

The research reported in this chapter has been made possible by a number of research grants but in particular UK Natural Environment Research Council grant number GR3 CO 030 and the European Union Framework 5 grant 'Development of an European Flood Forecasting System'.

References

Alsdorf, D., Melack, J.M., Dunne, T., Mertes, L.A.K., Hess, L.L. and Smith, L.C., 2000, Interferometric radar measurements of water level changes on the Amazon flood plain, *Nature*, **404**, 174–177.

Aronica, G., Bates, P.D. and Horritt, M.S., 2002, Assessing the uncertainty in distributed model predictions using observed binary pattern information within GLUE, *Hydrological Processes*, **16**, 2001–2016.

Aronica, G., Hankin, B. and Beven, K., 1998, Uncertainty and equifinality in calibrating distributed roughness coefficients in a flood propagation model with limited data, *Advances in Water Resources*, **22**, 349–365.

Bates, P.D., 2000, Development and testing of a sub-grid scale model for moving boundary hydrodynamic problems in shallow water, *Hydrological Processes*, **14**, 2073–2088.

Bates, P.D. and De Roo, A.P.J., 2000, A simple raster-based model for floodplain inundation, *Journal of Hydrology*, **236**, 54–77.

Bates, P.D. and Hervouet, J.-M., 1999, A new method for moving boundary hydrodynamic problems in shallow water, *Proceedings of the Royal Society of London, Series A*, **455**, 3107–3128.

Bates, P.D., Horritt, M. and Hervouet, J.-M., 1998b, Investigating two dimensional finite element predictions of floodplain inundation using fractal generated topography, *Hydrological Processes*, **12**, 1257–1277.

Bates, P.D., Horritt, M., Smith, C. and Mason, D., 1997, Integrating remote sensing observations of flood hydrology and hydraulic modelling, *Hydrological Processes*, **11**, 1777–1795.

Bates, P.D., Anderson, M.G., Baird, L., Walling, D.E. and Simm, D., 1992, Modelling floodplain flows using a two-dimensional finite element model, *Earth Surface Processes and Landforms*, **17**, 575–588.

Bates, P.D., Stewart, M.D., Siggers, G.B., Smith, C.N., Hervouet, J.-M. and Sellin, R.H.J., 1998a, Internal and external validation of a two dimensional finite element model for river flood simulation, *Proceedings of the Institution of Civil Engineers, Water Maritime and Energy*, **130**, 127–141.

Bechteler, W., Hartmaan, S. and Otto, A.J., 1994, Coupling of 2D and 1D models and integration into Geographic Information Systems (GIS), in W.R. White and J. Watts (eds), *Proceedings of the 2nd International Conference on River Flood Hydraulics* (Chichester: John Wiley & Sons), 155–165.

Beven, K.J., 1989, Changing ideas in hydrology: the case of physically based distributed models, *Journal of Hydrology*, **105**, 79–102.

Beven, K.J. and Binley, A.M., 1992, The future of distributed models: model calibration and predictive uncertainty, *Hydrological Processes*, **6**, 279–298.

Biggin, D.S. and Blyth, K., 1996, A comparison of ERS-1 satellite radar and aerial photography for river flood mapping, *Journal of the Chartered Institute of Water Engineers and Managers*, **10**(1), 59–64.

Chow, V.T., 1959, *Open Channel Hydraulics* (New York: McGraw-Hill).

Cobby, D.M., Mason, D.C. and Davenport, I.J., 2001, Image processing of airborne scanning laser altimetry for improved river flood modelling, *ISPRS Journal of Photogrammetry and Remote Sensing*, **56**, 121–138.

Cobby, D.M., Mason, D.C., Davenport, I.J. and Horritt, M.S., 2000, Obtaining accurate maps of topography and vegetation to improve 2D hydraulic flood models, *Proc. EOS/SPIE Symposium on Image Processing for Remote Sensing VI*, Barcelona, 25–29 Sept 2000, 125–136.

Costa, J.E., Spicer, K.R., Cheng, R.T., Haeni, F.P., Melcher, N.B., Thurman, E.M., Plant, W.J. and Keller, W.C., 2000, Measuring stream discharge by non-contact methods: a proof-of-concept experiment, *Geophysical Research Letters*, **27**, 553–556.

Defina, A., 2000, Two-dimensional shallow flow equations for partially dry areas, *Water Resources Research*, **36**, 3251–3264.

Defina, A., D'Alpaos, L. and Matticchio, B., 1994, A new set of equations for very shallow water and partially dry areas suitable to 2D numerical models, in P. Molinaro and I. Natale (eds), *Modelling Flood Propagation Over Initially Dry Areas* (New York: American Society of Civil Engineers), 72–81.

Engman, E.T. and Chauhan, N., 1995, Status of microwave soil moisture measurements with remote sensing, *Remote Sensing of the Environment*, **51**, 189–198.

Estrela, T. and Quintas, L., 1994, Use of GIS in the modelling of flows on floodplains, in W.R. White and J. Watts (eds), *Proceedings of the 2nd International Conference on River Flood Hydraulics* (Chichester: John Wiley & Lons), 177–189.

Fathi-Maghadam, M. and Kouwen, N., 1997, Nonrigid, nonsubmerged, vegetative roughness on floodplains, *Journal of Hydraulic Engineering, ASCE*, **123**, 51–57.

Feldhaus, R., Höttges, J., Brockhaus, T. and Rouvé, G., 1992, Finite element simulation of flow and pollution transport applied to a part of the River Rhine. in R.A. Falconer, K. Shiono and R.G.S. Matthews (eds), *Hydraulic and Environmental Modelling: Estuarine and River Waters* (Aldershot: Ashgate Publishing), 323–344.

Fuller, R.M. and Parsell, R.J., 1990, Classification of TM imagery in the study of land use in lowland Britain: practical considerations for operational use, *International Journal of Remote Sensing*, **11**, 1901–1917.

Fuller, R.M., Groom, G.B. and Jones, A.R., 1994a, The land cover map of Great Britain: an automated classification of Landsat Thematic Mapper data, *Photogrammetric Engineering & Remote Sensing*, **60**, 553–562.

Fuller, R.M., Groom, G.B. and Wallis, S.M., 1994b, The availability of Landsat TM images for Great Britain, *International Journal of Remote Sensing*, **15**, 1357–1362.

Gee, D.M., Anderson, M.G. and Baird, L., 1990, Large scale floodplain modelling, *Earth Surface Processes and Landforms*, **15**, 512–523.

Gomes Pereira, L.M.G. and Wicherson, R.J., 1999, Suitability of laser data for deriving geographical information: a case study in the context of management of fluvial zones, *ISPRS Journal of Photogrammetry and Remote Sensing*, **54**, 104–114.

Hey, R.D., 1979, Flow resistance in gravel-bed rivers, *Journal of the Hydraulics Division, ASCE*, **105**, 365–379.

Horritt, M.S., 1998, Enhanced flood flow modelling using remote sensing techniques. PhD thesis, University of Reading, UK.

Horritt, M.S., 1999, A statistical active contour model for SAR image segmentation, *Image and Vision Computing*, **17**, 213–224.

Horritt, M.S., 2000a, Calibration of a two-dimensional finite element flood flow model using satellite radar imagery, *Water Resources Research*, **36**, 3279–3291.

Horritt, M.S., 2000b, Development of physically based meshes for two dimensional models of meandering channel flow, *International Journal of Numerical Methods in Engineering*, **47**, 2019–2037.

Horritt, M.S. and Bates, P.D., 2001a, Predicting floodplain inundation: raster-based modelling versus the finite element approach, *Hydrological Processes*, **15**, 825–842.

Horritt, M.S. and Bates, P.D., 2001b, Effects of spatial resolution on a raster-based model of flood flow, *Journal of Hydrology*, **253**, 239–249.

Horritt, M.S. and Bates, P.D. 2002, Evaluation of 1-D and 2-D numerical models for predicting river flood inundation, *Journal of Hydrology*, **268**, 87–99.

Horritt, M.S., Mason, D.C. and Luckman, A.J., 2001, Flood boundary delineation from synthetic aperture radar imagery using a statistical active contour model, *International Journal of Remote Sensing*, **22**, 2489–2507.

Hwang, P.A., Krabill, W.B., Wright, W., Swift, R.N. and Walsh, E.J., 2000, Airborne scanning lidar measurement of ocean waves, *Remote Sensing of Environment*, **73**, 236–246.

King, I.P. and Roig, L.C., 1988, Two-dimensional finite element models for floodplains and tidal flats, in K. Niki and M. Kawahara (eds), *Proceedings of an International Conference on Computational Methods in Flow Analysis*, Okayama, Japan, 711–718.

Konikow, L.F. and Bredehoeft, J.D., 1992, Ground-water models cannot be validated, *Advances in Water Resources*, **15**, 75–83.

Kouwen, N., 1988, Field estimation of the biomechanical properties of grass, *Journal of Hydraulic Research*, **26**, 559–568.

Kouwen, N., 2000, Closure of 'effect of riparian vegetation on flow resistance and flood potential', *Journal of Hydraulic Engineering, ASCE*, **126**(12), 954.

Kouwen, N. and Fathi-Maghadam, M., 2000, Friction factors for coniferous trees along river, *Journal of Hydraulic Engineering, ASCE*, **126**, 732–740.

Kouwen, N. and Li, R.M., 1980, Biomechanics of vegetative channel linings, *Journal of the Hydraulics Division, ASCE*, **106**, 713–728.

Magnussen, S. and Boudewyn, P., 1998, Derivations of stand heights from airborne laser scanner data with canopy-based quantile estimators, *Canadian Journal of Forest Research*, **28**, 1016–1031.

Marks, K. and Bates, P.D., 2000, Integration of high resolution topographic data with floodplain flow models, *Hydrological Processes*, **14**, 2109–2122.

Marks, K.J., 2001, Enhanced flood hydraulic modelling using topographic remote sensing, PhD thesis, University of Bristol.

Mason, D.C., Cobby, D.M. and Davenport, I.J., 1999, Image processing of airborne scanning laser altimetry for some environmental applications, *Proc. EOS/SPIE Symposium on Image Processing for Remote Sensing V*, Florence, 20–24 Sept, vol. 3871, 55–62.

Mason, D.C., Cobby, D.M., Horritt, M.S. and Bates, P.D. 2003, Floodplain friction parameterisation in two-dimensional river flood models using vegetation heights derived from airborne scanning laser altimetry, *Hydrological Processes*, **17**, 1979–2000.

Naesset, E., 1997, Determination of mean tree height of forest stands using airborne laser scanner data, *ISPRS Journal of Photogrammetry and Remote Sensing*, **52**, 49–56.

Ormsby, J.P., Blanchard, B.J. and Blanchard, A.J., 1985, Detection of lowland flooding using active microwave systems, *International Journal of Remote Sensing*, **5**, 317–328.

Pearson, D., Horritt, M.S., Gurney, R.J. and Mason, D.C., 2001, The use of remote sensing to validate hydrological models. in M.G. Andeson and P.D. Bates (eds), *Model Validation: Perspectives in Hydrological Science* (Chichester: John Wiley & Sons), 163–194.

Richards, J.A., Woodgate, P.W. and Skidmore, A.K., 1987, An explanation of enhanced radar backscattering from flooded forests, *International Journal of Remote Sensing*, **8**, 1093–1100.

Romanowicz, R. and Beven, K.J., 1997, Dynamic real-time prediction of flood inundation probabilities, *Hydrological Sciences Journal*, **43**, 181–196.

Romanowicz, R., Beven, K.J. and Tawn, J., 1996, Bayesian calibration of flood inundation models, in M.G. Anderson, D.E. Walling and P.D. Bates (eds), *Floodplain Processes* (Chichester: John Wiley & Sons), 333–360.

Smith, C., Bates, P.D. and Anderson, M.G., 1997, Development of an inundation mapping capability using high resolution finite element modelling, *Final Technical Report to the United States Army*, Contract Number N681710-94-C-9109 R&D 7301-EN-01, European Research Office of the US Army, London.

Srokosz, M., 1997, Ocean surface currents and waves and along-track interferometric SAR, *Proceedings of a Workshop on Single-Pass Satellite Interferometry*, Imperial College, London, 22 July 1997.

Temple, D.M., 1987, Closure of 'velocity distribution coefficients for grass-lined channels', *Journal of Hydraulics Engineering, ASCE*, **113**, 1224–1226.

Töyrä, J., Pietroniro, A. and Martz, L.W., 2001, Multisensor hydrologic assessment of a freshwater wetland, *Remote Sensing of the Environment*, **75**, 162–173.

Ulaby, F.T., Dubois, P.C. and van Zyl, J., 1996, Radar mapping of surface soil moisture, *Journal of Hydrology*, **184**, 57–84.

Veitch, N., 2001, Novel use of aerial survey and remote sensing information on floods 2000, EA National Centre for Environmental Data and Surveillance, *Interim R&D Report*.

PART II

TERRESTRIAL SEDIMENT AND HEAT FLUX APPLICATIONS

Editorial: Terrestrial Sediment and Heat Fluxes

Nick Drake

Remote sensing has been used to either parameterize or validate a wide diversity of terrestrial environmental models. In this Part we consider the integration of remote sensing with models of river sediment and soil erosion and fire heat fluxes. These applications illustrate some of the diversity of modelling approaches that can be employed, the different ways by which models can be linked to remote sensing, the variety of scales at which they can be applied and the scaling problems that can be encountered when applying models at coarse scales.

Beginning with soil erosion, natural soil erosion rates are generally low, however, anthropogenic practices tend to increase erosion through factors such as overgrazing, agricultural intensification and the implementation of poor agricultural practices. Accelerated soil erosion leads to higher nutrient losses, a reduction in soil depth and thus over time a lower soil water holding capacity. These factors eventually lead to reduced biomass yields and can ultimately result in desertification. Erosion also has important off-site effects. For example, accelerated water erosion rates lead to increased sedimentation rates elsewhere that can adversely affect the ecology and biodiversity of aquatic systems. There is, therefore, a need to model soil erosion in order to predict the consequences of these actions.

Models of soil erosion by water were first developed in the 1940s by analysing the results of erosion plot studies (Zingg, 1940). This research led to the development of the universal soil loss equation (USLE) (Wischmeier and Smith, 1958), an empirical model that proved extremely popular with both managers and researchers and dominated the field for some time, particularly in America where it was developed. However, the problems of applying the model outside America, and the fact that it can only be used to estimate average annual erosion led to the development of alternatives. By the 1980s it was recognized that physically

---

Spatial Modelling of the Terrestrial Environment. Edited by R. Kelly, N. Drake, S. Barr.
© 2004 John Wiley & Sons, Ltd. ISBN: 0-470-84348-9.

based modes have the potential to overcome the limitations of empirical models and since then numerous physically based models have been developed and tested. However, the diverse array of factors that affect erosion means that the models are becoming increasingly complex and have thus, in turn, become more demanding on the number of parameters that they require. Furthermore, there is a lack of understanding of some of the key processes, such as soil crusting and rill initiation, that affect erosion dynamics and these processes are not usually parameterized in models. If they are, the equations used to represent them are derived from field or laboratory experiments, and are usually somewhat empirical. Consequently, many physically based models have empirical elements and, over time, a spectrum of models with varying amounts of theoretical rigour and empiricism has emerged.

Wind erosion modelling started with Bagnold (1941) who developed a theoretical model to explain the saltation of particles induced by wind. This model forms the basis of many of the physically based models used today. The development of empirical models such as the wind erosion equation (WEQ) (Woodruff and Siddoway, 1965) was attractive for resource managers as they could be readily implemented to provide an estimate of on-site soil loss. However, as was the case with water erosion models, now much effort is being spent developing physically based process models in order to overcome the limitations of their empirical counterparts. Thus, the physically based wind erosion prediction system (WEPS) has recently been developed as a replacement for the WEQ.

Early wind and water erosion modelling efforts were applied at the plot scale although it had long been recognized that the majority of eroded soil appeared to come from a limited number of sources and that spatial models were required to understand erosion at the regional scale. Remote sensing was an attractive solution in this regard but there was a large gulf between the accuracy of information that was required and what remote sensing could supply. Remote sensing was first used to study erosion by employing aerial photographs to interpret and map erosional and depositional landforms. Initially aerial photos simply enabled terrain mapping to be conducted over larger areas than was possible with localized field techniques. The next logical step was to apply these methods to monitor landform changes (for example, interpreting changes in coastal landforms in order to monitor coastal erosion). This spatial monitoring provided new insights. For example, Gay (1962) monitored the movement of Barchan Dunes by interpreting aerial photographs acquired in different years. He found that not only was the rate of movement proportional to the dune height, as Bagnold's (1941) model predicts, but the movement was also proportional to the dune width, and noted that this means that all Berhcan dunes, regardless of size, sweep equal areas at equal rates.

In the 1970s the advent of satellite remote sensing enabled the development of new applications. Sensors such as the coastal zone colour scanner were developed with wavelengths on them that can penetrate into the water column and allow monitoring of eroded sediments in near surface waters. Though this development provided new insights into near surface sediment transport and dispersal, it supplied little information on hillslope and river channel erosion. This was because the fine scale of hillslope processes is unresolved by the coarse spatial resolution of the sensors. Additionally, water erosion in these regions tends to occur when it is raining and, therefore, cloud obscures the view from optical sensors.

Developments in image processing methods and the ever-increasing diversity of sensors in space meant that by the 1980s it had become possible to derive many of the parameters that control erosion from remotely sensed imagery. For example, remote sensing provides numerous ways to map topography (e.g. photogrammetry, interferometry and LiDAR) and thus derive maps of slope and other topographic factors that control water erosion, while vegetation indices can be used to estimate vegetation cover, the most important parameter controlling resistance to erosion. Initially these parameters were combined in qualitative ways using GIS overlays to produce maps of erosion risk. However, they were soon used to implement spatial models of erosion.

Though these developments in remote sensing have enabled the estimation of numerous parameters of importance to soil erosion modelling, there is a gap between what the models require and what it is possible to obtain. As a result, in nearly all cases, some parameters need to be measured in the field and interpolated in a GIS for effective implementation of the model. It is often costly and time-consuming to frequently monitor such parameters in detail over large areas. Thus, there is a need for parsimonious models that utilize any parameters available from remote sensing.

The first chapter in this Part (chapter 6) is by Lane *et al.* and demonstrates how error can be managed in the application of digital photogrammetry to the quantification of river topography of large, braided, gravel-bed rivers. The chapter reviews the traditional treatment of error in digital elevation models (DEM), and then considers how the error can be identified, explained and corrected in this study in the context of a specific example. This is an important topic because many land erosion models rely on the parameterization of mass transfer processes using a DEM. The second two chapters of this Part illustrate the implementation of water and wind erosion models at coarse scales. In Chapter 7, Okin and Gilette outline the need for regional modelling of sand and dust emission, transport and deposition. They then implement a regional scale wind erosion model developed from Bagnold's equations and compare the results to ground measured observations in the Jornada Basin, New Mexico. They find that the model under-estimates wind erosion and dust flux throughout much of the basin because the parameterization using soil and landuse maps failed to capture the surface variability in the landscape. They conclude that there is a need to develop remote techniques to map the fine scale heterogeneity in the landscape that exerts a large control on wind erosion and dust flux.

In Chapter 8 Drake *et al.* implement a water erosion model for the catchment of Lake Tanganyika with the aim of quantifying the source areas within the catchment, the transfer of sediment to the lake and the sediment dispersal within the catchment. The model is applied using remote sensing to estimate spatial and temporal variations in vegetation cover, rainfall and the near surface sediment distribution within the lake. The erosion model results have been shown to be highly sensitive to scale, with increasingly coarser spatial resolutions causing a gradual reduction in predicted erosion rates. Scaling techniques have been employed in an effort to overcome these problems. It is concluded that validation of the model is needed but that this is problematic when applying coarse resolution data over large areas. These two chapters thus highlight the numerous problems that can be encountered when applying models to coarse resolution imagery but also provide methods that point towards solutions to certain aspects of the problem.

The final process that is considered in this section is fire. Fire affects humans, atmosphere, vegetation and soils, and successful fire modelling has the potential to further our understanding, prediction and management of this phenomenon. The first model of fire considered its spread and was developed by Fons (1946). Since then a large number of models have been developed to consider not only diffusion, but also fire properties and physical characteristics (e.g. surface, crown and ground fires). Similarly with approaches to erosion modelling, both empirical and physically based models have been developed, with the semi-empirical fire spread model of Rothermel (1972) attaining the most widespread use. More recently the secondary effects of fires on the atmosphere, hydrology, ecology and geomorphology of affected regions have started to be modelled, as has the economic impact. Many of these models are empirical (or contain significantly empirical elements) because our understanding of the fine detail of the way fire affects atmosphere, vegetation and soil systems is often poorly understood. Further research is needed into both process understanding and model development.

The development of GIS brought the first attempts to spatially predict fire growth, however, it was soon realized that the spatial data needed to parameterize such models were not available and would be extremely hard to collect. Developments in remote sensing have alleviated this problem to a certain extent and methods are being developed that allow fire detection, assessment of burnt areas, fuel load and fuel moisture content. This information has, in some cases, been integrated with models. For example, burnt area estimates derived from remote sensing have been used to provide important inputs into models that predict the quantity of carbon emitted by fires and the quantity of soil eroded after these events. Though such models provide a significant advance, it has been shown that factors such as errors in estimated fuel load can introduce a large amount of uncertainty. Therefore, new methods to accurately derive such parameters are needed. In the final chapter in this Part (Chapter 9), Wooster *et al.* introduce a new remote sensing method that holds great potential in directly measuring the amount of biomass combusted by the passage of a fire. They introduce a method that models the total amount of energy emitted by the fire, and show both empirically and theoretically that this can be related to the amount of biomass combusted and gases emitted. The method holds great promise in overcoming some of the current limitations to the implementation of the fire models outlined above over large areas.

References

Bagnold, R.A., 1941, *The Physics of Blown Sand and Desert Dunes* (New York: Methuen).

Fons, W.L., 1946, Analysis of fire spread in light forest fuels, *Journal of Agricultural Research*, **72**, 93–121.

Gay, S.P., 1962, Origen distribución y movimiento de las arenas eólicas en el área de Yauca a palpa, *Boletín de la Sociedad del Peru*, **27**, 37–58.

Rothermel, R.C., 1972, A Mathematical Model for Predicting Fire Spread in Wildland Fuels, USDA Forest Service Research Paper INT-115.

Wischmeier, W.H. and Smith, D.D., 1958, Rainfall energy and its relationship to soil loss, *Transactions of the American Geophysical Union*, **39**, 285–291.

Woodruff, N.P. and Siddoway, F.H., 1965, A wind erosion equation, *Proceedings of the Soil Science Society of America*, **29**, 602–608.

Zingg, A.W., 1940, Degree and length of slope as it affects soil loss runoff, *Agricultural Engineering*, **21**, 59–64.

6

Remotely Sensed Topographic Data for River Channel Research: The Identification, Explanation and Management of Error

Stuart N. Lane, Simon C. Reid, Richard M. Westaway and D. Murray Hicks

6.1 Introduction

The quality of Digital Elevation Model (DEM) data is all too often overlooked (Cooper, 1998; Lane, 2000) which can have serious implications for geo-morphological and hydrological study (Wise, 1998). There are a number of reasons why this can be the case. First, there is a growth in the availability of digital data sources, some of which have unknown or poorly specified data quality. A global measure of error (e.g. root mean square error, RMSE) can be misleading (1): if it is based upon test sites that bear little resemblance to the site of interest; and (2) if it fails to recognize the spatial structure of the error field. Second, when digital data are supplied that have been derived from a number of processes (e.g. photogrammetric generation of the initial topographic data, analysis to produce a DEM, digital contouring, construction of a DEM from digital contours), there is the possibility of propagation of error at each stage of the process. This can only be assessed through acquisition of the raw source data and repetition of the process. This raw data may not be available. This is especially problematic as research has shown that there can be substantial magnification of propagated error into DEM-derived geo-morphological and hydrological parameters (e.g. Wise, 1998, 2000; Lane *et al.*, 2000). Third, new methods of data generation have involved progressively greater levels of automation, with a corresponding increase in the amount of data whose quality must be determined (Lane, 2000) and a reduction in the amount of

Spatial Modelling of the Terrestrial Environment. Edited by R. Kelly, N. Drake, S. Barr.
© 2004 John Wiley & Sons, Ltd. ISBN: 0-470-84348-9.

manual quality control (e.g. Lane *et al.*, 1994). Finally, the advantage of automated data generation is also its nemesis. The volume of data generated means that the proportion of check data that can be used to assess error is reduced to the point of becoming unreliable (Lane *et al.*, 2000; Lane, 2000).

Central to the approach adopted in the chapter, and reflected in its structure, is the view that managing error in digital elevation models must be a three-stage process. First, the rate at which topographic data can be generated using remote sensing techniques requires robust methods for finding points that are in error and points that are not. Given the three-dimensional complexity of landform surfaces, this is not straightforward: there can be uncertainty as to whether points are in error, or a natural component of topography. Second, management of error must be based upon firm theoretical and empirical support for the correction procedures adopted. Where possible, this should involve re-collection of data having dealt with the root cause of the error. If this is not possible, then there should be a clear explanation for the cause of that error. This increases the confidence that error has been correctly identified, and also helps in informing the type of correction procedure that is required. Third, correction procedures must only affect points that are in error, and result in the removal of error as far as is possible, and should not introduce new types of error into either the points being collected or the neighbouring points.

With these issues in mind, the aim of this chapter is to demonstrate how we have managed error in the application of automated digital photogrammetry in the quantification of river topography in terms of large, braided, gravel-bed rivers. The focus of the case study is the Waimakariri River, South Island, New Zealand. This chapter explains the case study and methods used to generate the DEMs considered in this analysis, reviews the traditional treatment of error in DEMs, and then considers the manner in which error was identified, explained and corrected in this study. Although error management procedures were developed for a particular case study, they have attributes that may contribute to a more general error management procedure for natural landform surfaces.

6.2 The Waimakariri River Case Study

The method used to generate the data for this study is detailed in full in Westaway *et al.* (2003). This chapter develops this work by looking at the way in which error can be managed, and so only a summary of the data generation methods is provided here. The field site chosen for this study is a 3.3 km long reach of the lower Waimakariri River, South Island, New Zealand (Figure 6.1). The vertical relief in the study reach is small, generally less than 2 m, with a mixed gravel bed with a median grain size (D_{50}) of around 0.028 m (Carson and Griffiths, 1989). Monitoring of the lower Waimakariri River is currently undertaken by Environment Canterbury, the local regional council, who re-survey a number of cross-sections along the river, approximately every 5 years (Carson and Griffiths, 1989). In the 3.3 km study reach, there are four cross-section locations (17.81, 19.03, 19.93 (Crossbank) and 21.13 km from the river mouth), which indicate the spatial density of survey (approximately 1 cross-section per km in this reach).

Aerial photographs of the reach were taken in February and March 1999 and February 2000 by Air Logistics (NZ) Ltd, spaced to provide a 60% overlap which is desirable for photogrammetry. A feature of photogrammetric survey is the constant trade-off between

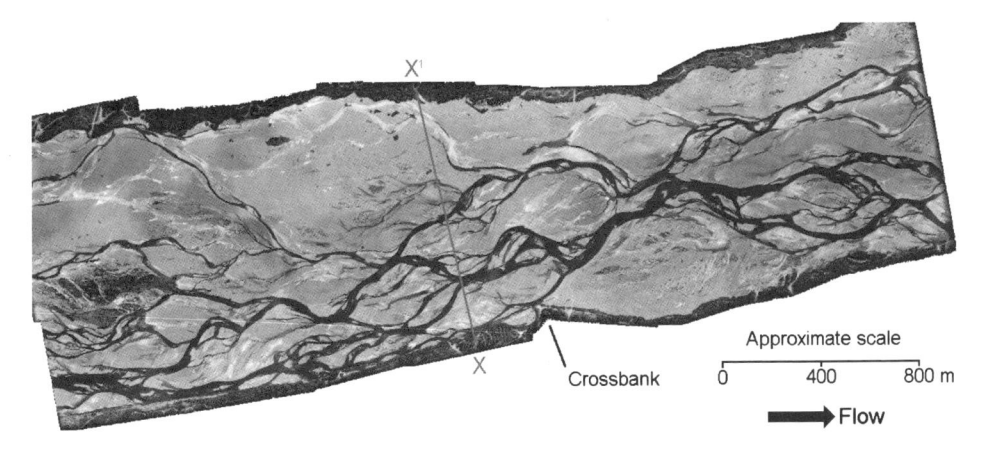

Figure 6.1 Mosaiced area image of the study area in the River Waimakariri, Christchurch, New Zealand

the increase in coverage that comes from a reduction in image scale and the improvement in point precision that results from an increase in image scale (Lane, 2000). This is particularly relevant with respect to large braided riverbeds, because of the low vertical relief (in the order of metres) compared to the spatial extent (in the order of kilometres). This imposes a minimum acceptable point precision for effective surface representation. In this study, a photographic scale of 1:5000 was chosen for the 1999 imagery, which, given a scanning resolution of 14 μm, results in an object space pixel size of 0.07 m (Jensen, 1996). Based on this, the lowest DEM resolution that can be produced is 0.37 m, approximately five times greater than the object space pixel size (Lane, 2000). Analysis of the 1999 imagery had been completed in time for the re-design of the airborne survey to acquire imagery at a larger scale (1:4000) in order to increase texture in the image and thereby improve stereo-matching performance. To achieve full coverage of the active bed, two flying lines were needed. Photo-control points (PCPs) were provided by 45 specially designed targets that were laid out on the riverbed prior to photography and positioned such that at least five (and typically 10–12) targets were visible in each photograph. Their position was determined to within a few centimetres using Trimble real-time kinematic (RTK) GPS survey.

DEMs of the study reach were produced using the OrthoMAX module of ERDAS Imagine installed on a Silicon Graphics UNIX workstation. DEMs were generated using the OrthoMAX default collection parameters with a 1 m horizontal spacing to give around four million (x, y, z) points on the riverbed. An individual DEM was generated for each photograph overlap: two flying lines of eight photographs each meant that in total 14 individual DEMs were produced for the 1999 surveys. The larger scale of the 2000 surveys increased the number of DEMs to 20. In addition, image analysis was conducted in order to obtain water depths for inundated areas of the bed. These methods are dealt with in full in Westaway *et al.* (2003). This chapter only addresses exposed areas of the riverbed.

To aid in the assessment of DEM quality, and hence to identify means of improving data collection, the generated DEMs were assessed using check point elevations measured by NIWA and Environment Canterbury field teams using a combination of Total Station and Trimble RTK GPS survey. An automated spatial correspondence algorithm was used to assign surveyed elevations to the nearest DEM elevation, provided the point was within a given search radius. The larger the search radius used, the more check data points are assigned DEM elevations. However, this reduces confidence in the resulting height discrepancies, as the average lateral distance between DEM and survey point is increased. In this study, the search radius was set to 0.5 m (to give a search diameter of 1 m), thereby matching the DEM grid spacing.

6.3 The Meaning of Error and the Treatment of Error in Digital Elevation Models

6.3.1 The Definition and Quantification of Error

Perhaps one of the most important aspects of error is a widespread variation in how it is defined and managed in the measurement of surface topography. By far the most common reference to surface quality is in terms of its accuracy (e.g. Davison, 1994; Neill, 1994; Gooch et al., 1999). However, in engineering surveying (e.g. Cooper, 1987; Cooper and Cross, 1988), it is very common to consider error more generically, in terms of the quality of topographic data, with accuracy describing one subset of data quality. There are then three types of error: systematic error, blunders and random errors (e.g. Cooper and Cross, 1988; Lane et al., 1994), and these are thought to control data accuracy, reliability and precision, respectively. Systematic errors occur when a measurement is used with an incorrect functional model (Cooper and Cross, 1988). Cooper and Cross give the example of how the use of the basic collinearity equations derived from the special case of a perspective projection (e.g. Ghosh, 1988) will result in systematic error as the functional model (the collinearity equations) does not include those effects that cause deviation from the perspective projection (e.g. sensor distortions). The functional model provides an inaccurate description and leads to systematic errors. Cooper and Cross (1988) note that there did not appear to be a widely accepted term to describe the quality of a dataset with respect to systematic errors and recommend the term *accuracy*. Blunders or gross errors arise from an incorrect measuring or recording procedure. Cooper and Cross noted that when measurements were made manually, it was relatively easy to identify and rectify gross errors through independent checks on measured data. However, with automation of the measurement process, this is more difficult, but still needs to be undertaken. The quality of a dataset in terms of blunders is defined in terms of its *reliability* (Cooper and Cross, 1988). Third, random errors relate to inconsistencies that are inherent to the measurement process, and cannot be refined by either development of the functional model or through the detection of blunders. The quality of a dataset in terms of random errors is defined as its *precision*.

Whilst these definitions largely relate to the engineering surveying approach to data quality, there is some difference in terms of the definitions used in statistical analysis. Everitt (1998) defines: (1) accuracy as the degree of conformity to some recognized standard value; (2) precision as the likely spread of estimates of a parameter in a statistical model; and introduces (3), bias, as the deviation of results or inferences from the truth. It is clear from these definitions that there is some confusion over the meaning of accuracy and of bias: they appear to be the same thing according to Everitt (1998). This implies that the distinction between accuracy and bias is a subtle one. Any observation may differ from its correct value, appearing to be inaccurate, but one observation does not allow us to decide whether this error is associated with random or systematic effects. If that observation were made many times, then there would be random variability in the observation, which strictly describes its precision, reflecting the inherent presence of error in all measurement systems. The mean value (i.e., a sample mean) of the observation may then be compared with the true value to decide on its bias, which is a measure of the extent to which the observation is free from systematic error, or is accurate. No measurement will be perfectly accurate so that accuracy has to be defined in terms of acceptability in relation to a particular study. Bias refers to the extent to which systematic error can be detected in a statistical sense, given the precision of a particular measurement. Most importantly, there is a general trend in the assessment of surface error to collapse accuracy and precision into a single measure of data quality (the RMSE) which is commonly labelled accuracy (e.g. Shearer, 1990; Davison, 1994; Neill, 1994; Zoej and Petrie, 1998). The RMSE is defined as:

$$\text{RMSE} = \sqrt{\frac{1}{n} \sum_{i=1}^{n} e_i^2} \qquad (1)$$

where n is the number of observations used to determine the RMSE; $ei = z_{ei} - z_{ci}$; z_{ei} is the estimated elevation of point i; and z_{ci} is the correct elevation of point i, measured independently. However, the standard engineering surveying definition of precision (Cooper and Cross, 1988) is the variance of the error in the derived parameter (in this case elevation):

$$\sigma_e^2 = \frac{1}{n-1} \sum_{i=1}^{n} (e_i - \overline{e_i})^2 \qquad (2)$$

where σ_e is the standard deviation of error (SDE). Equation (2) reduces to equation (1) if the mean error (ME) is zero. This distinction has been recognized (e.g. Gooch et al., 1999; Lane et al., 2000) and this emphasizes a basic problem with the RMSE as a measure of surface error: it only gives the precision of surface measurements if the ME is zero. In practice, we are interested in eliminating the ME and reducing the SDE.

If we return to the definitions of data quality provided by Cooper and Cross (1988), then if the correct functional model has been used (the surface is accurate), all blunders have been identified and corrected (the surface is reliable), and all data points have been obtained using the same data collection method, then the quality of the surface is only controlled by random errors, which may assumed to be spatially stationary. This can be expressed in terms of the surface precision defined in equation (2). If the surface also has a ME of zero, and there is no non-random error in the surface, equation (2) reduces to equation (1).

The determination of the SDE is important as a means of expressing data quality. It is also the first step in determining the quality of data derived from a surface. Commonly, geo-morphological research requires detection of change between two time periods, determination of the associated volume of material moved and evaluation of parameters such as slope. Lane et al. (2003) show that, with two surfaces (1 and 2) and under the assumption of spatially uniform precision in each surface and surfaces with independent error fields, the uncertainty in an elevation difference (σ_d) at any location is:

$$\sigma_d = \sqrt{(\sigma_{e1})^2 + (\sigma_{e2})^2} \qquad (3)$$

and that integration to the volume of change, with a regularly spaced DEM with $I \times J$ nodes spaced d apart, gives an uncertainty in volume estimates of:

$$\sigma_v = d^2 \left(IJ\left[\sigma_{e1}^2 + \sigma_{e2}^2\right]\right)^{0.5} \qquad (4)$$

A similar expression can be derived for slope estimates from a DEM where, under the assumption that the global SDE in elevation (σ_e) applies to every data point in a given DEM, the uncertainty in slope is:

$$\sigma_s = \frac{\sqrt{2}\sigma_e}{d} \qquad (5)$$

6.3.2 Local Versus Global Measurements of Data Quality

The above section emphasizes a major problem in the quantification of surface quality. Cooper and Cross (1988) make the distinction between local measures of precision that refer to a restricted number of parameters and global measures of precision that refer to all parameters using a single value. This is the essence of the problem of the description of surface quality using equation (2) and the propagation of surface error into derived parameters using equation (3) through equation (5). Strictly speaking, each data point has its own precision and a surface contains many data points, each with a unique measure of precision. However, in practice and as in Section 6.3.1, we substitute repeat measurement at a point on a surface with singular measurement of many data points distributed across a surface. Provided all of these data points have the same precision (i.e., the same random error), the variance of error in these spatially distributed data points tells us the precision of each data point in the surface. We can also determine the ME for the surface. Homogeneity of surface error then allows the use of global error descriptors. The problem here is the validity of the assumption that allows this spatial substitution for repeat measurement of the same point. Three issues emerge. First, it requires there to be no variation in precision as a function of space. The interaction between surface characteristics and the measurement techniques applied to those surfaces means that this is not always the case. For instance, in the case of digital photogrammetry, the precision of the stereo-matching process is controlled by surface texture, which will vary spatially across an image. Some data points may be matched with a greater precision than others. Similarly, where the rate of topographic change is greater, matching algorithms perform less effectively, and the precision of associated data points tends to be reduced (Lane et al., 2000). Second, it assumes that the error in each data point is purely random, and uncontaminated by both systematic error and blunders. This matters as, in theory, both systematic error and blunders may be reduced and possibly eliminated, either through improvement in the functional model used to relate measurements to parameters or through the identification and correction of blunders. Blunders are particularly problematic as they tend to be local (i.e., associated with individual points) and may be relatively small in number compared with the full dataset. Third, these issues mean that a global indicator of precision may mask localized and severe error, which can have particularly serious implications when parameters are derived from elevations (e.g. local surface slope or change between two dates), and error is propagated; the use of a global value of σ_e is not correct as locally $\sigma_{eij} > \sigma_e$ due to blunders and/or systematic error and/or spatial variation in random error.

This discussion implies that the description of surface quality needs very careful thought: global measures of error should only be applied over areas that have a globally homogeneous

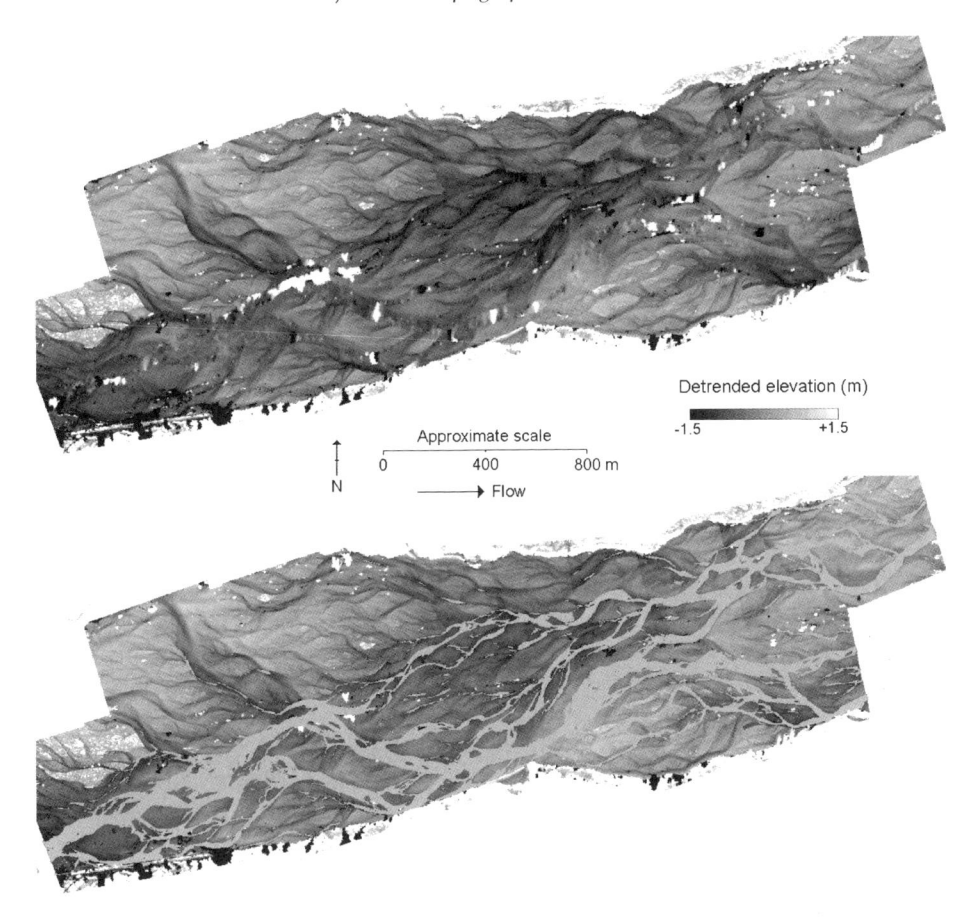

Figure 6.2 (a) The default photogrammetrically acquired Digital Elevation Model for the Waimakariri River at Crossbank in February 2000 and (b) with inundated and vegetated areas removed using unsupervised classification

precision. In some situations, it may be possible to classify different areas in a surface as having different precisions (see below) and apply equations (3)–(5) accordingly (i.e., regional estimates of precision). However, even then, methods must be developed that: (1) identify and correct localized error, which will essentially remove blunders; and (2) remove ME, which should deal with the effects of incorrect functional models, as these will commonly affect many points over a wide area, rather than individual points. If blunders and inaccuracies can be removed, then the precision of a region of a surface will be the same as individual data points that constitute that region, and regional measures of data quality become valid.

6.3.3 Error in the Waimakariri Study

Many of the points made in sections 6.3.1 and 6.3.2 are illustrated by considering a default DEM collected as part of the Waimakariri study (Figure 6.2(a)). This shows that there is clearly error in the surface. An obvious cause of photogrammetric failure is where the river

Table 6.1 Check data, photogrammetrically acquired points for elevations that were not inundated and the associated error statistics

DEM	Number of check points available	Number of photogrammetrically acquired data points	Check points as a percentage of dry points	Mean error (m)	Standard deviation of error (m)
February 1999	3700	2 451 166	0.15	0.542	2.113
March 1999	241	2 232 187	0.01	0.252	0.883
February 2000	1661	2 288 138	0.02	0.064	0.926

bed cannot be seen either because it is vegetated, or because it is inundated with water which, in the case of the Waimakariri, is commonly too turbid to allow through water photogrammetry (e.g. Westaway et al., 2000). However, even after erroneous data points are removed (Figure 6.2(b)), it is clear that many errors remain. To quantify the error in this surface, it is necessary to consider data points that have been stereo-matched and not those where matching has failed and which have been interpolated. Interpolated data points may be affected by incorrectly matched points. This will violate the use of equation (2) as it is possible that adjacent data points are not independent when surface derivatives such as slope are calculated.

Whilst the percentage of data points that could be checked is exceptionally low due to the relatively small size of the check dataset (Table 6.1), the magnitude and frequency of DEM errors are sufficient to give exceptionally poor data quality parameters (Table 6.1) given the total topographic relief of about 2 m. There is systematic error in the surface, associated with MEs significantly greater than zero, and poor surface point precision with very high-standard deviations of error. However, there are clearly areas of the riverbed which appear to be of a very high quality in visual terms and even with the errors present, there is considerable spatial structure present in the DEM. This aside, three issues emerge. First, many of the errors appear to be spatially concentrated along the main river channels. Similarly, there appears to be some sort of banding present in the default DEM, which may be due to some form of inaccuracy in the functional model (e.g. specification of the collinearity equations). Second, if these DEMs were used to calculate a mean bed level, there would be a systematic error of between 0.064 m (February 2000) and 0.542 m (February 1999). These are significantly greater than those that would arise from measurement using the conventional survey methods applied in the Waimakariri (levelling), except where bias is introduced due to the relatively large spacing between levelled cross-sections. In this study, these errors are too great to allow meaningful estimation of mean bed levels for river management purposes. Third, the number of data points used to represent the surface makes manual checking of each data point using stereo-vision unfeasible; a trained operator may check up to 500 points per hour, although this checking rate cannot always be sustained (Lane, 1998). This requires approximately 4900 hours to check each DEM. Similarly, the number of check points available for assessing each data point is a very low percentage of the number of data points generated photogrammetrically. Field data collection had to involve check data stratification to areas close to and along the channel edge, where it is suspected that photogrammetric error is likely to be greater (see below), so that these global estimates of surface quality are not necessarily reliable. Thus, we have the basic challenge that the remainder of this chapter seeks to address. We need: (1) to find out which data points are in error; and (2) to attempt to explain why they are in error to allow either data re-collection or to justify correction. This should lead to the removal of blunders and the reduction of systematic error, to produce a surface that is of a higher quality, where that quality is controlled by random error and defined by equation (2) above.

6.4 Error Identification

Central to the error identification methods that were evaluated and applied is automation in order to deal with the large number of data points that needed to be checked. First, image analysis using unsupervised two-way classification of rectified imagery was used to remove wet and vegetated data points (Figure 6.2(b)). Second, the stereo-matching process also produced a map identifying which data points had been unsuccessfully matched, and hence which needed to be interpolated. The result of these two components of post-processing was the distribution of data points shown in Figure 6.3 for February 2000. The automated identification of error needed to focus upon the individual point errors and banding shown in Figures 6.2(b) and 6.3.

6.4.1 The Identification of Localized Error

The photogrammetric data collection process included an element of post-processing to remove extreme point elevations. This involves a local variance based filter, based upon the Chauvenet principle (Taylor, 1997) with a point rejected if:

$$z_{ij} - \bar{z} => m\sigma_z \quad (6)$$

where z_{ij} is the point being considered, $\bar{z}$ is the average elevation defined for all data points

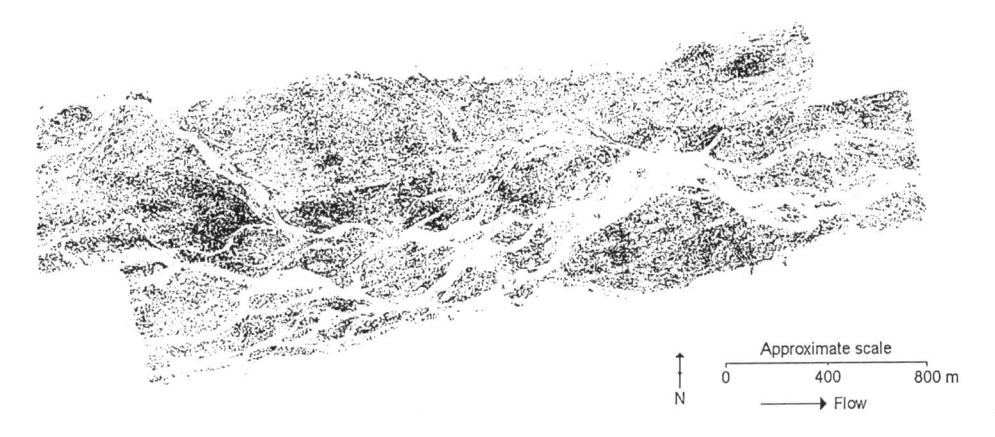

Figure 6.3 *The data points left in the surface after removing vegetated and inundated points shown in Figure 6.2(b)*

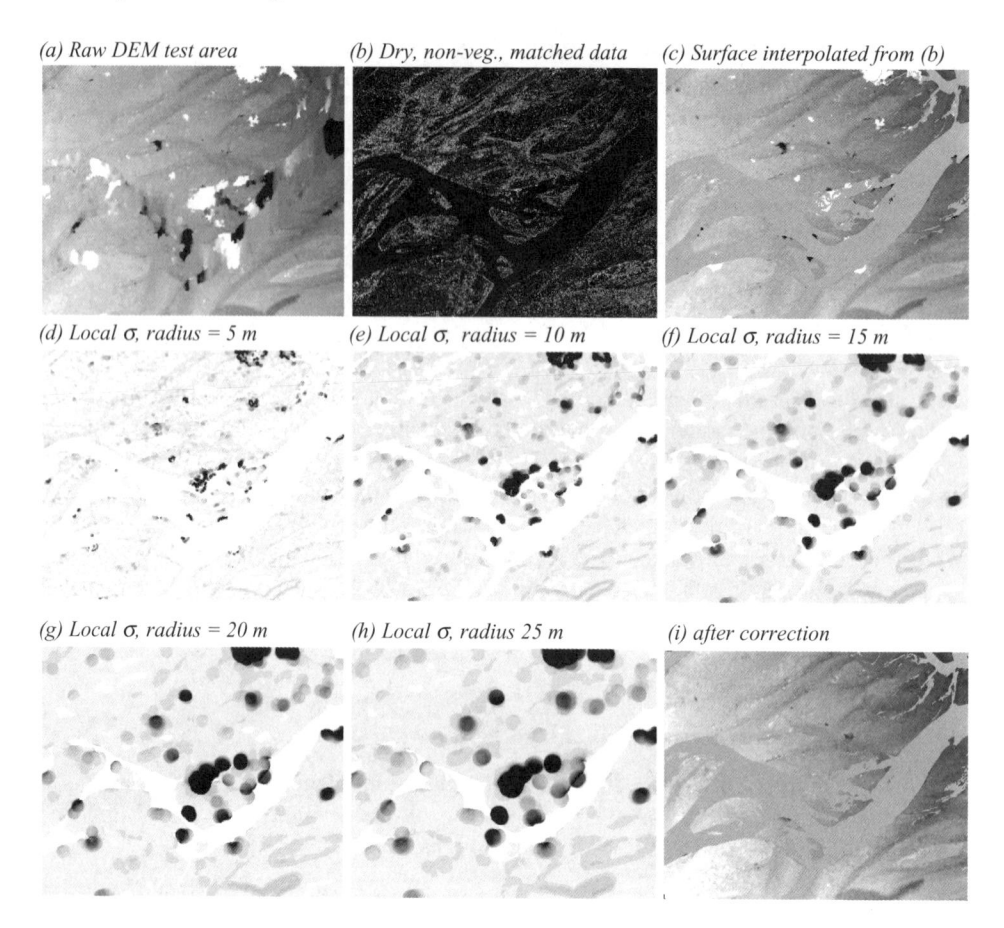

122 *Spatial Modelling of the Terrestrial Environment*

(a) Raw DEM test area   *(b) Dry, non-veg., matched data*   *(c) Surface interpolated from (b)*

(d) Local σ, radius = 5 m   *(e) Local σ, radius = 10 m*   *(f) Local σ, radius = 15 m*

(g) Local σ, radius = 20 m   *(h) Local σ, radius 25 m*   *(i) after correction*

Figure 6.4 *Diagrams to illustrate application of standard deviation-based filtering of error (approach 1) to the test area. Part (b) shows the data points that were matched, in non-inundated and non-vegetated areas and part (c) the surface interpolated from these. Parts (d) through (h) show the effects of calculating point standard deviation for different search radii. Part (i) shows the final accepted combination of search radius (20 m) and filter tolerance (0.5 m), demonstrating that many of the spikes and pits in part (c) have now been removed*

neighbouring and including the pixel (i.e., a 3 × 3 matrix), σ_z is the SD of those elevations used to calculate, and m is a user-specified value. This is a local topography-based filter that will clearly be effective in the removal of isolated spikes and dips. However, it will break down where there are clusters of points in error, reflected in the retention of significant error in Figure 6.2(a). The effect of equation (6) was explored for a test area (Figure 6.4(a)), which contains isolated points in error (Figure 6.4(c)) after filtered out of unmatched, dry and vegetated points (Figure 6.4(b)). This was applied in a manner similar to Felicísimo (1994) for grid-based data but used in this case with point data. A local SD was calculated for circles of varying search radii. A point was eliminated if the calculated SD centred on a point was greater than a SD tolerance. Note that if the method had been applied to the full-gridded DEM, then it would have been ineffective because point errors would have been magnified due to propagation of error through incorrectly matched points or interpolated points. This would raise σ_z and make the incorrect elevations more difficult to find: it is vital to remove blunders before a surface is fitted to a dataset.

The user has to choose two main parameters: (1) the radius of the search error (r); and (2) the SD tolerance (t). In the case of search radius, 5 values were explored: $r = 5$ m, $r = 10$ m, $r = 15$ m, $r = 20$ m and $r = 25$ m. The effects of search area upon SD is shown in Figure 6.4(d) through (h) and it is clear that there is some relationship between SD and the error that is clear in Figure 6.4(c). The use of a search radius means that a larger area can be considered than the 3×3 elevation matrix used in the photogrammetric software's application of equation (6). This is important as it increases the reliability of derived standard deviations due to the inclusion of a larger number of data points. In practice then, the actual area chosen should reflect the topography of the surface being considered. The greater the local rate of elevation change (i.e., with less local stationarity at order 2), the smaller the search area that may be used for reliable SD estimation. However, there is a trade-off here as with larger numbers of data points, the ability to detect individual points that are in error, as opposed to clusters of data points, is reduced: highly localized errors may not be detected by this method. Thus, we explore the effects of a number of different search ranges to identify an optimal one for the dataset used here.

Three tolerance levels were tested (2 m, 1 m and 0.5 m). There is an *a priori* reason for the choice of tolerance levels based upon the expected SD and the characteristics topography of this type of river: relatively flat bar surfaces with breaks of slope between bar surfaces of different elevation. The maximum possible expected standard deviation, for a dataset within a given area, will occur when the break of slope falls exactly along the diameter of each search radius, with 50% of elevations at the maximum value for the surface and 50% of elevations at the minimum value for the surface. This results in an area-independent justification of all three tolerance values. If the surface topographic range is 2 m, then the maximum possible expected SD is 1 m, which was set as the median tolerance value considered. The 2 m tolerance value was chosen to recognize that the SD is almost certainly higher than this because of grain and bedform scale related elevation variation on both high- and low-bar surfaces. The 0.5 m tolerance was chosen under two assumptions: (1) the removal of too much data would not be a problem in a surface that already contained so much data; and (2) that the 1 m tolerance is essentially the maximum possible standard deviation.

Figure 6.5 shows the test results, plotted for different radii of curvature and tolerance levels, as compared with check point data, for the test area shown in Figure 6.4. As expected, there is strong interaction between search radius and tolerance. The search radius of 5 m retains a large number of points and ME and SDE both remain high. Once the 10 m search radius is reached, with the exception of the 2 m tolerance level, the SDE falls significantly, to below 1 m. Optimal results are obtained with a 15 m or 20 m search radius. After this, the SDEs for the 0.5 m and 2 m tolerances begin to rise again. The observations reflect the point made above: as the search radius is increased, the SD becomes more reliable as more good points will be included. However, at large search radii, the effects of points in error upon the SD become too small for even the tolerances used here. Thus there is a link between search radius and tolerance. This link will be surface dependent (in relation to topographic variability) and error dependent (according to the level of point clustering). The 0.5 m tolerance level consistently produces the lowest SDE for all radii of curvature

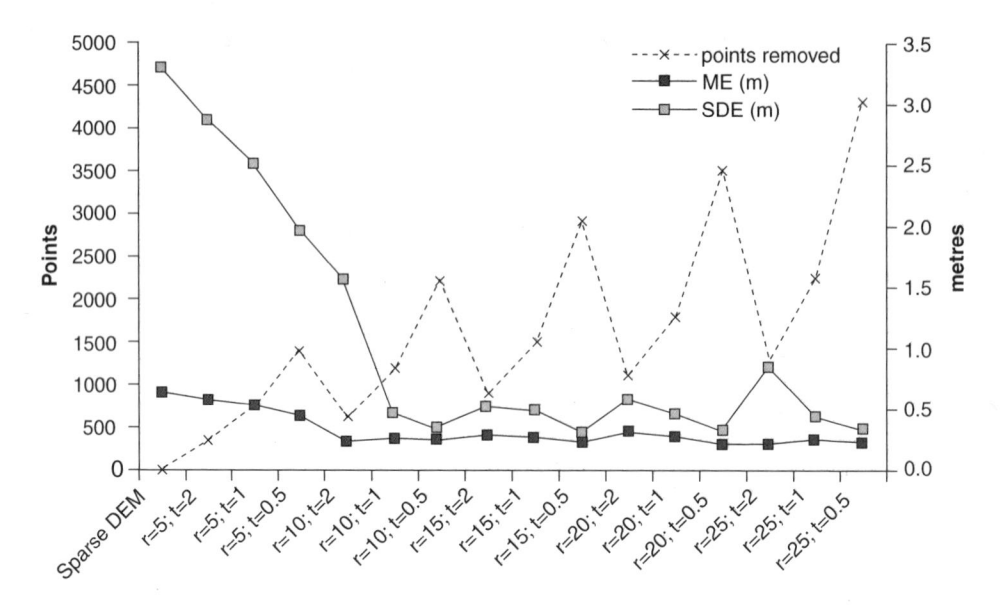

Figure 6.5 *The effects of different values of search radius (r) and tolerance (t) upon the number of points, mean error and standard deviation of error, using approach 1 for the test area shown in Figure 6.4(a)*

and this is reflected in the significantly larger number of data points removed with the 0.5 m tolerance as compared with 1 m and 2 m tolerances. This emphasizes the basic problem with any type of thresholding like this: as tolerance is reduced, smaller reductions in SDE occur for proportionately larger numbers of points lost. There is a tendency towards the loss of greater amounts of potential 'signal' as the boundary between signal and 'noise' is approached. On the basis of the theoretical reasoning above, given the relatively low relief of this surface in relation to the large amount of available data, it was decided to adopt a radius of curvature of 20 m with a SD tolerance of 0.5 m (Figure 6.4(i)). It is clear that some, but not all, of the error in Figure 4(c) has been removed by using a local SD filter.

The main limitation of the approach above is that it is not as topographically sensitive as it could be. The SD is locally determined, but there is no variation in tolerance level in relation to the expected standard deviation. As a result, there is a greater probability of removal of correct points (e.g. along near-vertical channel banks) or retention of erroneous points in areas of greater rate of topographic change than in those where the surface topography is smoother. Thus, as a generic method for the identification of DEM error, the success of the method will be spatially variable in the presence of spatial variation in natural topographic variability. It is likely to fail over very rough surfaces where spatial structure in surface variability is the norm. A second method sought to address this problem and is based upon the observation that DEMs collected at coarser resolutions using digital photogrammetry tend to be smoother. The method also reflects the hierarchical nature of the stereo-matching used here. The stereo-matching was area-based, the algorithm produces elevation averages weighted over a given template. Thus, coarser resolution DEMs use a large template in the image (i.e., image area) to determine elevation. This tends to increase the precision of the

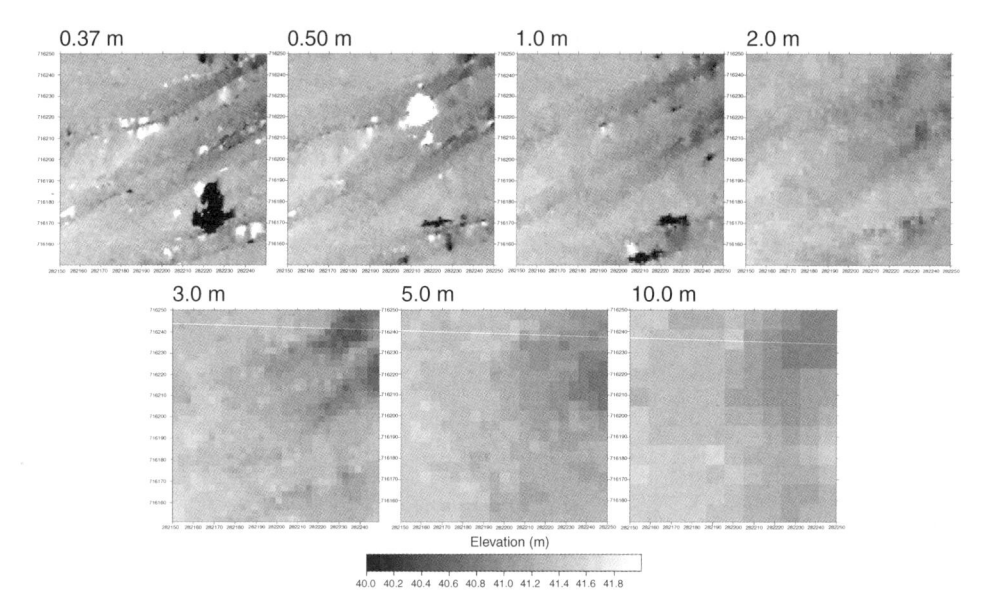

Figure 6.6 The effects of changing the resolution at which DEM data were collected upon surface representation

matching process as there is a greater probability of variation in image texture over larger areas and hence a greater probability that a coarser resolution DEM is matched successfully and correctly. In practice, surface representation is reduced, as DEM resolution is increased. Figure 6.6 shows that as DEM resolution is coarsened to 2 m, many of the errors obtained at finer resolutions have disappeared. However, significant topographic structure (i.e., bar forms) is still apparent in the DEM. Thus, the coarser resolution DEMs contain relatively error-free topographic data that may be used to interrogate finer resolution DEMs, allowing introduction of topographic variability into the filtering process that was not available in the SD filtering method (approach 1).

Thus, three DEMs were generated of the test area shown in Figure 6.6, using 2.5 m, 5 m and 10 m resolutions (Plate 2(a)), and differenced with the 1 m DEM (Plate 2(b)). This shows a progressive increase in the number of points identified as differing between surfaces as resolution is coarsened, reflecting the interaction between greater spatial averaging and poorer surface representation in coarser DEMs. To identify point errors, a tolerance level has to be introduced that recognizes this trade-off: points that exceeded a given elevation difference would need to be removed as some difference between points is inevitable as a result of natural topographic variability. This deals implicitly with the problem of the SD-based approach: the filter here is photogrammetrically averaged elevation, determined in the absence of error, and this is being allowed to vary spatially. To test the model, three tolerance values were tested: 2 m, 1 m and 0.5 m. The elevation range expected for this surface was 2 m. As the filter is photogrametrically averaged elevation, tolerances were expected to lie within this range.

Figure 6.7 shows the effects of different DEM resolutions and elevation tolerance values upon the ME, SDE and number of points retained. It is clear that a DEM resolution of about

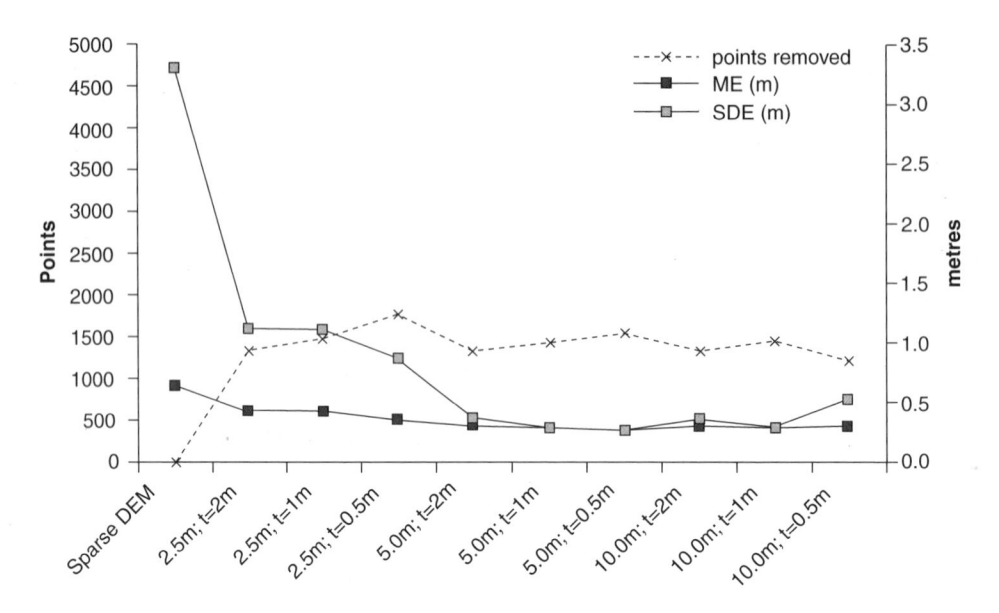

Figure 6.7 *The effects of different DEM resolutions and elevation tolerance values using approach 2, for the test area shown in Figure 6.4(c)*

5 m is required to reduce the SDE to acceptable levels. As compared with Figure 6.5, this method is interesting as if either the 5 m or 10 m resolution DEM is used, there is a relatively low sensitivity to the tolerance level. This suggests that point errors manifest themselves as local deviations from local topography and demonstrates the importance and the effectiveness of a topography-based filter criterion. There is some increase in SDE with a very low tolerance and a 10 m resolution DEM, and this may be the point at which natural topographic variability is being removed due to the filtering DEM being measured at too coarse a resolution and the tolerance threshold being set too low. It appears that the 5 m resolution DEM with a 1 m or 0.5 m tolerance gives the best results. The 1 m tolerance was adopted.

Figure 6.8 compares approaches 1 and 2 after bilinear interpolation. This shows that method 2 produced visually better results, although the SDEs for the two correction methods (0.308 m in method 1 and 0.291 m in method 2) were not significantly different. This is because the SDEs are based upon retained points only, whereas the surfaces shown in Figure 6.8 are obtained after interpolation. Method 2 removes fewer points (1444) as compared with method 1 (3500), thus reducing the level of smoothing in the interpolation process and maintaining a better representation of relief.

6.4.2 The Identification of Systematic Surface Error: Banding

It was clear from Figures 6.2(b) and 6.3 that there was some sort of banding present in the DEM which was observed in DEMs obtained from a braided river from a laboratory flume (Stojic et al., 1998). The fact that it affects large areas of the acquired data suggests that this is some form of systematic error, probably associated with some error in the assumed functional model derived as a consequence of the bundle adjustment. The most notable trait

(a) *(b)*

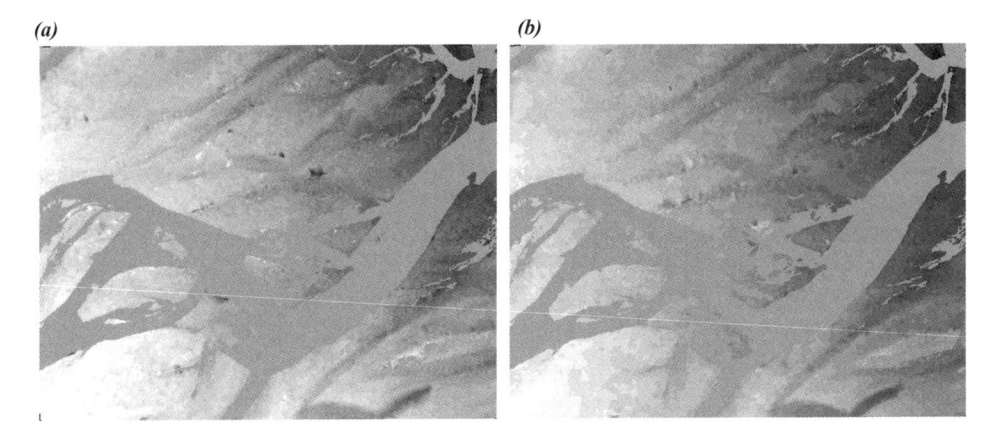

Figure 6.8 *A visual comparison of applying method 1 (search radius = 20 m, tolerance = 0.5 m) and Method 2 (coarse DEM resolution = 5 m, tolerance = 1 m)*

of the banding was that it was in the areas where the individual DEMs overlapped with one another. Initially, after application of the point post-processing described in section 6.4.1, the DEMs were stitched together by combining all dry points that had been deemed acceptable, and then using bilinear interpolation. The effects of this could be explored by comparing the surfaces interpolated from the individual DEMs where they overlapped. In this analysis, it is not possible to determine which independent DEM surface is correct: any discrepancies imply that either or both of the datasets are in error. The results are shown in Plate 3(a) and in schematic in Plate 3(b). The effect of the patterns shown in Plate 3(a) and (b), after bilinear interpolation of the full dataset, is shown in schematic in Plate 3(c). Systematic error in both surfaces increases the mottling with distance from the midpoint of the overlap (cf. Plate 3(a)); downstream data points cause more mottling upstream with distance from the centre of the overlap and upstream data points cause more mottling downstream with distance from the centre of the overlap. The problem here is that we do not know which if either surface is in error. It is in the areas of overlap that camera calibration tends to have greatest effect as the overlap areas will be based upon image pixels that are furthest from the principal point of symmetry, and where the precision of the camera calibration parameters will be poorest. The geometry of stereo-imagery in areas of DEM overlap is such that both DEMs will involve image data that is slightly away from the principal point.

6.5 Explanation of Errors

Too many of the attempts to manage error in DEMs are based upon an incomplete understanding of the nature of that error. A classic example of this is the application of smoothing functions, such as those based upon a 3×3 variance filter, as these simply propagate error from incorrect points into adjacent points without necessarily removing all of the error in the incorrect points. One of the crucial steps in an error analysis is the explanation of the causes of that error, so that the data collection process can be improved and the error

Table 6.2 The response of data quality parameters to the exclusion of data points in their derivation. Note that the changes in ME and SDE are statistically significant at the 95% level when compared with the raw data case

	Raw data	Excluding points within 1 m of a wetted channel	Excluding points within 2 m of a wetted channel	Excluding points within 5 m of a wetted channel
ME (mean error), m	0.197	0.014	0.051	0.063
SDE (standard deviation of error), m	0.313	0.195	0.193	0.258
Maximum error, m	0.828	0.231	0.231	0.231
Minimum error, m	0.265	−0.265	−0.265	−0.265

eliminated without recourse to error removal once the data has been generated. However, this is an idealistic view. In this section we explore the causes of the errors identified in section 6.4 and demonstrate that even if data re-collection is not possible, explaining the cause of DEM error goes some way to justifying the correction methods adopted.

6.5.1 Causes of Localized Error

The methods described in Section 6.4.1 allowed the identification of erroneous pits and spikes in the dataset. The first step in attempting to explain these errors involved visualizing and quantifying their spatial distribution. Plate 4 shows a contour plot of error superimposed upon a wet–dry classification of an orthorectified image. This immediately shows that the errors are associated with proximity to wet areas, with some concentration of error along the water edges. To assess the extent to which this pattern could be supported by quantitative analysis, edge detection was applied to an orthorectified image of the area shown in Plate 4. Data points in the raw photogrammetric DEM were then removed from the error analysis according to their distance from the water's edge.

Table 6.2 confirms the observation in Plate 4 that errors were significantly greater along the channel margin. Exclusion of points within a 1 m radius results in the largest reduction in ME and a major reduction in the SDE. With 2 m exclusion, similar patterns are found. With 5 m exclusion, data quality starts to worsen, albeit marginally as compared with the raw data case. It is also interesting to note that the minimum error is insensitive to point exclusion as compared to maximum error. In this case, positive error is where check point elevation is higher than photogrammetric point elevation, suggesting that photogrammetrically acquired data that are close to the channel margin appear to be too low compared to their correct elevation. As we are not considering interpolated data points here, only those that have been matched successfully using the photogrammetric method, there are two hypotheses for this effect: (1) structure of river bed relief; and (2) the stereo-matching process. Large discrepancies between check data and photogrammetric data might be expected in the vicinity of breaks of slope, where the river banks are steep. This relates to the resolution of data collection; with a 1 m data collection interval, elevations are averaged over a 1 m^2 area due to the template based stereo-matching. However, the check data refer to point samples that are not areally weighted in any way. This means that larger errors would be expected

Table 6.3 *The correct camera calibration parameters as compared to camera parameters with error simulated under a normal distribution as defined by camera parameter standard errors obtained from the bundle adjustment*

	Image 1 Correct	Image 1 Simulated	Difference	Image 2 Correct	Image 2 Simulated	Difference
Xo (m)	283909.26	283909.12	−0.147	283483.46	283483.35	−0.105
Yo (m)	716100.37	716100.33	−0.037	716013.21	716013.13	−0.075
Xo (m)	615.53	615.35	−0.187	615.53	615.53	0.000
omega (degrees)	0.1970	0.1942	−0.0029	0.2507	0.2464	−0.004
kappa (degrees)	10.5958	10.5985	0.0026	9.7322	9.7287	−0.0035
phi (degrees)	0.3520	0.3442	−0.0078	0.3988	0.3985	−0.0003

wherever a grid cell fell across a break of slope. Assuming the stereo-matching process is operating properly, there is an equal probability of the photogrammetric data points being too high (where the grid cell is predominantly dry) or too low (where the grid cell is predominantly wet). The results in Table 6.3 suggest that the errors are systematic. The second hypothesis relates to the operation of the stereo-matching process. Breaks of slope will have an effect upon the viewing angle of the two images. Using imagery of a similar scale and breaks of slope associated with a landform of similar relative relief, Lane *et al.* (2000) showed that the stereo-correlating algorithm had a tendency to produce erroneous matches due to dead ground problems. In summary, this demonstrates that determination of the spatial distribution of error is effective in helping to understand the causes of that error, and where error is most heavily concentrated. The analysis described in section 6.4.1 could potentially be improved through a more intelligent approach to error correction, in which image information content (e.g. water edges) is used to inform the error search process. However, the finding that errors are greatest at the water edges raises issues of surface representation, as edge data will be required to capture breaks of slope adequately. The selective nature of the point removal in section 6.4.1, rather than using a 1 m channel edge exclusion zone, is clearly preferable.

6.5.2 Causes of the Banding

Initial inspection of the banding suggested that this was probably associated with error in the triangulation. To assess the extent to which this could be the case, a simple simulation exercise was undertaken (Figure 6.9(a)). This involved taking a flat DEM, similar in platform extent to an individual DEM. The associated matrix of object space co-ordinates was then applied to the collinearity equations used to generate that DEM. This produced a set of correct image space co-ordinates. The correct image space co-ordinates were then re-applied to the collinearity equations but with camera positions and orientations perturbed under a normal distribution according to the SD of each parameter defined during triangulation. This produces a DEM containing simulated error due to random parameter perturbation. In addition, it is possible to simulate the effects of matching upon the DEM. We do this by adding a random perturbation to the image co-ordinates to reflect the non-perfect matching process. This is defined as having a SD of 1 image pixel.

130 Spatial Modelling of the Terrestrial Environment

```
┌─────────────────────────────────────┐
│  A DEM of a flat plane similar in planform │
│       dimensions to the study area       │
└─────────────────────────────────────┘
```

$$x = \frac{-c[m_{11}(X-X_o)+m_{12}(Y-Y_o)+m_{13}(Z-Z_o)]}{[m_{31}(X-X_o)+m_{32}(Y-Y_o)+m_{33}(Z-Z_o)]}$$

$$y = \frac{-c[m_{21}(X-X_o)+m_{22}(Y-Y_o)+m_{23}(Z-Z_o)]}{[m_{31}(X-X_o)+m_{32}(Y-Y_o)+m_{33}(Z-Z_o)]}$$

Perturb estimated camera position and orientation under standard deviations defined during bundle adjustment

A set of 'correct' image co-ordinates

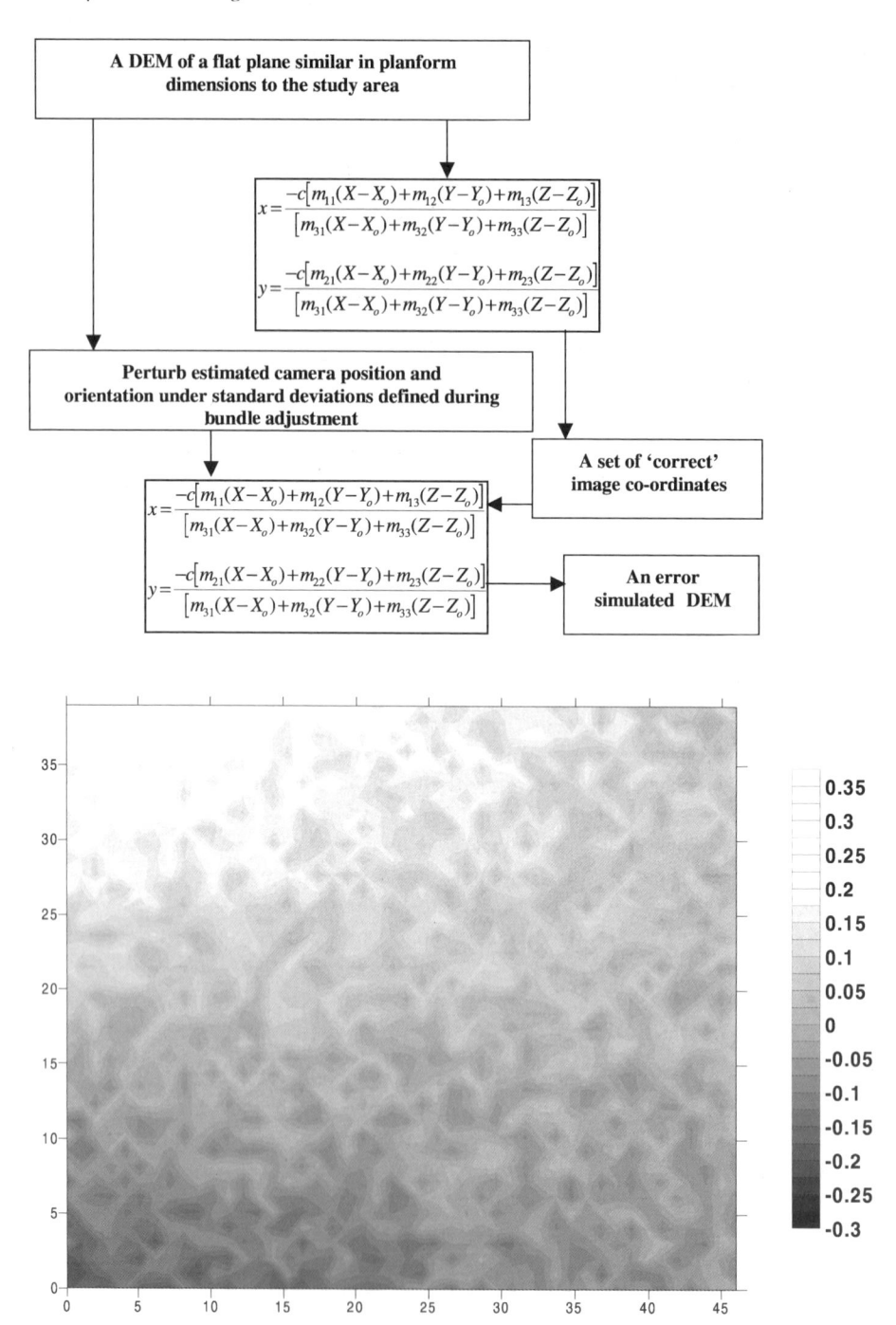

$$x = \frac{-c[m_{11}(X-X_o)+m_{12}(Y-Y_o)+m_{13}(Z-Z_o)]}{[m_{31}(X-X_o)+m_{32}(Y-Y_o)+m_{33}(Z-Z_o)]}$$

$$y = \frac{-c[m_{21}(X-X_o)+m_{22}(Y-Y_o)+m_{23}(Z-Z_o)]}{[m_{31}(X-X_o)+m_{32}(Y-Y_o)+m_{33}(Z-Z_o)]}$$

An error simulated DEM

Figure 6.9 The simulation exercise used to assess the effects of random error in camera parameters upon an idealized planar DEM (a) and the results of simulation of DEM error due to random perturbation of camera positions and orientations and the effects of random error in the stereo-matching process (b)

Table 6.4 The effects of introduction of tie points upon the average standard deviation of exposure stations. Figures in brackets give the change in parameter due to tie point addition with respect to the no tie point case

Photogrammetric block		Standard deviation of unit weight	Mean exposure station residual standard deviation X (m)	Mean exposure station residual standard deviation Y (m)	Mean exposure station residual standard deviation Z (m)
No tie points					
Waimakariri— 1999	Feb	0.55	±0.088	±0.084	±0.061
Waimakariri— 1999	March	0.98	±0.097	±0.102	±0.081
Waimakariri— 2000	Feb	0.89	±0.110	±0.119	±0.052
With tie points					
Waimakariri— 1999	Feb	1.04(+0.490)	±0.082(−0.006)	±0.088(0.004)	±0.061(0.000)
Waimakariri— 1999	March	1.13(+0.150)	±0.071(−0.026)	±0.076(-0.026)	±0.059(−0.022)
Waimakariri— 2000	Feb	0.94(+0.050)	±0.080(−0.030)	±0.091(−0.028)	±0.045(−0.007)

Figure 6.9(b) demonstrates two important characteristics. First, systematic error is evident in the dataset. This is significant with an elevation range of 0.50 m. With two DEMs, and banding in the same direction, the total banding could be as much as 1 m, and commensurate with the scale of banding shown in Figures 6.2(b) and 6.3. This suggests that a major cause of the banding that is commonly observed in studies of low relief surfaces when DEMs are stitched together (e.g. Stojic *et al.*, 1998) could be the propagation of random error in camera positions and orientations into locally systematic error in the DEM. The banding is only about 0.13% of the flying height and likely to be a problem when the feature of interest has relief of the same order as the banding.

The main problem with banding is how to deal with it. The predominant cause of banding was thought to be the propagation of errors in camera position and orientation into the derived DEM surfaces. This implies that improvement in the precision of camera position and orientation should result in less error in the DEM surface. There was no additional ground control available. However, it was possible to add additional tie points into the bundle adjustment used to determine camera position and orientations. Tie points are based upon the principle that the co-location of a point of unknown object space co-ordinates upon n images results in the creation of $2n$ equations to determine 3 unknowns. Thus, it introduces $2n$-3 units of redundancy for each point, which can be used to improve the precision of camera positions and orientations. This has appeal because it addresses the hypothesized root causes of the surface error. Thus, points that could be visually identified on both overlapping images were used as tie points (e.g. Stojic *et al.*, 1998; Chandler *et al.*, 2001). This was performed in a systematic manner, by adding additional rows of tie-points between existing rows of ground control points. In total, 94, 113 and 88 tie-points were added to the February 1999, March 1999 and February 2000 Waimakariri blocks, respectively.

In photogrammetric terms, addition of tie-points appears to have two main effects. First, those SD values that were especially high are reduced to more acceptable, albeit still large, values (Table 6.4). This is especially the case in z, and this reflects a problem common to

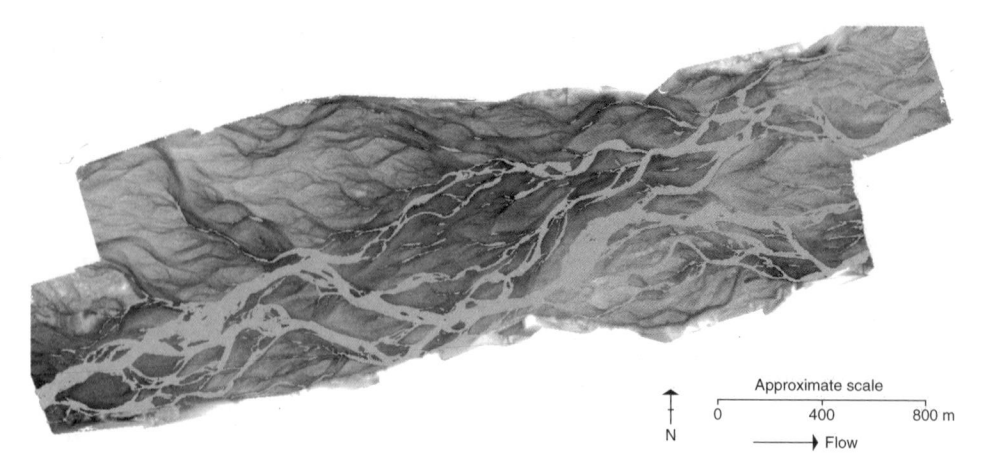

132 *Spatial Modelling of the Terrestrial Environment*

Figure 6.10 *The dry area DEM for February 2000 after removal of wet and vegetated points, post-processing to remove local error and distance-weighted stitching*

applications of photogrammetry to relatively flat surfaces of low relief where the imagery is taken near-vertical. Second, in two cases (February 1999 and February 2000) the SD of unit weight approaches 1 after tie-point addition, which is the ideal value, implying that the specified precision of the ground control points is commensurate with the determined precision of the bundle adjustment.

Even after the addition of tie-points, the banding problem remained. The second and third approaches sought to deal with this problem in a more empirical fashion, but under the justification that some banding would be expected given the analysis presented in Figure 6.9. The first approach is illustrated in Figure 6.10, and is based upon refinement of the stitching process. A distance-weighted algorithm is used to stitch the DEMs. This involves averaging corresponding points on individual DEM surfaces in areas of overlap, but assigning weights to the averaging that are a function of distance from the centre of the overlap. In this case, if we use distance-linear weights (i.e., 90% to the left of the centre takes 90% of the elevation from DEM 1 and 10% from DEM 2), then the systematic error was less visible (Figure 6.10).

Although Figure 6.10 shows that much of the banding effect appears to have been removed following the refined-stitching approach, the extent to which this is the case was checked. The remaining banding effect is most apparent when consecutive DEMs are compared by differencing them. Such comparison is of considerable geomorphological significance, as it is the first step in the identification of the spatial patterns of river channel change (e.g. Lane, 1998) and the estimation of the distributed patterns of sediment transport due to channel change (e.g. Lane *et al.*, 1995). Plate 5 shows that the estimated patterns of erosion and deposition from DEM comparison are sufficiently large on an annual comparison for the banding effect to almost have disappeared (Plate 5(a)). However, when evaluating river response to a storm event (Plate 5(b)), the banding effect is still very much in evidence.

The final method developed to deal with the banding problem was based upon using the ground control points. If a single stereo-pair is used, the bundle adjustment should honour

Table 6.5 The effects of ground control point discrepancy removal upon DEM elevations

DEM	Pre-discrepancy removal: mean error (m)	Pre-discrepancy removal: standard deviation of error (m)	Post-discrepancy removal: mean error (m)	Post-discrepancy removal: Standard deviation of error (m)
February 1999	0.174	2.113	0.084	0.261
March 1999	0.154	0.260	0.010	0.261
February 2000	0.110	0.141	0.088	0.131

Table 6.6 Check data, photogrammetrically acquired points for elevations that were not inundated, corrected photogrammetrically acquired points for elevations that were not inundated, and the associated error statistics

DEM	Number of check points available (raw data)	Raw mean error (m)	Raw standard deviation of error (m)	Corrected mean error (m)	Corrected standard deviation of error (m)	Theoretical standard deviation of error (m)
February 1999	3700	0.542	±2.113	0.084	0.261	±0.070
March 1999	241	0.252	±0.883	0.010	0.261	±0.070
February 2000	1661	0.064	±0.926	0.088	0.131	±0.056

all the ground control points used in its determination. However, if multiple stereo-pairs are being used in a common bundle adjustment, there is a growing possibility that there will be discrepancies between the derived DEM elevations and the correct ground control point elevations. This should provide a useful means of detecting systematic DEM errors in situations where there is banding, provided the random contribution to those errors is small. To assess this, discrepancy maps were contoured and an example is shown in Plate 6. for February 2000. Although this was a DEM for which little banding was thought to exist, Plate 6 shows that there are quite considerable discrepancies for some, but not all, of the photocontrol data points. These discrepancy maps were then used to remove the effects of banding by subtracting the discrepancies from the stitched DEMs. This resulted in a statistically significant reduction of ME in all three cases, with, as expected, negligible changes in SDE (Table 6.5). This demonstrated that this aspect of post-processing resulted in a significant reduction in systematic error.

6.6 Summary of Data Quality

Table 6.6 shows the results of the correction methods developed for and applied to this dataset. In all cases, the SDE has been reduced significantly. In the two cases where there was a serious ME, this has also been reduced. This demonstrates the potential of automated data correction methods for improving surface quality to levels that are more acceptable. Lane et al. (2003) propagated the standard deviations of error shown in Table 6.6 through to surfaces of difference to identify where there had been statistically significant change and also through to estimates of volumes of change. This showed that these errors were reduced to such a level that the DEMs could be used to detect erosion and deposition patterns and to determine volumes of material eroded and deposited. Table 6.6 also shows the theoretical estimates of precision that the design of this study was based upon, using the scale of imagery acquired and the scanning resolution used to create digital imagery from diapositives. This shows how the theoretical precision is seriously downgraded in practice as a result of the effects of surface structure upon the data collection process. If a certain data quality is critical, room for degradation of the theoretical precision is required. Table 6.6 also shows how the final data quality improved more than expected when the image scale was increased as compared with the improvement in theoretical precision. This emphasizes the sensitivity of the data collection method used in this study to surface texture, and this will apply to any study where stereo-matching is used for surface representation.

6.7 Conclusion

This chapter has described the way in which we developed systems for managing error in a complex digital photogrammetric project. The need to develop these systems arose from an increasingly common problem for remote sensing in general: the growth of automated methods for data generation reduce the traditional interpretation of individual data points; whilst this benefits data collection processes by allowing data to be acquired over a much larger area and much more frequently, problems arise if there is degradation in the quality of results obtained. In relation to environmental modelling, this is of particular concern, because the parameters that determine earth surface processes tend to be derivatives of a surface (e.g. slope, difference between two surfaces, surface curvature), and not the surface itself. As equations (3)–(5) show, errors will propagate into these derivatives so that subtle errors can have quite severe consequences at the point at which remotely sensed data is used for process modelling. It also follows that process modellers must become closely involved in the data generation and interpretation process in order to make sure that avoidable decisions that affect data quality adversely (e.g. certain types of point filtering algorithms) are not taken.

The process of error management we used in the Waimakariri case example was based upon the identification of three types of error: systematic error, blunders and random error. The work shows that these sorts of errors are manifest and need to be treated in different ways. Most notably, error correction methods needed to reflect: (1) how the method used to generate the data introduces error (e.g. the banding that emerged during DEM stitching, associated with uncertainties in sensor position and orientation); and (2) how the surface determines the nature of error that arises (e.g. problems associated with sudden breaks of slope close to the water edge). In this case, the most important progress was associated with developing methods for finding spatially-localized error using a form of smoothing based upon the stereo-matching algorithm adopted. This approach has the potential to be of more widespread use in terms of other surface types and terrain collection methods, wherever there are a relatively small number of seriously erroneous data points within a surface that is generally good. In principle, this means that an approximate surface representation may be achieved as a means of identifying isolated data points that are in error. In practice, research is required to assess the best means of obtaining both an approximate surface representation and the threshold at which a point becomes erroneous when judged with respect to the approximate representation. As different surfaces will have different degrees of topographic variability, these issues may be surface-specific. However, the development of *a priori* recommendations (e.g. expected topographic variability over a given spatial scale) and/or the combination of different data sources (e.g. using map-derived elevations to construct an approximate terrain model which is used to filter data derived using higher resolution remote sensing) is worth further exploration.

Finally, a major theme in this chapter has been the detection and management of locally systematic error. Section 6.3.2 emphasizes the importance of local versus global estimates of data quality. When a study is interested in distributed process modelling, and process rates are defined by local gradients in surface topography, getting those gradients right is crucial. In turn, this requires individual data points to be correct, which emphasizes the importance of using local topographic variability in the data correction procedure, rather than the black box smoothing typical of some DEM products. However, it is equally important to design topographic surveys that consider how a particular data collection method interacts with a particular surface of interest. If this cannot be done using *a priori* reasoning, then it should be done through appropriate pilot survey, so that data collection parameters are optimized before the expense of a full survey. This will prevent potential users of remotely sensed topographic data from being disappointed by its apparent failure when the failure arises simply because the survey has not been designed properly, and the data have not been processed adequately.

Acknowledgements

The study was partially funded by the New Zealand Foundation for Research, Science and Technology under contracts CO1818 and CO1X0014. Maurice Duncan, NIWA, assisted with field data collection and bathymetric data analysis. B. Fraser, D. Pettigrew and W. Mecchia, Environment Canterbury, assisted with field surveys. Air photography was by Air Logistics, Auckland, while AAM Geodan, Brisbane, undertook the airborne laser scanning. G. Chisholm, Trimble Navigation N.Z. Ltd., assisted with GPS equipment set-up. RMW was supported by a NERC studentship. SNL was supported by the Royal Society and the University of Leeds.

References

Carson, M.A. and Griffiths, G.A., 1989, Gravel transport in the braided Waimakariri River-transport mechanisms, measurements and applications, *Journal of Hydrology*, **109**, 201–220.

Chandler, J.H., Shiono, K., Rameshwaren, P. and Lane, S.N., 2001, Automated DEM extraction for hydraulics research, *Photogrammetric Record*, **17**, 39–61.

Cooper, M.A.R., 1987, *Control Surveys in Civil Engineering* (London: Collins).

Cooper, M.A.R., 1998, Datums, coordinates and differences, in S.N. Lane, K.S. Richards and J.H. Chandler (eds), *Landform Monitoring, Modelling and Analysis* (Chichester: John Wiley & Sons), 21–36.

Cooper, M.A.R. and Cross, P.A., 1988, Statistical concepts and their application in photogrammetry and surveying, *Photogrammetric Record*, **12**, 637–663.

Davison, M., 1994, Heighting accuracy test from 1:1000 scale aerial photography, *Photogrammetric Record*, **16**, 922–925.

Everitt, B.S., 1998, *The Cambridge Dictionary of Statistics* (Cambridge: Cambridge University Press).

Felicísimo, A.M., 1994, Parametric statistical-method for error detection in digital elevation models, *ISPRS Journal of Photogrammetry and Remote Sensing*, **49**, 29–33.

Ghosh, S.K., 1988, *Analytical Photogrammetry*, 2nd edition (New York: Pergamon Press).

Gooch, M.J., Chandler, J.H. and Stojic, M., 1999, Accuracy assessment of digital elevation models generated using the ERDAS Imagine OrthoMAX digital photogrammetric system, *Photogrammetric Record*, **16**, 519–531.

Jensen, J.R., 1996, Issues involving the creation of digital elevation models and terrain corrected orthoimagery using soft-copy photogrammetry, in C.W. Greve (ed.), *Digital Photogrammetry: An Addendum to the Manual of Photogrammetry* (Bethesda: American Society of Photogrammetry and Remote Sensing), 167–179.

Lane, S.N., 1998, The use of digital terrain modelling in the understanding of dynamic river channel systems, in S.N. Lane, K.S. Richards and J.H. Chandler (eds), *Landform Monitoring, Modelling and Analysis* (Chichester: John Wiley & Sons), 311–342.

Lane, S.N., 2000, The measurement of river channel morphology using digital photogrammetry, *Photogrammetric Record*, **16**, 937–957.

Lane, S.N., James, T.D. and Crowell, M.D. 2000. Application of digital photogrammetry to complex topography for geomorphological research, *Photogrammetric Record*, **16**, 793–821.

Lane, S.N., Richards, K.S. and Chandler, J.H., 1994, Developments in monitoring and terrain modelling of small-scale riverbed topography, *Earth Surface Processes and Landforms*, **19**, 349–368.

Lane, S.N., Richards, K.S. and Chandler, J.H., 1995, Morphological estimation of the time-integrated bed-load transport rate, *Water Resources Research*, **31**, 761–772.

Lane, S.N., Westaway, R.M. and Hicks, D.M., 2003, Estimation of erosion and deposition volumes in a large gravel-bed, braided river using synoptic remote sensing, *Earth Surface Processes and Landforms*, **28**, 249–271.

Neill, L.E., 1994, The accuracy of heighting from aerial photography, *Photogrammetric Record*, **14**, 917–922.

Shearer, J.W., 1990, The accuracy of digital terrain models, in G. Petrie and T.J.M. Kennie (eds), *Terrain Modelling in Surveying and Civil Engineering* (London: Whittles), 315–336.

Stojic, M., Chandler, J.H., Ashmore, P. and Luce, J., 1998, The assessment of sediment transport rates by automated digital photogrammetry, *Photogrammetric Engineering and Remote Sensing*, **64**, 387–395.

Taylor, J.R., 1997, *An Introduction to Error Analysis: The Study of Uncertainties in Physical Measurements*, 2nd edition (Sausalito, CA: University Science Books).

Westaway, R.M., Lane, S.N. and Hicks, D.M., 2000, Development of an automated correction procedure for digital photogrammetry for the study of wide, shallow gravel-bed rivers, *Earth Surface Processes and Landforms*, **25**, 200–226.

Westaway, R.M., Lane, S.N. and Hicks, D.M., 2003, Remote survey of large-scale braided rivers using digital photogrammetry and image analysis, *International Journal of Remote Sensing*, **24** 795–816.

Wise, S., 1998, The effect of GIS interpolation errors on the use of DEMs in geomorphology, in S.N. Lane, K.S. Richards and J.H. Chandler (eds), *Landform Monitoring, Modelling and Analysis* (Chichester: John Wiley & Sons), 139–164.

Wise, S., 2000, Assessing the quality for hydrological applications of digital elevation models derived from contours, *Hydrological Processes*, **14**, 1909–1929.

Zoej, M.D.V. and Petrie, G., 1998, Mathematical modelling and accuracy testing of SPOT Level 1B stereopairs, *Photogrammetric Record*, **16**, 67–82.

7
Modelling Wind Erosion and Dust Emission on Vegetated Surfaces

Gregory S. Okin and Dale A. Gillette

7.1 Introduction

Erosion, transportation and deposition of material by wind are fundamental abiotic processes in the world's drylands. Atmospheric dust created in deserts by wind erosion and subsequently transported around the globe is a vital component of the entire Earth system. Mineral particulates from local or regional sources can dramatically impact air quality. Despite this, there is often little information about where, precisely, in the landscape wind erosion and dust emission occur. This is largely because the conditions on the surface that determine the extent and degree of wind erosion tend to vary across both space and time and relatively complex interactions between soil and vegetation conspire to allow or suppress wind erosion. Although the mathematical understanding of wind erosion and dust emission has progressed to a point where current models can accommodate a fair degree of complexity, few studies to date have integrated the models in such a way that spatially explicit estimates of wind erosion and dust flux can be made. This technological step is required if we are to start to understand the distribution of atmospheric dust observed from remote sensing platforms or are to predict dust events that can affect local human health or global biogeochemical cycles.

The purpose of the present research is to develop a modelling framework in which spatially explicit estimates of wind erosion and dust emission for vegetated surfaces can be derived. The model that is developed here allows land surface-based estimates of dust emission at the regional scale and is intended to be complementary to global scale estimates of dust emission that can be derived from modelling (e.g. Tegen *et al.*, 1996). Coarse resolution remote sensing observations of the atmosphere using sensors such as the Total

---

Spatial Modelling of the Terrestrial Environment. Edited by R. Kelly, N. Drake, S. Barr.
© 2004 John Wiley & Sons, Ltd. ISBN: 0-470-84348-9.

Ozone Mapping Spectrometer (TOMS) (e.g. Prospero, 1999) and meteorological satellites such as METEOSAT (Legrand *et al.*, 1989) also provide useful general information on source areas. While global scale estimates of atmospheric dust distributions are routinely produced using these sensors, land surface-based wind erosion estimation remains in its infancy, particularly when applied at the regional scale. This chapter is intended to describe the state of the science of regional scale modelling, outline the implementation of a regional scale model and identify future areas for research.

7.1.1 The Importance of Atmospheric Dust

Atmospheric mineral dust can affect the global energy balance by changing the radiative forcing of the atmosphere through its ability to scatter and absorb light (Sokolik and Toon, 1996; Tegen and Lacis, 1996). Atmospheric mineral aerosols may also provide surfaces for reactions, change the concentration of other aerosols in the atmosphere and affect cloud nucleation and optical properties (Dentener *et al.*, 1996; Levin *et al.*, 1996; Dickerson *et al.*, 1997; Wurzler *et al.*, 2000). The iron in dust is thought to play a major role in ocean fertilization and oceanic CO_2 uptake-thereby affecting the global carbon budget (Duce and Tindale, 1991; Coale *et al.*, 1996; Piketh *et al.*, 2000). Dust transported to downwind terrestrial ecosystems can play a major role in soil formation and nutrient cycling (Chadwick *et al.*, 1999; Reynolds *et al.*, 2001) and present serious health concerns (Griffin *et al.*, 2001).

On local and regional scales, anthropogenic disturbance can have consequences on wind erosion and plant community composition and nutrient cycling in arid and semiarid environments. Human land use practices can reduce vegetation cover and expose wind-erodible soils to mobilization (Stockton and Gillette, 1990; Tegen and Fung, 1995; Lancaster and Baas, 1998; Okin *et al.*, 2001a). For example, the Dust Bowl resulted from a decade-scale drought and poor land use practices and showed that throughout the central United States, wind erosion remains a potentially important ecosystem and geomorphic process (Schlesinger *et al.*, 1990; Forman *et al.*, 1992; Rosenzweig and Hillel, 1993; Schultz and Ostler, 1993; Brown *et al.*, 1997; Alward *et al.*, 1999).

Recent results from the Jornada Basin in south-central New Mexico, part of the National Science Foundation's Long-Term Ecological Research (LTER) network, indicate that severe wind erosion can induce both plant community change and disruption of the soil nutrient cycle (Okin *et al.*, 2001b). A site was cleared of vegetation in 1990 for an experiment aimed at measuring dust flux from the loamy sand soils common at Jornada (Figure 7.1). In the 11 years since its establishment, the site and the area directly downwind of it have undergone major changes in nutrient budget and community composition (Okin *et al.*, 2001b). Here, removal of vegetation has triggered wind erosion by enhancing particle saltation and suspension processes. Saltation of sand-sized grains leads to negative physical consequences for vegetation such as burial and abrasion. Since plant nutrients are concentrated on suspension-sized particles, suspension leads to the permanent removal of nearly all soil nutrients in the surface soils in approximately 10 years. Thus, saltation kills mature vegetation while suspension inhibits the growth of new plants, leading to changes in plant community by differentially affecting shrub and grass species. With their biomass largely below the saltation layer, grasses are abraded and buried by sand particles, whereas shrubs are largely spared this fate. Growth of new plants (including annuals) that rely on nutrients

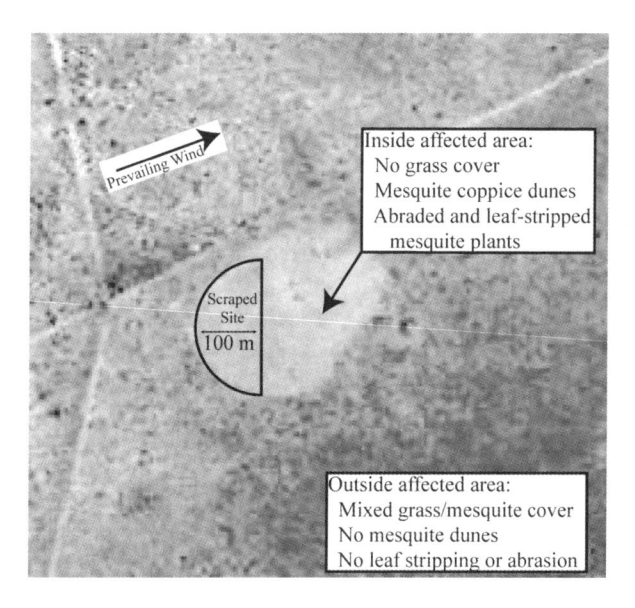

[Figure: aerial image with annotations — "Prevailing Wind"; "Scraped Site 100 m"; "Inside affected area: No grass cover; Mesquite coppice dunes; Abraded and leaf-stripped mesquite plants"; "Outside affected area: Mixed grass/mesquite cover; No mesquite dunes; No leaf stripping or abrasion"]

Figure 7.1 *The Jornada 'scraped site' shown here in a 1999 low-altitude airborne visible/infrared spectroradiometer (AVIRIS) image with resolution of ~5 m. This site proves that wind erosion can affect ecosystem stability in this ecosystem. It has experienced an average deflation rate of 1.8 cm yr^{-1} (Gillette and Chen, 2001) and plant-available N and P have been reduced by 89% and 78%, respectively (Okin et al., 2001b). No vegetation has regrown on the scraped site itself. The vegetation community downwind of the site has changed from a grassland to a shrubland due to burial, abrasion and leaf stripping from saltating particles from the scraped site itself (Okin et al., 2001a) and plant-available N and P in surface soils in the downwind area have been reduced by 82% and 62%, respectively (Okin et al., 2001b)*

in the topmost soil layer is inhibited. The seedbank may also be depleted in the winnowed surface soils (A. van Rooyen, personal communication). Thus perennial grasses are killed, new vegetation growth is suppressed and shrubs, where established, gain a competitive edge. Creation of an erosive discontinuity in a mixed shrub/grass ecosystem accelerates the conversion of grassland to shrubland.

7.1.2 Where Does the Dust Come from?

Despite the importance of desert dust, it is often unclear in detail where it is produced and what role humans play in mediating its production. Sensors such as TOMS can be used to directly observe dust in the atmosphere using methods such as the TOMS aerosol index (e.g. Prospero, 1999), but current technologies do not allow unique identification of the loci due to their coarse resolution and, in the case of TOMS, because of problems in observing the lower levels of the atmosphere. Because of this, controversy remains about the extent to which land use contributes to the atmospheric mineral dust. Recent work by Prospero *et al.* (2002) and Ginoux *et al.* (2001) suggests that the overwhelming majority of desert dust comes from closed basins in arid areas related to now-dry or ephemeral lakes. They argue further that humans do not significantly perturb the dust cycle, a conclusion supported by

140 *Spatial Modelling of the Terrestrial Environment*

Guelle *et al.* (2000). This point of view contrasts sharply with that of Tegen and Fung (1995) who suggest that land use may in fact dramatically affect the amount of dust emitted in desert regions.

7.1.3 Aims and Objectives

In summary, all of the studies which have examined dust emission in order to disentangle the natural and land use signals have either used coarse scale modelling or remote sensing data where there are many potential sources in each pixel that could produce the observed atmospheric distribution. In order to resolve the question of the contribution of land use and land cover to desert dust, local or regional scale models that incorporate data from the Earth's land surface are required to identify dust sources. But dust emission depends on several soil and vegetation parameters (e.g. soil texture, crown height and plant density) that can vary in both space and time. New techniques which allow modelling of wind erosion and dust emission based on known distributions of vegetation and soils in small well-studied areas are a necessary first step in achieving the goal of land surface-based models of wind erosion. The aim of the present chapter is to do just this for the Jornada Basin in south-central New Mexico.

7.2 Study Site Description

The Jornada del Muerto basin lies approximately 30 km northeast of Las Cruces, NM, in the Chihuahuan Desert ecosystem (Figure 7.2). It is bounded by the San Andres mountains on the east and by the Rio Grande valley and the Fra Cristobal-Caballo mountain complex on the west. Elevation above sea level varies from 1180 to 1360 m. The Jornada Plain consists

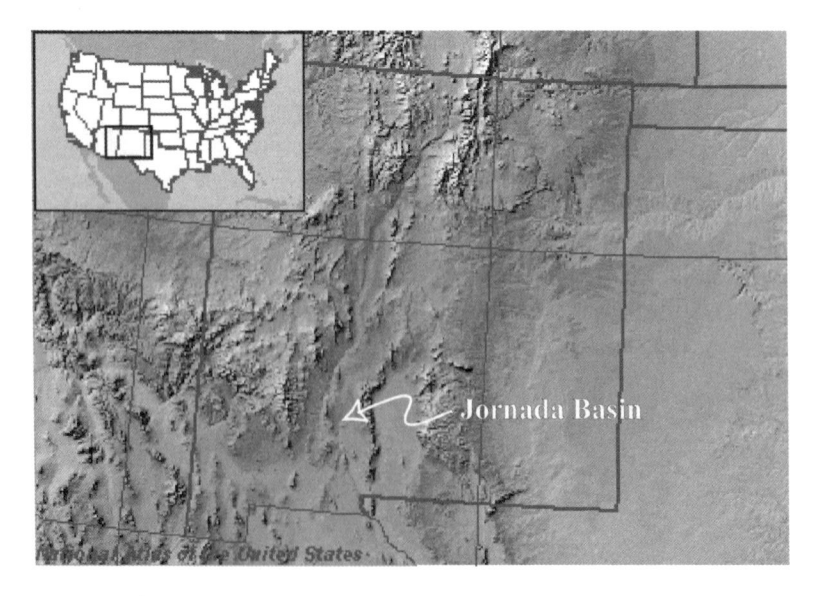

Figure 7.2 Location of the Jornada Basin, New Mexico. Maps from National Atlas of the United States, October 29, 2003, www.nationalatlas.gov

of unconsolidated Pleistocene detritus. This alluvial fill from the nearby mountains is 100 m thick in places and the aggradation process is still active. Coarser materials are found near foothills along the eastern part of the study area. The topography of the study area consists of gently rolling to nearly level uplands, interspersed with swales and old lakebeds (Buffington and Herbel, 1965).

The climate of the area is characterized by cold winters and hot summers and displays a bimodal precipitation distribution. Winter precipitation usually occurs as low-intensity rains or occasionally as snow and contributes to the greening of shrub species in the basin in the early spring. Summer monsoonal precipitation, usually in the form of patchy, but intense, afternoon thunderstorms, is responsible for the late-summer greening of grasses. The average annual precipitation between 1915 and 1962 in the basin was 230 mm, with 52% falling between July 1 and September 30 (Paulsen and Ares, 1962). The average maximum temperature is highest in June, when it averages 36°C and lowest in January, when it averages 13°C (Buffington and Herbel, 1965).

The principal grass species in the study area are *Scleropogon brevifolius* (burrograss), *Hilaria mutica* (tobosa grass) and several species of *Aristida* while major shrubs are *Larrea tridentata, Prosopis glandulosa,* and *Florensia cernua.* Soils in the basin are quite complex but generally range from clay loams to loamy fine sands, with some areas being sandy or gravelly (Soil Conservation Service, 1980).

The Jornada Basin is an area of intense research as part of the Jornada LTER programme of the National Science Foundation. The primary reason for the broad scientific interest in the Jornada Basin is because it is the premier site in which widespread conversion of grassland communities to shrublands over the past 100 years (Buffington and Herbel, 1965; Bahre, 1991) has been studied. Grazing and other anthropogenic disturbances are commonly invoked to explain this trend (Bahre and Shelton, 1993). Extensive work on soil structure and plant community composition at the Jornada LTER site and the adjacent New Mexico State University (NMSU) Ranch strongly supports this hypothesis (Schlesinger et al., 1990).

7.3 Review of Basic Equations Relating to Wind Erosion Modelling

Saltation mass flux has been related by a large number of authors using different approaches. Bagnold (1941) approached the problem first and provided a physically reasonable solution that was later confirmed by the analysis of Owen (1964). The saltation equation has been verified by wind tunnels, field experiments and alternate theoretical derivations (Shao and Raupach, 1993):

$$Q_{\text{Tot}} = A \frac{\rho}{g} \sum_{u_*} u_* (u_*^2 - u_{*t}^2) \Delta T, \qquad (1)$$

where Q_{Tot} is the horizontal (saltation) mass flux (g cm^{-1} s^{-1}), A is a unitless parameter (usually assumed to be equal to 1) related to supply limitation (Gillette and Chen, 2001), ρ is the density of air (g cm^{-3}), g is the acceleration of gravity (cm s^{-1}), u_* is the wind shear velocity and u_{*t} is the threshold shear velocity (both cm s^{-1}). Q_{Tot} is envisioned as the mass flowing past a pane one length unit wide, perpendicular to both the wind and the ground. Vertical mass flux (dust flux), F_a (g cm^{-2} s^{-1}), is linearly related to Q_{Tot} by a constant K

Table 7.1 Variations in F_a/Q_{Tot} as a function of soil texture (from Gillette et al. (1997))

Texture	Log of F_a/Q_{tot}(cm^{-1}) Average	Minimum	Maximum	Number of samples
Clay	−6.4	−6.5	−6.3	2
Loam	−5.7	−5.7	−5.7	1
Sandy loam	−3.7	−4.1	−3.5	2
Loamy sand	−4.5	−5.9	−4.2	7
Sand	−5.7	−6.6	−5.1	28

©1997 *American Geophysical Union*, reproduced with permission

(cm^{-1}) that is typically of order of 10^{-4}–10^{-5} cm^{-1} that varies with soil texture (see Table 7.1) (Gillette et al., 1997). F_a is envisioned as the mass leaving the surface per unit time.

The shear velocities, u_* and u_{*t}, are related to wind speed at height z, $U(z)$ (cm s^{-1}), by:

$$U(z) = \frac{u_*}{k} \ln\left(\frac{z-D}{z_0}\right), \qquad (2)$$

where k is von Karmann's constant (unitless), z_0 is the roughness height (cm) and D is the displacement height (cm). If $D > 0$, then turbulent flow in wakes predominates and no wind erosion can occur.

Kaimal and Finnigan (1994, p. 68) have suggested that D is approximately related to canopy height, h_c (cm) by:

$$D = 0.75h_c. \qquad (3)$$

While this relationship may hold when considering boundary layer phenomena that take place above the canopy, equation (3) dictates a non-zero D and therefore no wind erosion, for all vegetated surfaces (surfaces with $h_c \neq 0$). Modelling approaches to wind erosion must assume that $D = 0$, so that wind erosion is allowed on vegetated surfaces and roughness height, z_0, must be assumed to be the determinant of the relationship between $U(z)$ and u_*.

Several authors have proposed relations to predict z_0 as a function of surface roughness. Lettau (1969) and Wooding et al. (1973) have suggested parameterization of z_0 by:

$$z_0 = 0.5h_c\lambda, \qquad (4)$$

where λ is a unitless parameter known as 'lateral cover' (Raupach et al., 1993) or 'roughness density'(Marticorena et al., 1997). Lateral cover is equal to the average frontal area of plants multiplied by their number density, N (cm^{-2}). Number density is related to fractional cover of vegetation, C (unitless) by:

$$N = \frac{C}{A_p}, \qquad (5)$$

where A_p is the average footprint of individual plants (cm^2).

The linear relation in equation (4) should hold until λ reaches ~ 0.1. Above this value, z_0/h_c remains approximately constant. Marticorena et al. (1997) have adjusted equation (4) to incorporate the levelling-off of the linear relationship with increasing λ:

$$z_0 = \begin{cases} (0.479\lambda - 0.001)h_c & \text{for } \lambda \leq 0.11 \\ 0.005 h_c & \text{for } \lambda > 0.11 \end{cases} \quad (6)$$

Thus, using either equation (4) or (6), one can predict z_0 based on knowledge of vegetation parameters and assuming randomly distributed vegetation.

For conditions of partial, randomly distributed vegetation cover, u_{*t} is related to the threshold shear velocity for an unvegetated surface u_{*ts} by:

$$u_{*t} = u_{*ts}\sqrt{(1-\sigma\lambda)(1+\beta\lambda)}, \quad (7)$$

where σ (unitless) is equal to the ratio of the average basal area to average frontal area of individual plants, β (unitless) is the ratio of the drag coefficient of an isolated plant to the drag coefficient of the ground surface in the absence of the plant and is of order 10^2.

The wind erosion and dust flux model discussed above (equations (1)–(3)) assumes randomly and isotropically spaced vegetation. In desert areas, particularly in areas undergoing severe wind erosion, this may not be the case. Okin and Gillette (2001) have shown using standard geostatistical techniques and 1-m resolution digital orthophotos that vegetation can be distributed anisotropically which can enhance dust emission. They found that in mesquite dunelands in the northern Chihuahuan Desert, vegetation is oriented in elongated areas whose direction nearly parallels the direction of the prevailing wind. These landscapes also display horizontal mass fluxes (Q_{Tot}) several times those observed in adjacent lands dominated by other vegetation. The vegetation-free areas elongated in the direction of the dominant eroding winds that accounted for the increased wind erosion were named 'streets'. It is likely that the streets are the primary loci of wind erosion and that in areas without streets, or at times when the wind is not aligned with the streets, the wind erosion from the area is negligible.

7.4 The Spatially Explicit Wind Erosion and Dust Flux Model

Significant work over the past half-century has led to the development of parameterized wind erosion models that are based on theoretical consideration, field experiments and wind-tunnel experiments. A small number of parameters controls wind erosion and dust flux in areas with randomly distributed partial cover. The model parameters, in turn, are closely related to characteristics of the soil (e.g. surface texture) and vegetation (e.g. crown size and geometry, plant number density, plant distribution anisotropy) (Table 7.2).

In the present research, we have created a spatially explicit wind erosion and dust flux model (SWEMO) that allows estimation of wind erosion and dust flux across a landscape by incorporating spatial distributions of important parameters. Users can thus see how in the landscape soil and vegetation parameters interact to create patterns of wind erosion and dust emission. This approach provides a powerful basis for trying to understand where in landscapes the strongest dust sources are and is therefore applicable to trying to understand the most important or persistent dust sources in an area. At its heart are flux equations that represent the state of the art in wind erosion and dust flux modelling, discussed above,

144 Spatial Modelling of the Terrestrial Environment

Table 7.2 Relations between wind erosion model parameters and vegetation/soil parameters

Model parameter		Vegetation/Soil parameter
Threshold shear velocity of soil	u_{*ts}	Soil grain size, crusting, disturbance
Displacement height	D	Assumed to be zero
Roughness height	z_0	Plant height, radius, and density
Basal/Frontal area ratio	σ	Plant height and radius
Drag coefficient ratio	β	Approx. constant ($\sim$100)
Lateral cover	λ	Plant height, radius and number density
Fractional cover	C	Plant radius and density
Number density	N	Fractional cover and plant radius

which are similar to those used by Marticorena *et al.* (1997) in their simulation of Saharan dust sources.

SWEMO uses maps of soil texture and vegetation, in addition to knowledge of vegetation cover and size parameters, to create derived maps of threshold shear velocity (with vegetation), u_{*t} and z_0. Values u_{*ts} are derived by extracting median values of u_{*ts} for each soil texture class from Gillette *et al.* (1997). Soil crusting and disturbance were not included in this version of the model. For each cell in the model, a histogram of shear velocity is derived from a histogram of wind speed at one height using the value of z_0 at that cell and equation (2). Equation (1) is then evaluated for each cell to derive an estimate of total horizontal flux, Q_{Tot}. A soil-texture based value of F_a/Q_{Tot} (see Table 7.1) is then used to calculate the amount of vertical flux, F_a, for each cell. The processing stream for SWEMO is depicted in Figure 7.3.

7.4.1 Issues in the Integration of Parameters for SWEMO

Wind erosion depends on several parameters that vary as a function of soil and vegetation cover. By using maps of dominant vegetation type and soil texture, SWEMO is able to impose spatial variability by allowing the main parameters that determine wind erosion and dust flux (see Table 7.3) to vary according to the specific soil and vegetation found at any location. Thus, the primary constraint on the use of SWEMO in natural landscapes is the availability of overlapping vegetation and soil maps that provide information at scales of interest.

However, even if categorical maps of soil texture and vegetation type, with polygons labelled, for example, 'sandy loam' and 'creosote', are 100% accurate, they do not represent the full variability of the landscape: among other things, the size and spacing of plants vary even among areas with the same polygon labels. Thus, although SWEMO is able to evaluate the wind erosion equations (equations (1) and (2)) using different parameters for different soils and community types, those parameters cannot vary in the model within classes. In addition, the parameters used to evaluate the wind erosion equations must be derived from literature values or field measurements and there is no guarantee that the values chosen for integration into SWEMO are representative over the entire area of interest.

Finally, and probably most important, the majority of wind erosion may occur in small areas not well represented by local averages. A small hole in vegetation, such as a natural

Table 7.3 SWEMO input parameters for soils of the Jornada Basin

Texture	Sandy loam	Loamy sand	Clay loam	Gravelly sand
u_{*ts} (cm s^{-1})	29	34	68	28
$Log(F_a/Q_{tot}$ (cm^{-1}))	−3.7	−4.5	−5.7	−5.7
A	1.0	1.0	1.0	1.0

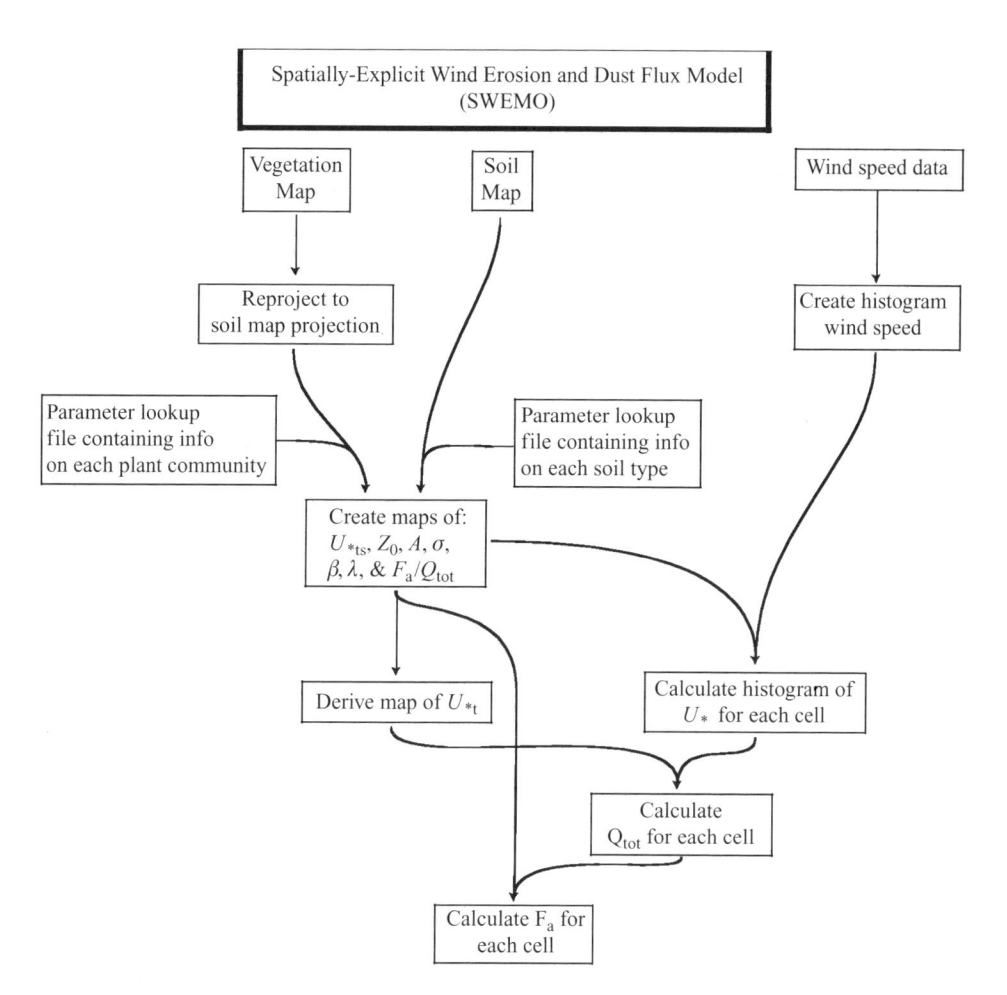

Figure 7.3 Processing stream for the spatially explicit wind erosion and dust flux model (SWEMO)

disturbance, a road, a dry river or a dry lake, may account for the majority of dust emitted in an area, but be insignificant on the scale at which most maps are produced.

In the present study, a site with all requisite data (soil maps, vegetation maps, wind data and a wealth of ongoing ecological research) was chosen to test SWEMO. The Jornada

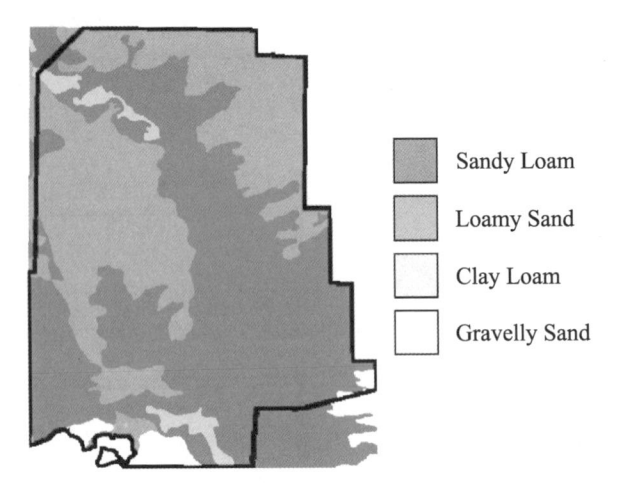

146 Spatial Modelling of the Terrestrial Environment

Figure 7.4 Soil texture map of the Jornada Basin derived from the soil survey of Doña Ana County, New Mexico (Soil Conservation Service, 1980). The area outlined in black is the region where the soil and vegetation maps overlap and is the study area for the present research

Basin in south-central New Mexico is a part of the National Science Foundation's LTER network and as such provides a wealth of required and ancillary data. Ongoing research into dust emissions at this site by one of the authors (Gillette) also allows us to compare modelled fluxes with measured values.

7.5 Using SWEMO to Predict Wind Erosion and Dust Flux at the Jornada Basin

7.5.1 Data Sources and Model Inputs

Soil Data. Portions of the Jornada Basin have been mapped by the US Soil Conservation Service (1980) (Figure 7.4). Each polygon in this soil map is labelled with a soil texture of the dominant soil type in that polygon, which allows estimation of u_{*ts} and F_a/Q_{Tot} (Table 7.3). The particle-limitation coefficient, A, is assumed to be 1.0 for the entire study area. Values of u_{*ts} used in this study were derived from mean values provided by Gillette (1988). Values of F_a/Q_{Tot} were estimated using data from Gillette *et al.* (1997) (see Table 7.2).

Vegetation Data. A vegetation map of portions of the Jornada Basin was made available to these study through the Jornada LTER project (digital data produced by R. Gibbens, R. McNeely and B. Nolen). This map contained information on spatial distribution of the dominant plant communities in the basin (see Table 7.4 and Figure 7.5): grassland, mesquite (*Prosopis glandulosa*), creosote (*Larrea tridentata*), tarbush (*Flourensia cernua*), snakeweed (*Xanthocephalum* spp.), other shrubs and no vegetation. Fractional cover for the grassland and snakeweed cover types and plant diameter and height for all vegetation

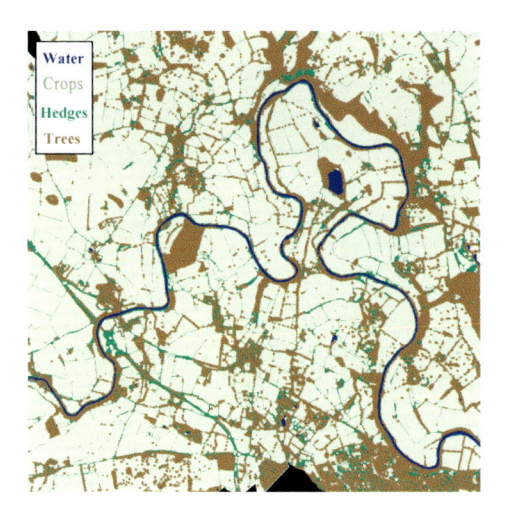

Plate 1 Connected regions identified in the segmentation. Short vegetation is light brown, hedges are green, tall vegetation dark brown and water regions blue

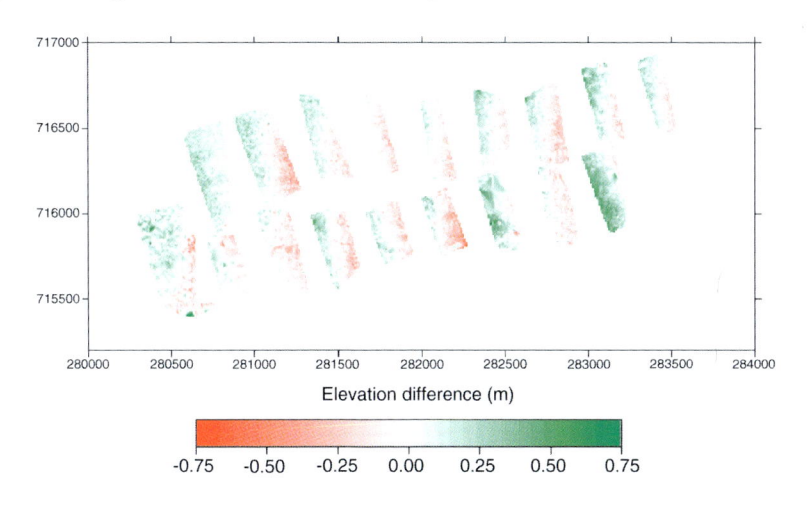

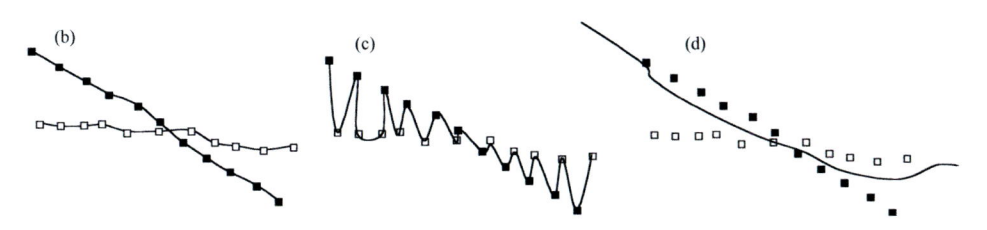

Plate 3 The problem of systematic surface error in areas of DEM overlap. Part (a) shows the results of the DEM overlap; (b) shows this effect in schematic terms; (c) shows the effect of bilinear interpolation as applied to the full dataset without a more sophisticated treatment of overlap areas; and (d) shows DEM combination using distance weighting, considered in section 6.6. Black points are the upstream DEM and white points are the downstream DEM

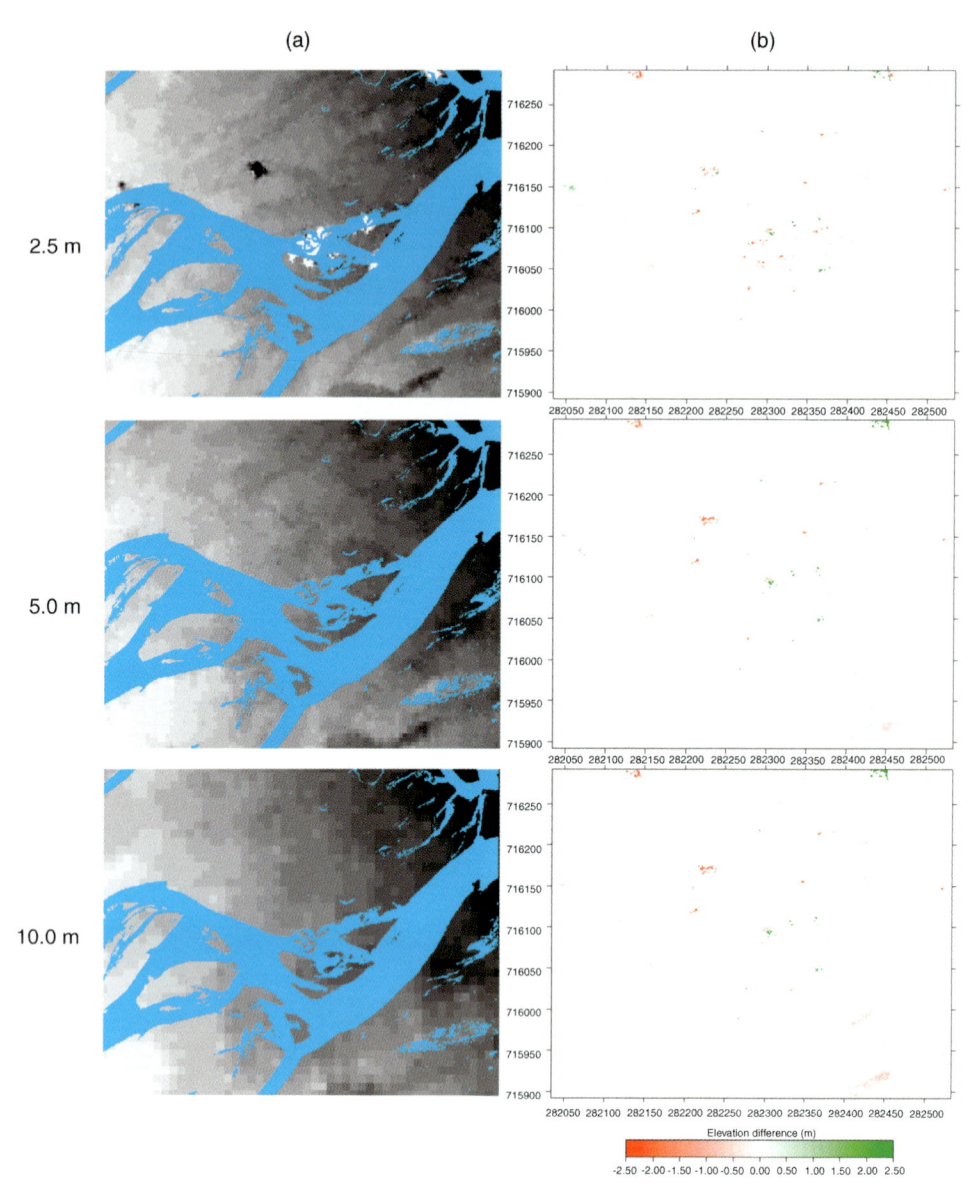

Plate 2 (a) DEMs of the test area collected at three different resolutions and (b) compared with a 1 m DEM

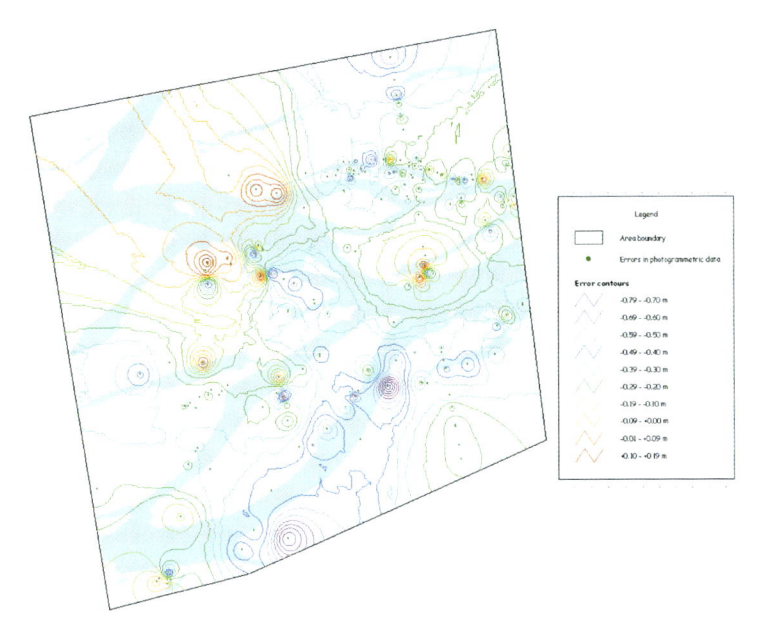

Plate 4 Contoured error surface for an intensively sampled sub-area. It shows that the errors are mainly concentrated around channel margins

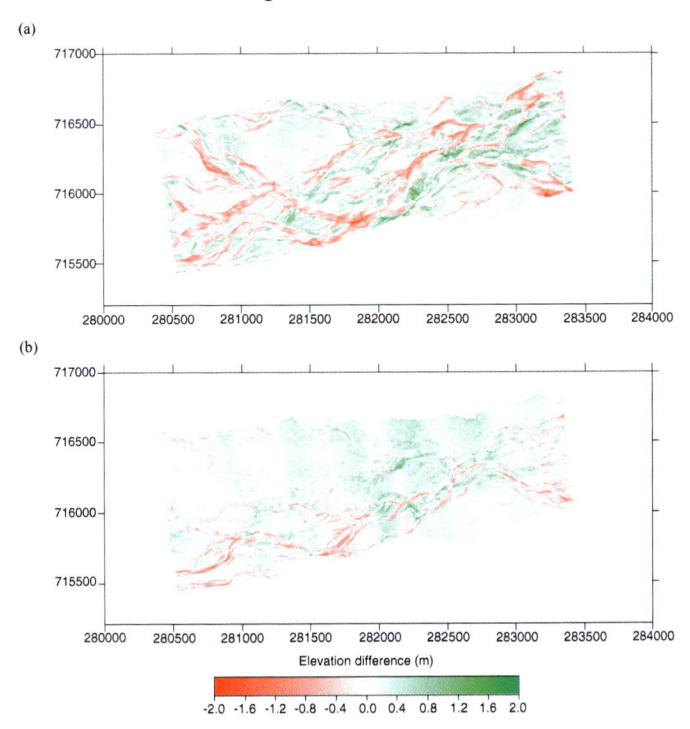

Plate 5 Comparison of the February 1999 DEM with the February 2000 DEM. (a) annual and with the March 1999 DEM (b) storm event, to demonstrate that when erosion and deposition changes are small, the banding effect can still be clearly identified

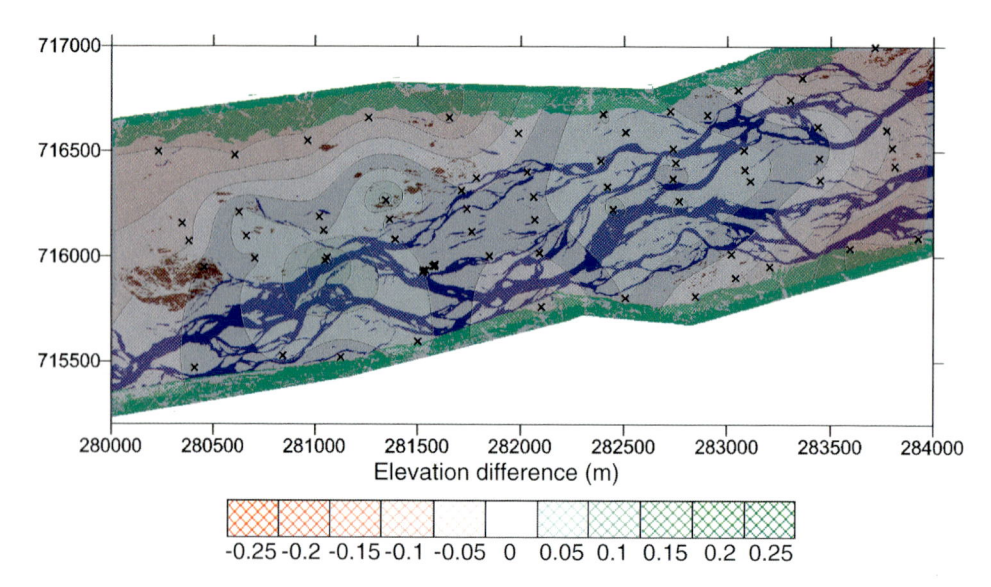

Plate 6 Contoured map of height discrepancies between the post-processed DEM surfaces and ground control point elevations for the February 2000 DEM. Zones of no shading are within ±0.0025 m of surveyed PCP position. The location of individual ground control points is marked

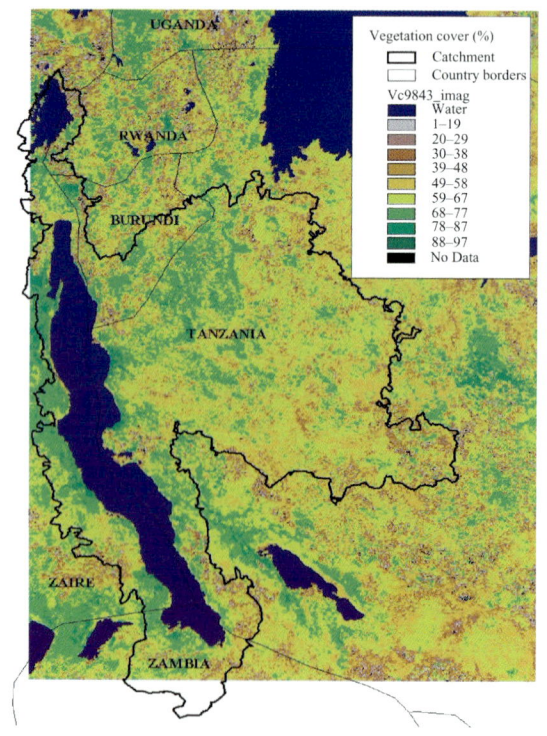

Plate 8 Vegetation cover image (1 km NDVI) for April 1998. Data from the March and May were used to fill the gaps in areas of persistent cloud cover

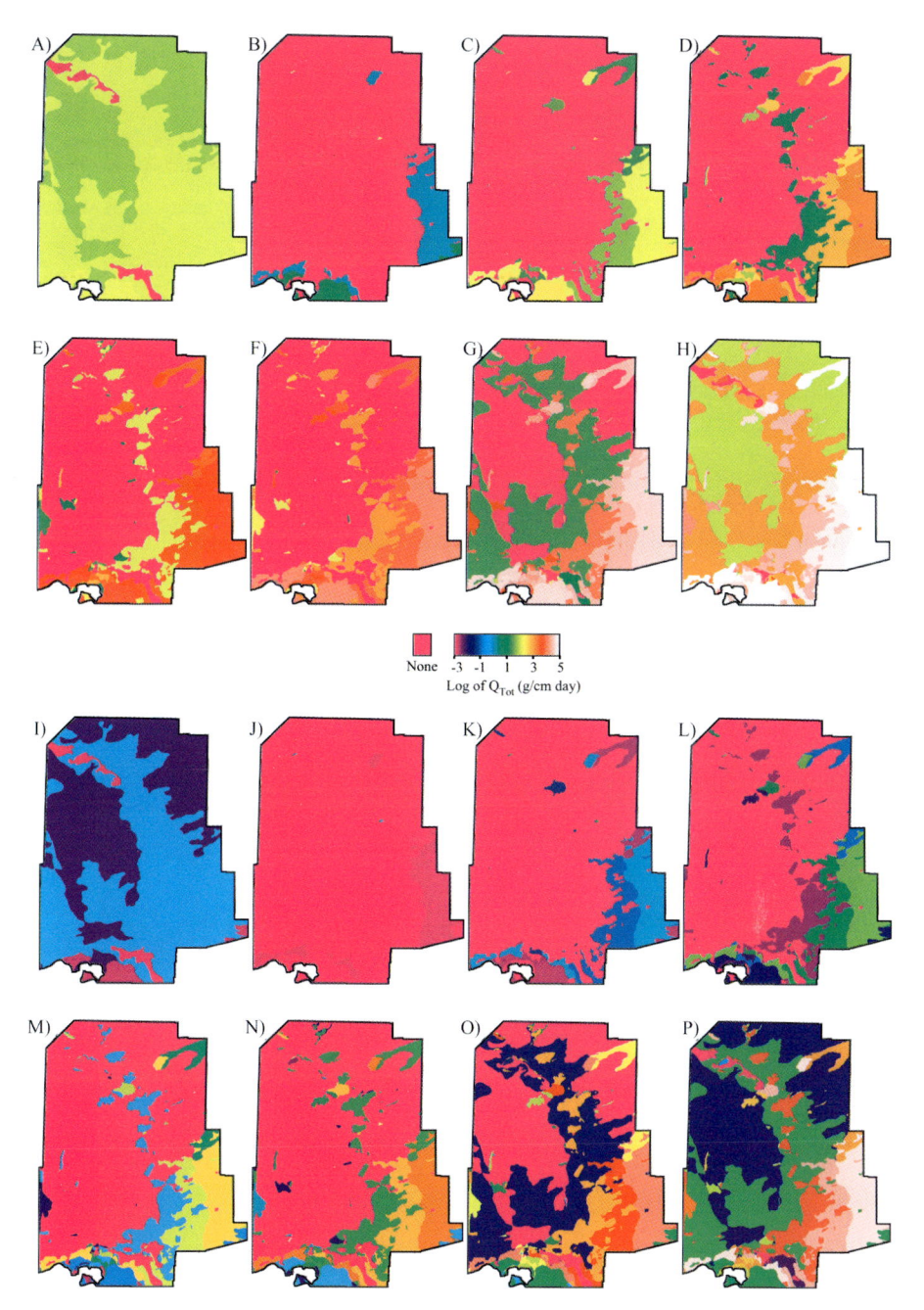

Plate 7 Model estimates of total horizontal flux Q_{Tot} and vertical flux F_a for several scenarios. Wind speeds observed at Jornada, Spring 2000: (a and i) no vegetation (using u_{*t}); (b and j) vegetation (using u_{*t}); (c and k) vegetation (using u_{*t}) using 1.25 × wind speed; (d and l) vegetation (using u_{*t}) using 1.5 × wind speed; (e and m) vegetation (using u_{*t}) using 1.75 × wind speed; (f and n) vegetation (using u_{*t}) using 2.0 × wind speed; (g and o) vegetation (using u_{*t}) using 2.5 × wind speed; (h and p) vegetation (using u_{*t}) using 3.0 × wind speed

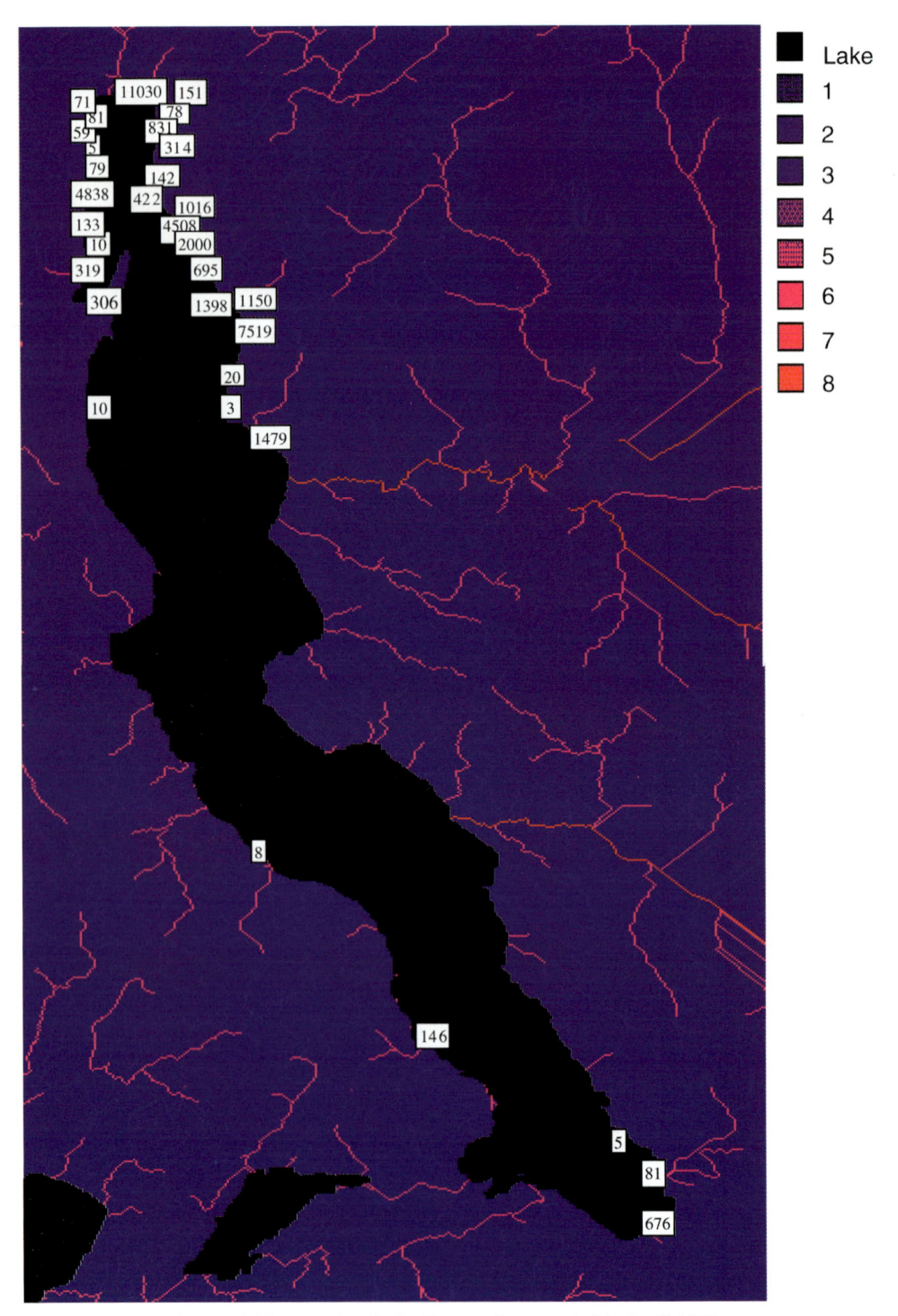

Plate 9 Total sediment yield (tonnes) to Lake Tanganyika recorded in April 1998

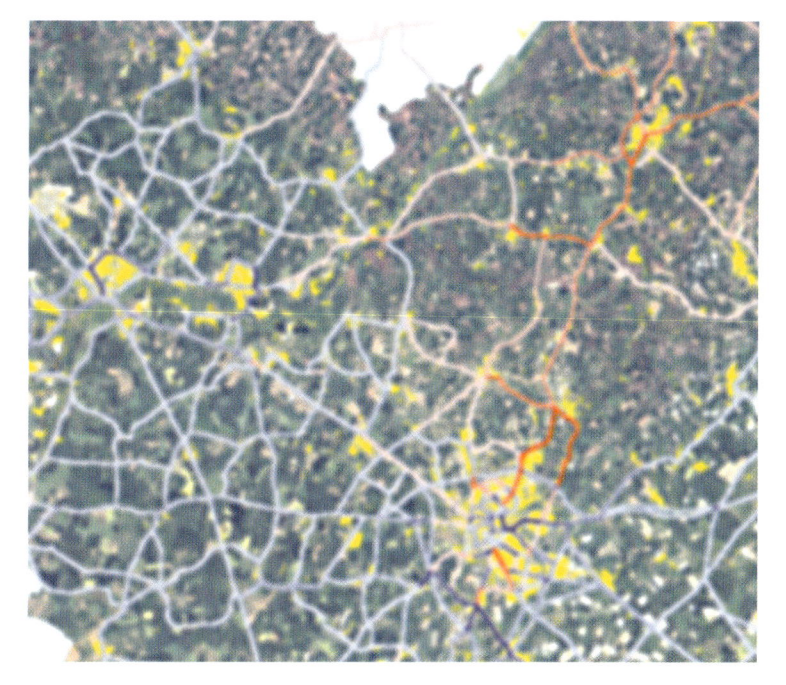

Plate 10 Change in emissions concentrations between the reference scenario and the policy scenario. Dark blue indicates a large decrease in concentration, light blue a small decrease, light pink a small increase and dark pink a large increase. Settlement is picked out in yellow

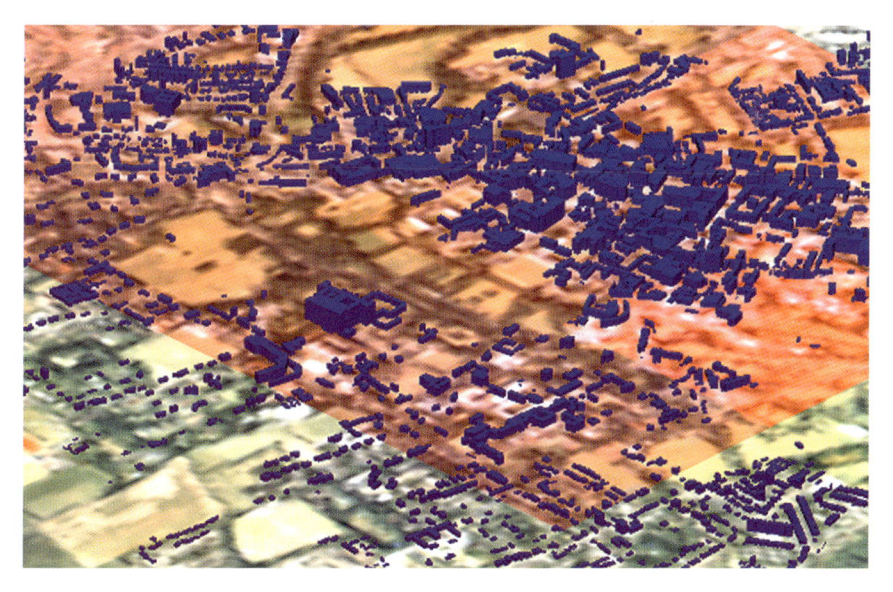

Plate 11 Visualization of emissions concentrations for the Policy Case in West Cambridge. Buildings have been modelled from airborne LiDAR data. Concentration increases with increasing opacity and increasing shades of red

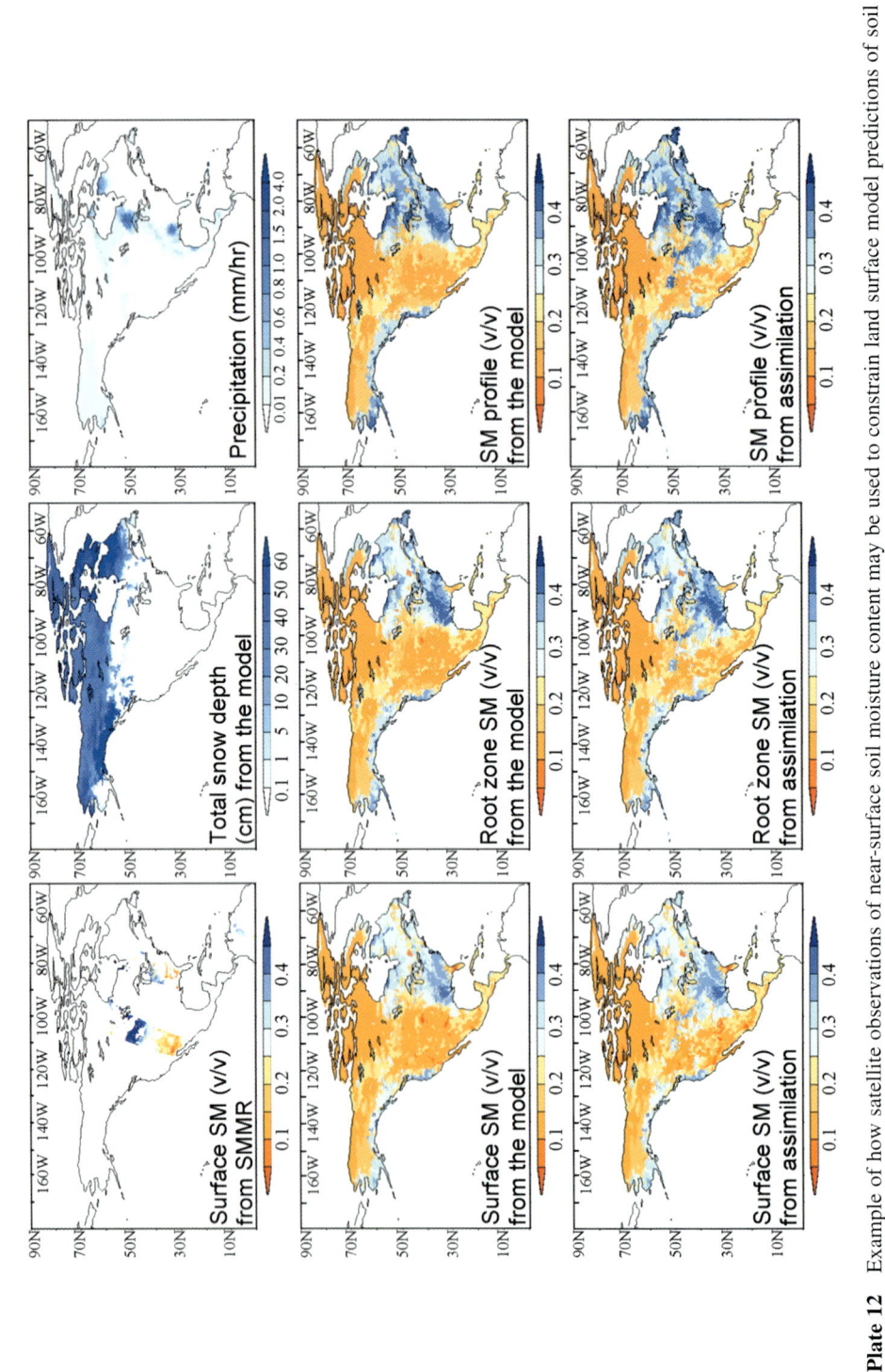

Plate 12 Example of how satellite observations of near-surface soil moisture content may be used to constrain land surface model predictions of soil moisture throughout the soil profile using data assimilation. (Adapted from Walker and Houser 2001)

Table 7.4 SWEMO input parameters for vegetation of the Jornada Basin

Vegetation type	Grass	Mesquite	Creosote	Tarbush	Snakeweed	Other shrubs	Bare
Fractional cover	0.25	0.21	0.17	0.12	0.29	0.17	0.001
Basal area (cm^2)	900	5100	10 800	13 200	100	8500	0.001
Profile area (cm^2)	1000	18 500	9200	13 800	100	8000	0.001
β	100	100	100	100	100	100	1
z_0 (cm)	3.8	4.1	8.2	4.6	1.8	5.6	0.04

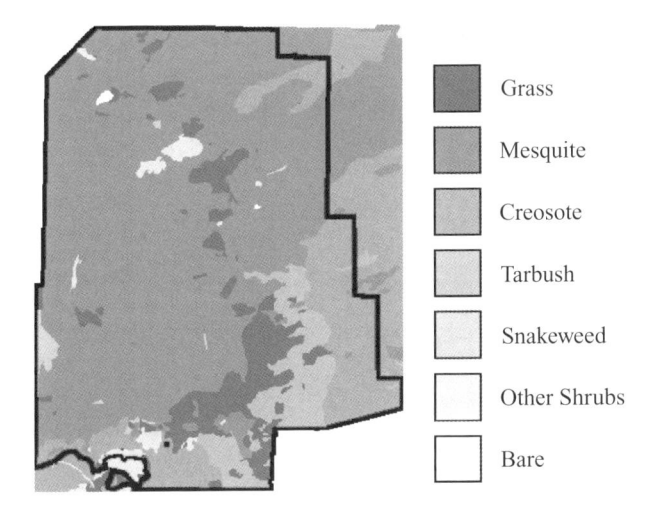

Figure 7.5 Vegetation map of the Jornada Basin derived from mapping efforts at the Jornada Experimental Range and the NMSU College Ranch. The area outlined in black is the region where the soil and vegetation maps overlap and is the study area for the present research

cover types were derived from ongoing vegetation monitoring data as part of the biodiversity Vegetation Transect at Jornada. Fractional cover for the creosote and tarbush cover types was derived from ongoing vegetation monitoring data as part of the Small Mammal Exclosure experiment at Jornada. Fractional cover for the creosote vegetation cover type was taken from Okin and Gillette (2001). Plants were assumed to be cylindrical in shape with basal area equal to π(plant radius)^2 and profile area equal to (plant diameter)(plant height). The drag ratio, β, was taken from Raupach *et al.* (1993) and assumed to be constant for all cover types. Roughness height, z_0, was calculated using equation (4) based on vegetation height, profile area and number density.

Wind Data. Wind monitoring by one of the present authors (Gillette) has been ongoing at several sites in the Jornada for many years. Data from one windy season (28 March 2000–10 July 2000) was used in this study (Figure 7.6). Although wind direction was also

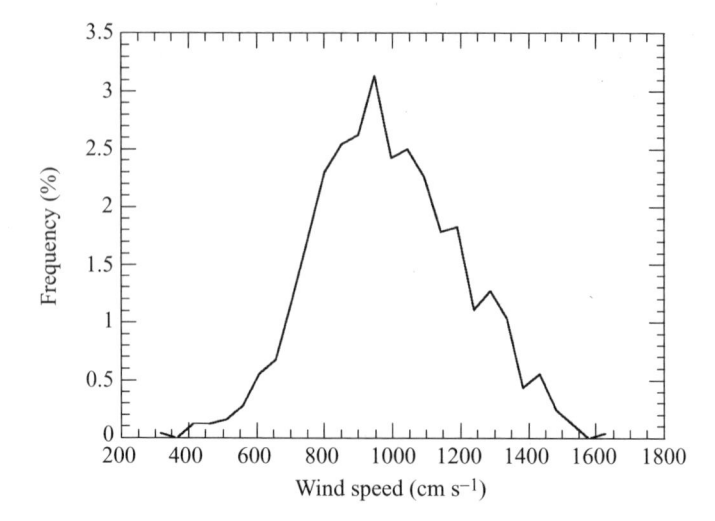

Figure 7.6 Frequency of wind speeds at one monitoring site at Jornada. Measurements were taken at 1490 cm from the period 28 March 2000 to 10 July 2000. Wind speeds below 300 cm s^{-1} were not recorded. Frequency is reported as percent of entire 105-day wind record

recorded, direction is assumed not to have an impact on the magnitude of wind erosion and dust flux.

7.5.2 Running the Model

SWEMO uses soil texture and vegetation maps as its primary inputs, and from these produces intermediate maps of parameters required for evaluating the equations (1) and (2). Thus, the first step in generating spatially explicit estimates of wind erosion and dust flux, is the production of these intermediate products (Figures 7.7–7.10). Figures 7.7–7.9 are generated in a straightforward way by determining vegetation type or soil texture in each model cell and then looking up the appropriate value of u_{*ts}, z_0 or k from associated look-up table files (see Tables 7.3 and 7.4). Figure 7.10 is generated by determining vegetation type and soil texture for each model cell, looking up values of u_{*ts} as well as plant structural parameters in the associated look-up table files and evaluating equation (7) for each cell.

A record of wind speed is imported into the model and compressed by creating a histogram of the data. This histogram is then used to calculate a histogram of u_* using equation (2) and values of z_0 for each cell (Figure 7.8). The histogram approach to working with wind speed data is the most computationally efficient way to evaluate equation (1) for each cell, because it allows estimation of horizontal flux for a few wind speed bins, as opposed to each entry in an entire wind speed record.

Once the intermediate maps and u_* histograms have been created, equation (1) is evaluated for each cell to generate estimates of Q_{Tot} for each cell. Data from Figure 7.9 are then used to convert the Q_{Tot} map into maps of dust flux, F_a.

In the present research, Q_{Tot} and F_a were evaluated for the entire study area for eight scenarios:

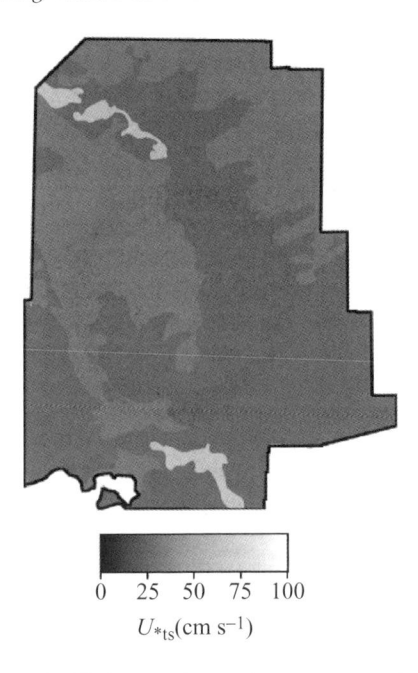

Modelling Wind Erosion and Dust Emission on Vegetated Surfaces 149

Figure 7.7 Map of the threshold shear velocity for unvegetated soil, u_{*ts}, derived from the soil texture map and literature values for u_{*ts}

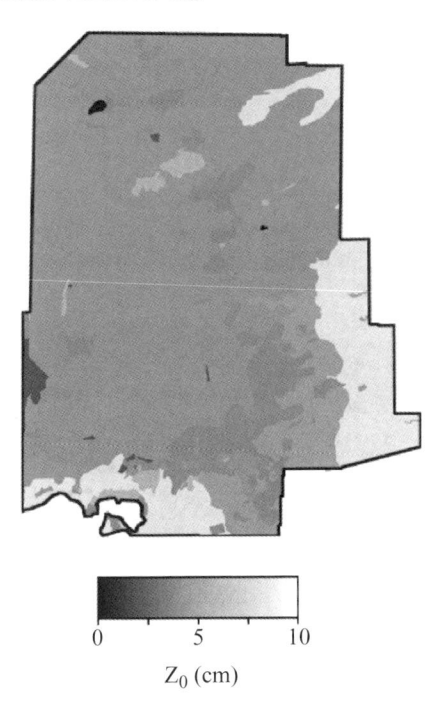

Figure 7.8 Map of the roughness height, z_0, derived by evaluating equation (4) for each vegetation height based on its measured height (see Table 7.4)

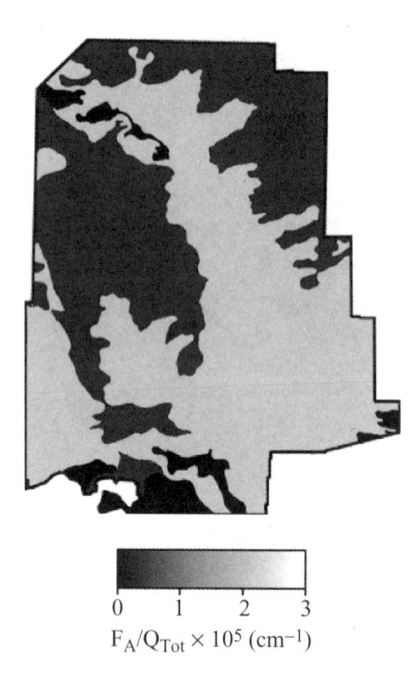

150 Spatial Modelling of the Terrestrial Environment

Figure 7.9 Map of the ratio of vertical flux to horizontal flux, F_a/Q_{Tot}, derived from the soil texture map and literature values

Figure 7.10 Map of the threshold shear velocity for vegetated soil, u_{*t}, derived by evaluating equation (7) for the entire study area

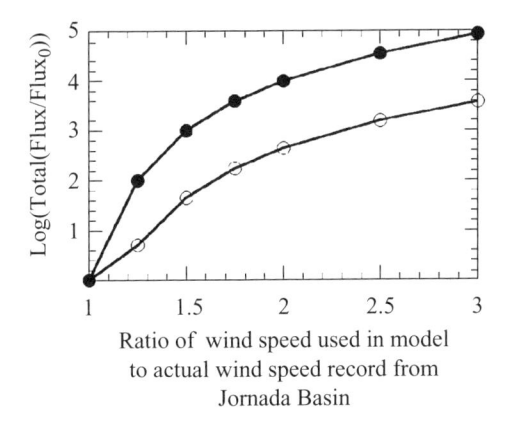

Figure 7.11 Log of total flux with elevated wind speed divided by total flux with measured wind speed. Flux is either Q_{Tot} or F_a for elevated wind speed. Flux$_0$ is either Q_{Tot} and F_a for measured wind speed. Results for Q_{Tot} are shown by close circles and results for F_a are shown by open circles

1. No vegetation, wind record from Jornada (Plates 7(a) and 7(i))
2. Mapped vegetation (Figure 7.5), wind speed record from Jornada (Plates 7(b) and 7(j); wind speed record shown in Figure 7.6)
3. Mapped vegetation, Jornada wind speed record multiplied by 1.25 (Plates 7(c) and 7(k))
4. Mapped vegetation, Jornada wind speed record multiplied by 1.5 (Plates 7(d) and 7(l))
5. Mapped vegetation, Jornada wind speed record multiplied by 1.75 (Plates 7(e) and 7(m))
6. Mapped vegetation, Jornada wind speed record multiplied by 2.0 (Plates 7(f) and 7(n))
7. Mapped vegetation, Jornada wind speed record multiplied by 2.5 (Plates 7(g) and 7(o))
8. Mapped vegetation, Jornada wind speed record multiplied by 3.0 (Plates 7(h) and 7(p))

By increasing the wind speeds used in the model over the measured wind speeds at Jornada, we were able to evaluate the effect that windier conditions would have at the site. Calculating the total of Q_{Tot} and F_a for the entire area, the overall effect of increased wind speed for the area was evaluated (Figure 7.13).

7.5.3 Discussion of Model Results

Values of Q_{Tot} for the study area (Plate 7) compare favourably with measured values of dust flux in several cover types within the Basin for a period encompassing the period of the wind record (Table 7.5). However, the model predicts no wind erosion or dust flux for much of the Basin, most notably the mesquite areas which display the highest measured

Table 7.5 Summary of measured values of Q_{Tot} for different vegetation types in the Jornada Basin for the period 24 July 1998 to 19 April 2001

Vegetation type	\multicolumn{6}{c}{$\text{Log}(Q_{Tot}(\text{g cm}^{-1}/\text{day}))$}
	Grassland	Mesquite	Creosote	Tarbush	Playa	Bare
Average	−1.1	0.0	−1.0	−1.0	−0.8	1.5
Minimum	−3.0	−0.6	−3.0	−2.3	−1.9	0.7
Maximum	−0.3	0.9	−0.2	−0.4	−0.3	2.3

Figure 7.12 Oblique aerial photograph taken in 1975 during a dust storm in West Texas, USA. Notice that plumes are formed over a small fraction of the total amount of agricultural land (from Gillette (1999), reproduced with permission)

dust flux among vegetated surfaces. This effect is largely due to the high-threshold shear velocity calculated for combinations of vegetation and soil (Figure 7.7) and as a result, the area of no modelled wind erosion is seen to dwindle with increasing wind speed (Plate 7).

Gillette (1999) has suggested that the bulk of wind erosion and dust flux that occurs often occurs as small 'hot spots' within the landscape (Figure 7.12). These areas may be roads, small disturbance, or as Okin and Gillette (2001) have pointed out, self-organizing 'streets' in the landscape. Streets are elongated unvegetated areas, usually aligned with the wind, down which wind blows, thus effectively reducing u_{*t} over small scales. Okin and Gillette have suggested that these streets may be responsible for the observed high-flux rates in mesquite dunelands. SWEMO assumes homogenous vegetation cover and distribution within each vegetation type. Thus, our results suggest that the assumption of homogenous vegetation cover is not adequate for spatial modelling of wind erosion. More complicated models that incorporate these erosion hot spots are required to reproduce vegetation with high fidelity.

Despite this, the modelled increase in Q_{Tot} and F_a with increasing wind speed confirms the predictions of Gregory et al. (1999) that wind erosion and dust flux increase non-linearly with increasing wind speed. As Plate 7 shows, this effect is the result of two factors: (1) increased number of areas subject to wind erosion (areas with $u_* > u_{*t}$); and (2) increased flux from areas undergoing wind erosion (increased $u_*^2 - u_{*t}^2$). Because wind erosion is controlled in large part by the difference between u_* and u_{*t} for conditions where $u_* > u_{*t}$, and u_{*t} is largely controlled by variable vegetation cover (equation (7)), evaluating Q_{Tot} and F_a for elevated wind speeds also serves as a proxy for conditions when vegetation cover may be decreased due to low vegetation conditions, due, say, to drought.

7.6 Conclusion

The mismatch between observed patterns of wind erosion and dust flux and the patterns modelled using SWEMO most likely arises from the fact that in natural environments, small wind erosion hot spots are responsible for the vast majority of flux. New modelling techniques need to be developed which can incorporate these small-scale spatial phenomena into large-scale dust flux models. In addition, new techniques must be developed which can identify, quantify and map these erosion hot spots in the landscape.

A potential method to map hot spots, in addition to providing more detailed information about surface properties, is to derive estimates of the relevant parameters from remote sensing data. Some attempt has been made to do this in previous studies, although not for the original purposes of wind erosion modelling. Okin et al. (2001c) and Okin and Gillette (2001), for example, have derived estimates of surface cover in areas susceptible to wind erosion using spectral mixture analysis and geostatistics of aerial photographs, respectively. Other authors have tried to develop remote sensing tools that estimate vegetation number density and size using the Li-Strahler optical canopy model (see, for example, Li and Strahler, 1985; Franklin and Strahler, 1988; Li and Strahler, 1992; Scarth and Phinn, 2000).

Spatial modelling of wind erosion remains in its infancy. Because large-scale dust emission often depends on very small-scale heterogeneities on the land surface, the SWEMO model described here is only a first step in developing a full modelling approach to this fundamental geomorphological process that dominates the world's arid and semi-arid regions – 30% of the Earth's land surface.

Acknowledgements

Data was provided for this work by NSF grant DEB 00-84012 as a contribution to the Jornada (LTER) programme.

References

Alward, R.D., Detling, J.K. and Milchunas, D.G., 1999, Grassland vegetation changes and nocturnal global warming, *Science*, **283**, 229–231.
Bagnold, R.A., 1941, *The Physics of Blown Sand and Desert Dunes* (New York: Methuen).
Bahre, C.J., 1991, *A Legacy of Change: Historic Human Impact on Vegetation in the Arizona Borderlands* (Tucson, Arizona: University of Arizona Press).
Bahre, C.J. and Shelton, M.L., 1993, Historic vegetation change, mesquite increases and climate in southeastern Arizona, *Journal of Biogeography*, **20**, 489–514.

Brown, J.H., Valone, T.J. and Curtin, C.G., 1997, Reorganization of an arid ecosystem in response to recent climate change, *Proceedings of the National Academy of Science of the United States of America*, **94**, 9729–9733.

Buffington, L.C. and Herbel, C.H., 1965, Vegetational changes on a semidesert grassland range from 1858 to 1963, *Ecological Monograph*, **35**, 139–164.

Chadwick, O.A., Derry, L.A., Vitousek, P.M., Huebert, B.J. and Hedin, L.O., 1999, Changing sources of nutrients during four million years of ecosystem development, *Nature*, **397**, 491–497.

Coale, K.H., Johnson, K.S., Fitzwater, S.E., Gordon, R.M., Tanner, S., Chavez, F.P., Ferioli, L., Sakamoto, C., Rogers, P., Millero, F., Steinberg, P., Nightingale, P., Cooper, D., Cochlan, W.P., Landry, M.R., Constantinou, J., Rollwagen, G., Trasvina, A. and Kudela, R., 1996, A massive phytoplankton bloom induced by an ecosystem-scale iron fertilization experiment in the equatorial Pacific Ocean, *Nature*, **383**, 495–501.

Dentener, F.J., Carmichael, G.R., Zhang, Y., Lelieveld, J. and Crutzen, P.J., 1996, Role of mineral aerosol as a reactive surface in the global troposphere, *Journal of Geophysical Research-Atmospheres*, **101**, 22869–22889.

Dickerson, R.R., Kondragunta, S., Stenchikov, G., Civerolo, K.L., Doddridge, B.G. and Holben, B.N., 1997, The impact of aerosols on solar ultraviolet radiation and photochemical smog, *Science*, **278**, 827–830.

Duce, R.A. and Tindale, N.W., 1991, Atmospheric transport of iron and its deposition in the ocean, *Limnology and Oceanography*, **36**, 1715–1726.

Forman, S.L., Goetz, A.F.H. and Yuhas, R.H., 1992, Large-scale stabilized dunes on the high plains of Colorado: understanding the landscape response to Holocene climates with the aid of images from space, *Geology*, **20**, 145–148.

Franklin, J. and Strahler, A.H., 1988, Invertible canopy reflectance modelling of vegetation structure in semiarid woodland, *IEEE Transactions on Geoscience and Remote Sensing*, **26**, 809–825.

Gillette, D.A., 1988, Threshold friction velocities for dust production for agricultural soils, *Journal of Geophysical Research*, **93**, 12645–12662.

Gillette, D.A., 1999, A qualitative geophysical explanation for 'hot spot' dust emission source regions, *Contributions to Atmospheric Physic*, **72**, 67–77.

Gillette, D.A. and Chen, W.A., 2001, Particle production and aeolian transport from a 'supply-limited' source area in the Chihuahuan desert, New Mexico, United States, *Journal of Geophysical Research*, **106**, 5267–5278.

Gillette, D.A., Fryrear, D.W., Gill, T.E., Ley, T., Cahill, T.A. and Gearhart, E.A., 1997, Relation of vertical flux of particles smaller than 10 μm to aeolian horizontal mass flux at Owens Lake, *Journal of Geophysical Research*, **102**, 26009–26015.

Ginoux, P., Chin, M., Tegen, I., Prospero, J.M., Holben, B., Dubovik, O. and Lin, S.J., 2001, Sources and distributions of dust aerosols simulated with the GOCART model, *Journal of Geophysical Research-Atmospheres*, **106**, 20255–20273.

Gregory, P.J., Ingram, J.S.I., Campbell, B., Goudriaan, J., Hunt, L.A., Landsberg, J.J., Linder, S., Stafford Smith, M. and Sutherst, R.W., 1999, Managed production systems, in B. Walker, W. Steffen, J. Canadell, and J. Ingram (eds), *The Terrestrial Biosphere and Global Change: Implications for Natural and Managed Ecosystems*, Vol. 4 (Cambridge: Cambridge University Press), 229–270.

Griffin, D.W., Garrison, V.H., Herman, J.R. and Shinn, E.A., 2001, African desert dust in the Caribbean atmosphere: microbiology and public health, *Aerobiologia*, **17**, 203–213.

Guelle, W., Balkanski, Y.J., Schulz, M., Marticorena, B., Bergametti, G., Moulin, C., Arimoto, R. and Perry, K.D., 2000, Modeling the atmospheric distribution of mineral aerosol: comparison with ground measurements and satellite observations for yearly and synoptic timescales over the North Atlantic, *Journal of Geophysical Research-Atmospheres*, **105**, 1997–2012.

Kaimal, J.C. and Finnigan, J.J., 1994, *Atmospheric Boundary Layer Flows* (Oxford: Oxford University press).

Lancaster, N. and Baas, A., 1998, Influence of vegetation cover on sand transportation by wind: field studies at Owens Lake, California, *Earth Surface Processes and Landforms*, **23**, 69–82.

Lettau, H.H., 1969, Note on aerodynamic roughness-parameter estimation on the basis of roughness element description, *Journal of Applied Meteorology*, **8**, 828–832.

Levin, Z., Ganor, E. and Gladstein, V., 1996, The effects of desert particles coated with sulfate on rain formation in the eastern Mediterranean, *Journal of Applied Meteorology*, **35**, 1511–1523.

Li, X. and Strahler, A.H., 1985, Geometric-optical modelling of a conifer forest, *IEEE Transactions on Geoscience and Remote Sensing*, **23**, 705–721.

Li, X. and Strahler, A.H., 1992, Geometric-optical bidirectional reflectance modelling of the discrete crown vegetation canopy: effect of crown shape and mutual shadowing, *IEEE Transactions on Geoscience and Remote Sensing*, **30**, 276–291.

Mahowald, N.M., Zender, C., Luo, C. and del Corral, J. (submitted), Understanding the 30-year Barbados desert dust record, *Journal of Geophysical Research*.

Marticorena, B., Bergametti, G., Aumont, B., Callot, Y., N'Doumé, C. and Legrand, M., 1997, Modeling the atmospheric dust cycle: 2. Simulation of Saharan sources, *Journal of Geophysical Research*, **102**, 4387–4404.

Okin, G.S. and Gillette, D.A., 2001, Distribution of vegetation in wind-dominated landscapes: implications for wind erosion modelling and landscape processes, *Journal of Geophysical Research*, **106**, 9673–9683.

Okin, G.S., Murray, B. and Schlesinger, W.H., 2001a, Degradation of sandy arid shrubland environments: observations, process modelling and management implications, *Journal of Arid Environments*, **47**, 123–144.

Okin, G.S., Murray, B. and Schlesinger, W.H., 2001b, Desertification in an arid shrubland in the southwestern United States: process modeling and validation, in A. Conacher (ed.), *Land Degradation: Papers Selected from Contributions to the Sixth Meeting of the International Geographical Union's Commission on Land Degradation and Desertification, Perth, Western Australia, 20–28 September 1999* (Dordrecht: Kluwer Academic Publishers), 53–70.

Okin, G.S., Okin, W.J., Murray, B. and Roberts, D.A., 2001c, Practical limits on hyperspectral vegetation discrimination in arid and semiarid environments, *Remote Sensing of Environment*, **77**, 212–225.

Owen, P.R., 1964, Saltation of uniform grains in air, *Journal of Fluid Mechanics*, **20**, 225–242.

Paulsen, H.A., Jr. and Ares, F.N., 1962, Grazing values and management of black gramma and tobosa grasslands and associated shrub ranges of the Southwest, *U.S. Forest Service*.

Piketh, S.J., Tyson, P.D. and Steffen, W., 2000, Aeolian transport from southern Africa and iron fertilization of marine biota in the South Indian Ocean, *South African Journal of Geology*, **96**, 244–246.

Prospero, J.M., Ginoux, P., Torres, O. and Nicholson, S.E. (2002), Environmental characterization of global sources of atmospheric soil dust derived from the NIMBUS-7 absorbing aerosol product, *Reviews of Geophysics*, **40**, 2.1–2.31.

Raupach, M.R., Gillette, D.A. and Leys, J.F., 1993, The effect of roughness elements on wind erosion threshold, *Journal of Geophysical Research*, **98**, 3023–3029.

Reynolds, R., Belnap, J., Reheis, M., Lamothe, P. and Luiszer, F., 2001, Aeolian dust in Colorado Plateau soils: nutrient inputs and recent change in source, *Proceedings of the National Academy of Sciences of the United States of America*, **98**, 7123–7127.

Rosenzweig, C. and Hillel, D., 1993, The dust bowl of the 1930s: analog of greenhouse effect in the Great Plains, *Journal of Environmental Quality*, **22**, 9–22.

Scarth, P. and Phinn, S., 2000, Determining forest structural attributes using an inverted geometric-optical model in mixed eucalypt forests, Southeast Queensland, Australia, *Remote Sensing of Environment*, **71**, 141–157.

Schlesinger, W.H., Reynolds, J.F., Cunningham, G.L., Huenneke, L.F., Jarrell, W.M., Virginia, R.A. and Whitford, W.G., 1990, Biological feedbacks in global desertification, *Science*, **247**, 1043–1048.

Schultz, B.W. and Ostler, W.K., 1993, Effects of prolonged drought on vegetation association in the Northern Mojave Desert, *Wildland Shrubs and Arid Land Restoration Symposium*, Las Vegas, Nevada, 228–235.

Shao, Y. and Raupach, M.R., 1993, Effect of saltation bombardment on the entrainment of dust by wind, *Journal of Geophysical Research*, **98**, 12719–12726.

Soil Conservation Service, 1980, *Soil Survey of Doña Ana County, New Mexico* (United States Department of Agriculture, Soil Conservation Service).

Sokolik, I.N. and Toon, O.B., 1996, Direct radiative forcing by anthropogenic airborne mineral aerosols, *Nature*, **381**, 681–683.

Stockton, P.H. and Gillette, D.A., 1990, Field measurement of the sheltering effect of vegetation on erodible land surfaces, *Land Degradation and Rehabilitation*, **2**, 77–85.

Tegen, I. and Fung, I., 1995, Contribution to the atmospheric mineral aerosol load from land surface modification, *Journal of Geophysical Research*, **100**, 18707–18726.

Tegen, I. and Lacis, A.A., 1996, Modeling of particle size distribution and its influence on the radiative properties of mineral dust aerosol, *Journal of Geophysical Research*, **101**, 19237–19244.

Wooding, R.A., Bradley, E.F. and Marshall, J.K., 1973, Drag due to regular arrays of roughness elements of varying geometry, *Boundary-Layer Meteorology*, **5**, 285–308.

Wurzler, S., Reisin, T.G. and Levin, Z., 2000, Modification of mineral dust particles by cloud processing and subsequent effects on drop size distributions, *Journal of Geophysical Research*, **105**, 4501–4512.

8

Near Real-Time Modelling of Regional Scale Soil Erosion Using AVHRR and METEOSAT Data: A Tool for Monitoring the Impact of Sediment Yield on the Biodiversity of Lake Tanganyika

Nick Drake, Xiaoyang Zhang, Elias Symeonakis, Martin Wooster, Graeme Patterson and Ross Bryant

8.1 Introduction

Evaluation and management of sediment inputs into lakes require an understanding of all the sediment sources within the basin, the rate of transfer of these sediments from the hillslopes to the channels, from the channels to the lake and the pattern of dispersal within the lake itself. This is problematic for large lakes because they tend to have very large catchments and thus both lake and catchment require monitoring over large areas. Because of this, our understanding of sediment transfers within large lake systems is often limited.

Lake Tanganyika is a good example of an important large lake ecosystem that is sensitive to sedimentation problems and requires management. It contains around one-sixth of the world's liquid freshwater and is almost one and a half kilometres at its deepest point. The present lake basin has been water-filled for at least 10 million years and some sediments date back twice this period (Patterson and Mackin, 1998). In addition to these almost unparalleled physical characteristics, the lake is also a truly unique ecosystem; the second most biologically diverse lake on Earth. There are at least 300 endemic fish species in the lake and more are continually being discovered. Six of these species occur in vast numbers and are an important source of food and income for the local population.

Sediment inputs from accelerating land erosion caused by deforestation appears to be the current main environmental threat to the lake's unique ecosystem, far outweighing that from organic pollution (Vandelannoote et al., 1996). Cohen et al. (1993) showed how the species richness of fish and ostracods was significantly reduced in regions of the lake subject to sedimentation from land erosion, with species richness in some cases being 65% lower. The majority of these species are endemic to the lake, or even to specific regions of the lake, and have highly particular habitat requirements that are likely to be severely impacted by increases in suspended sediment concentration brought about by erosion within the catchment and sediment transport into the lake.

The importance of sediment pollution requires that the erosion of the entire lake drainage basin be investigated in order to determine the sediment sources and sinks within it. This study forms the first step in the development of this capability, which combines remote sensing data collection and analysis with spatial modelling of the erosion at the catchment scale, modelling the transport of the eroded soil particles into Lake Tanganyika and monitoring of sediment movement within the lake. This research was conducted as part of the Lake Tanganyika Biodiversity Project (LTBP) Special Sediment Study. The LTBP is assisting the four riparian states (Burundi, Democratic Republic of Congo, Tanzania and Zambia) in producing an effective and sustainable system for managing and conserving the unique biodiversity of Lake Tanganyika. This study forms part of the development of this capability and focuses on evaluating the feasibility of using the model for real-time erosion monitoring and prediction that could form part of an operational monitoring system.

One of the main problems with modelling erosion and deposition in large catchments is obtaining the necessary data over such large areas. We have overcome this problem by using remote sensing data where possible and by choosing models that can be readily implemented using such data but are otherwise parsimonious in their data requirements. Our approach to monitoring sediment dynamics is composed of three components. First, we have implemented a soil erosion model of the Lake Tanganyika catchment and the surrounding area at the regional scale (1–8 km pixel size), on a decadal time step. This is achieved by using an on-site Advanced Very High Resolution Radiometer (AVHRR) receiver at Kigoma (Tanzania) and access to the internet in order to obtain the US Aid Famine Early Warning System (FEWS) METEOSAT-based rainfall estimates (Herman et al., 1997). This modelling system can be used to monitor erosion, predict source areas, and define areas within the catchment where it is increasing. Second, we have developed a method of routing erosion in order to estimate sediment inputs to the lake and their likely source areas. Thus the model has the potential to identify river mouths that are affected by high-sediment discharge rates, the areas in the catchment responsible for this, and the likely cause of the phenomena (e.g. extreme rainfall, reduced vegetation cover or highly erodible soils). Finally, we have used AVHRR and Along Track Scanning Radiometer (ATSR-2) data to map sediment plumes emanating from these rivers and monitor their dynamics within the lake. The latter imagery is also being retrieved from the internet.

We applied the monitoring system outlined above for April 1998, the wettest month in the 1998 wet season. This month was used because such wet periods are likely to produce the most erosion. It is interesting to note that the region experienced extremely high rainfall throughout the 1998 wet season that can be attributed to the El Niño conditions of that year, thus rainfall was even higher than usual.

8.2 Methods

Methods of implementation of the three components of the sediment monitoring system are discussed in turn below.

8.2.1 Soil Erosion Modelling

The following soil erosion model was applied to the Lake Tanganyika catchment at the regional scale (1–8 km pixel size) on a decadal time step (Thornes 1985, 1989):

$$E = kOF^2 s^{1.67} e^{-0.07v} \tag{1}$$

where E is the erosion (mm/decad), k is a soil erodibility coefficient, OF is the overland flow (mm/decad), s is the slope(m/m) and v is the vegetation cover (%). The model predicts soil detachment by rainsplash and the ability of overland flow to transport this material. However, it does not consider where the sediment is transported to or any subsequent deposition.

A problem is posed by the choice of the spatial scale of the model because soil erosion predictions are sensitive to the spatial resolution of the model input parameters (Drake et al., 1999). For example, as the scale of the digital elevation model (DEM) is reduced, the slope derived from it is also reduced because the averaging inherent in the coarser resolution data means that high altitudes become lower and low altitudes higher (Zhang et al., 1999). The lower slopes result in lower predicted erosion. To overcome this problem we have developed a way of calculating slope independent of scale using a fractal approach. Most landscapes can be considered to be fractal and it can be shown (Zhang et al., 1999) that the percentage slope (s) is related to s at any scale d by the equation:

$$s = \alpha d^{1-D} \tag{2}$$

where α is the fractal constant and D is the fractal dimension. Slope in this study was estimated from the GTPO30 global 1 km digital elevation model (DEM) by calculating α and D from the DEM and then using equation (1) with a d of 30 m in order to estimate the slope at this scale. Zhang et al. (1999) have shown that the method is effective at the scales being considered here. They tested the method using the same DEM, calculated slope for a d of 200 m and 30 m, compared the slope estimates to two DEMs of this resolution, and found RMS errors of 3.92% at 30 m and 8.63% at 200 m.

To estimate vegetation cover we used LAC AVHRR data obtained from the Kigoma AVHRR receiving station. Imagery was calibrated, geocoded using information in the file header, cloud masked, converted into NDVI and then calibrated to vegetation cover. The cloud masking technique of Saunders and Kriebel (1988) was employed. The method uses

12 μm channel of AVHRR because clouds have a large optical depth at this wavelength and utilizes the fact that cloud tops tend to be cold due to their altitude. Thus, a simple brightness temperature threshold of this channel can separate cloud from other features in the image. Thresholds were relatively easy to determine over Lake Tanganyika since the water surface temperature varies by only 6° C over the year. A threshold of between 283 and 289 K, depending on the particular image, was found to be effective.

NDVI was then calculated for the cloud-free areas in each image and maximum decadal composites were then produced. We applied this method to the April imagery because such wet periods are likely to produce the most erosion. However, because of these adverse weather conditions, cloud was always a problem and all three maximum value composites contained many areas that were cloud masked for the entire decad. We overcame this by implementing the above-mentioned procedure for March and May and filling the cloud-masked areas with the average NDVI for that location from the preceding and following decad. If there were still regions with no NDVI information, then this process of averaging is continued until they are filled. However, there were so many areas with no information during April so it was decided that decadal composites were not practical at the height of the wet season. Thus, a monthly maximum value composite for June was computed. This image still contained continuously cloud-covered areas in a few mountainous locations in Rwanda and Burundi where cloud cover exists for the entire month needing data from all of March and April to fill them. Thus, it was only possible to parameterize the model in terms of monthly vegetation cover during such a wet period.

This maximum monthly NDVI image was then converted into vegetation cover using the calibration outlined in equation (3) see Plate 8. Numerous authors have found a relationship between either total or green vegetation cover (v) and NDVI and these data have been collated by Drake *et al.* (1995) and Zhang (1999) (Figure 8.1). It is clear that there

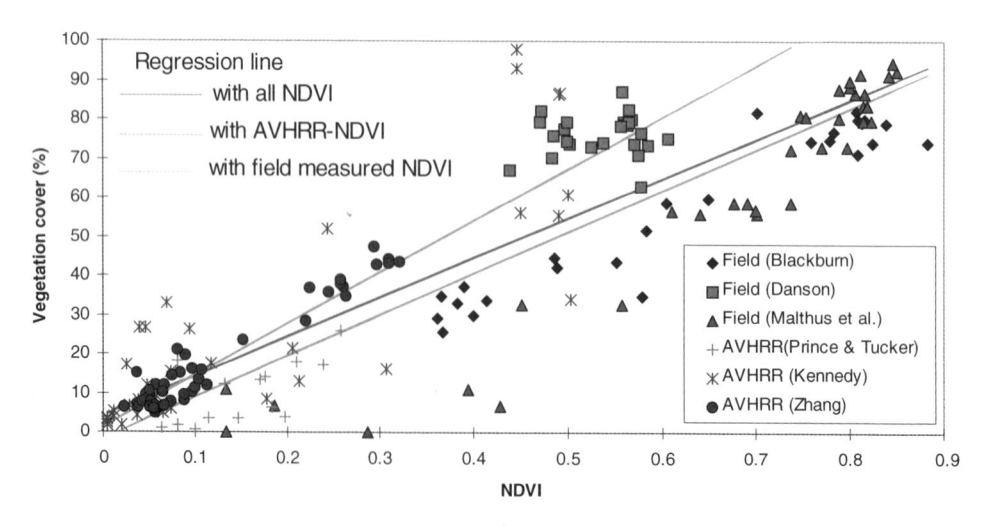

Figure 8.1 *Scattergram showing the relationship between percent vegetation cover (v) and NDVI data found by previous researchers. NDVI data were calculated from ground-based spectroradiometer (field) or satellite-based (AVHRR) radiometric measurements*

is a strong relationship between v and NDVI, however, this relationship varies, with field measurements using radiometers and spectroradiometers providing higher NDVI values than those obtained from ground measurements correlated to AVHRR imagery. This is due to atmospheric effects that affect the satellite-based measurements but not the spectroradiometer data. Because no atmospheric correction was applied to the AVHRR imagery a regression relationship was calculated using only the AVHRR-derived NDVI and the vegetation cover information:

$$V_c = 1.333 + 131.877\ NDVI \qquad (3)$$

Fit statistics: $r^2 = 0.73$, $F = 241.63, p = 4.26 \times 10^{-27}$.

It is important to note that the regions of persistent cloud cover could introduce errors in vegetation cover estimation and thus potentially introduce uncertainties for soil erosion estimation. In cloud-covered areas it is likely to be raining for much of the time, and these regions will tend to suffer high-overland flow and erosion. However, they are also the areas where the actual vegetation cover is at its most uncertain and errors in vegetation cover can produce a considerable change in predicted erosion as it is the most sensitive parameter in the model (Drake et al., 1999).

Scale also poses a problem when using coarse resolution vegetation cover estimates to parameterize regional soil erosion models. A reduction in predicted erosion occurs as the spatial resolution of the vegetation cover is reduced (Drake et al., 1999). This is caused by a loss of variance and is a result of the way erosion is parameterized in the model. The negative exponential relationship between cover and erosion (equation (1)) means that erosion is very high in bare areas but very low once cover exceeds around 40%. If we consider an image of vegetation patches (i.e., fields) with 100% cover, surrounded by bare fields, erosion is very high in bare areas and non-existent in the vegetated fields. When the spatial resolution of the image is reduced to a point where it exceeds the field size, predicted erosion is reduced because some of the vegetation cover of the vegetated field is assigned to the bare areas. To overcome this scaling problem we have implemented the Polya function method developed by Zhang et al. (2002). This technique represents spatial variation of vegetation cover in a pixel as a histogram of cover and uses eight neighbouring pixels to estimate the sub-pixel histogram at a specified spatial resolution (we use 30 m). Thus the output of the Poyla function is an image with a degraded spatial resolution (8 km) that contains a histogram of vegetation cover variation within each pixel. The erosion model is then run for each column of the histogram and the results summed to obtain erosion for the decad in question.

Soil erodibility is computed from maps of soil properties. No detailed soil maps of the region are available and thus it was estimated from data contained within the 8 km resolution UNEP/GRID Digital FAO Soil Map of the World. We employed the table of Mitchel and Bubenzer (1980) (Table 8.1) and Boolean algebra to convert maps of texture and organic matter data into spatially distributed K values. Because the soils are clay rich, soil erodibility within the catchment is generally quite low.

The SCS model (Soil Conservation Service, 1972) has been used to estimate overland flow. The model was selected not only because it is a catchment scale model and thus should not suffer spatial scaling problems, but also because it has similar parameter requirements to the soil erosion model. The only additional parameters needed are landuse and rainfall.

Table 8.1 The relationship between soil texture, organic matter and soil erodibility

Texture Class	0.875 < %OM	0.875 ≤ OM < 1.625	1.625 ≤ OM < 2.5	2.5 ≤ OM < 3.5	OM ≥ 3.5
Sand	0.05	0.04	0.03	0.025	0.02
Loamy sand	0.12	0.11	0.1	0.09	0.08
Sandy loam	0.27	0.255	0.24	0.215	0.19
Loam	0.38	0.36	0.34	0.315	0.29
Silt loam	0.48	0.45	0.42	0.375	0.33
Silt	0.6	0.56	0.52	0.47	0.42
Sandy clay loam	0.27	0.26	0.25	0.23	0.21
Clay loam	0.28	0.265	0.25	0.23	0.21
Sandy clay	0.14	0.135	0.13	0.125	0.12
Silty clay	0.25	0.24	0.23	0.21	0.19
Clay	0.13	0.17	0.21	0.25	0.29

The model is defined as:

$$OFp = \frac{(r_i - Ia)^2}{(r_i + 0.8S)} \quad (4)$$

where OFp is the overland flow and r_i is a rainfall event in the catchment, Ia is the initial abstraction and is equal to $0.2S$ where S is the the potential retention of water in the catchments and is calculated by:

$$S = \frac{25400}{CN} - 254 \text{ (mm)} \quad (5)$$

where CN is the runoff curve number. Curve numbers can be estimated from published tables (e.g. Rawls et al., 1993) and describe the ability of the pixel to produce runoff, with high curve numbers providing ideal conditions for runoff generation. We have used these tables to estimate curve numbers with information on soil texture derived from the FAO Soil Map of the World, vegetation cover derived from NDVI, and landcover information from the International Geosphere and Biosphere Program (IGBP) Global 1 km Landuse Map. These curve number maps are dynamic on a decadal basis since they change according to variations in vegetation cover.

Famine Early Warning System (FEWS) precipitation estimates were used to derive overland flow from the curve number maps (equation (4)). They are based upon a combination of METEOSAT Cold Cloud Duration estimates and Global Telecommunications System (GTS) raingauge data (Herman et al., 1997). First, a preliminary estimate of accumulated precipitation is made using the GOES Precipitation Index (GPI) (Arkin and Meisner, 1987). The GPI uses the hourly duration of cold cloud tops during a decad for the determination of accumulated rainfall, 3 mm of precipitation being allocated to each hour that cloud top temperatures are less than 235 K. The GPI estimate is corrected using a bias field that is calculated by incorporating the GTS observational data and fitting the biases to a grid using optimal interpolation producing an estimate of convective rainfall. Over regions in which precipitation is due to orographic lifting and the clouds are relatively warm, an additional procedure is used which incorporates the local terrain features with numerical model analyses of meteorological parameters. This process for warm cloud precipitation estimation takes into account surface wind direction, relative humidity and terrain. The combined technique incorporates rainfall from both the convective and the stratiform cloud types, producing a final estimate of total accumulated precipitation.

The main problem of implementing the SCS model with the FEWS rainfall data is that the model is only applicable for rainfall events whereas the rainfall data are on a decadal timestep. To overcome this, the model was adjusted to run on a decadal basis where $OF(i)$ is calculated in the following manner (Zhang et al., 2002):

$$OF(i) = \sum_{i=1}^{n} \Delta r OFpJ \tag{6}$$

OFp is overland flow per n different classes of rainfall intensity, $i = 1, \ldots, n$,

$$\Delta r = \frac{(r_{max} - I_a)}{n} \tag{7}$$

r_{max} is the maximum rainfall per rain day which is assumed to be 500 mm, and J is the rain day frequency density function which is assumed to be:

$$J = \frac{J_0}{r_0} e^{-\frac{r_i}{r_0}} \tag{8}$$

where J_0 is the total number of rain days per month, r_i is the rainfall per rain day and r_0 is the mean rainfall per rain day (mm). This methodology allows the use of FEWS data in the overland flow model; however, a map of the number of rain days per decad is required. We produced this by interpolating the daily GTS raingauge data using indicator Kriging (Symeonakis, 2001). Indicator Kriging maps the probability of a pixel being wet. By defining a probability level of 50% this field was transformed to a binary one, with ones where rain occurred and zeros where it did not for each day of the month. These maps were then summed into decads and used to implement the overland flow model.

8.2.2 Sediment Transport and Routing

The process of soil erosion by water includes soil detachment, transport and deposition. So far we have only modelled soil detachment and availability for transport but not transport itself or when and where it will be deposited. To estimate sediment input into the lake we have to consider these factors.

Eroded soil can be divided into two types: (i) sediment deposited when the sediment concentration is higher than the transport capacity or (ii) sediment transported into the lake when the concentration of sediment in the overland flow is less than the transport capacity. Sediment transport capacity can be calculated using models such as the European Soil-Erosion Model (EUROSEM), Water Erosion Prediction Project (WEPP) and the sediment continuity equation (Kothyari et al., 1997). However, such models require numerous input parameters, many of which cannot be readily derived at the regional scale. Therefore, such models are not currently suitable for regional scale modelling, particularly in Africa where input data are relatively sparse.

An alternative way to model deposition is to estimate the sediment delivery ratio (Dr), the ratio of sediment yield to erosion. The delivery ratio of each cell in a catchment can be estimated using 'upland theory' (Boyce, 1975: American Society of Civil Engineering (ASCE), 1975), which argues that steep headwater areas are the main sediment-producing zones of a basin and that sediment production per unit area decreases as average slope decreases with increasing basin size. Indeed, a power relationship has been demonstrated between Dr and catchment size (ASCE, 1975):

$$Dr = C_3 A^{C_2} \tag{9}$$

where C_2 and C_3 are empirical coefficients and A is the area of a catchment (km^2). There is little information in Africa as to what the values of the coefficients should be. The exponent ranges between -0.01 and -0.25 but has been found to be applicable with a value of -0.125 in the USA (ASCE, 1975) and this value is used here. There is less information available about the value of C_3, however, a value of C_3 of 0.46 was determined by Richards (1993) and this value is used here.

When sediment moves through a catchment, the route is primarily controlled by topography. The effects of slope on sediment delivery can be used to decide on the drainage direction within the cell on the basis of the steepest descent. The routed delivery ratio (Dr'_i) in ith cell is controlled by the contributing area in the following manner:

$$Dr'_i = Dr_i . Dr_{i-1} . Dr_{i-2} \ldots Dr_{i-N} \tag{10}$$

where N is the number of pixels in the upstream catchment. Therefore, the routed delivery ratio in a raster image can be expressed as:

$$Dr'_i = C_3 \left(\sum_{j=1}^{i} A_j \right)^{C_2} \tag{11}$$

where i represents ith cell, A_j is the cell size. Once the delivery ratio is calculated, the sediment yield E_y can be estimated using:

$$E_y = \sum_{i=1}^{n} Dr'_i E_i \tag{12}$$

where E_i is the soil erosion in ith cell, and n is the number of cells upstream.

8.2.3 Estimation of Lake Sediment Concentrations

Visible and infrared remote sensing can be used to estimate the near surface sediment concentration of natural waters due to the increase in reflectance caused by scattering from particles of sediment. Sediment plumes emanating from catchment rivers will be detected if they stay near the surface; however, if they rapidly sink, remote sensing will be of little use for detecting and monitoring them as light penetration into water is limited. The buoyancy of plumes is largely controlled by differences in density between them and the lake water and density is primarily controlled by temperature and salinity. These factors coupled with the problem of cloud cover will control, to a large extent, the effectiveness of remote sensing for monitoring sediment dispersal within the lake.

Due to persistent cloud only four largely cloud-free AVHRR images were acquired over Lake Tanganyika in April 1998. In order to obtain as much cloud-free imagery as possible, the ATSR-2 sensor on-board the ESA ERS-2 satellite was also used and two largely cloud-free images of the lake were found in the archive. Both sensors have a spatial resolution of approximately 1 km, however, ATSR-2 images have a much smaller scene size than AVHRR (512 × 512 km compared to 2700 km × 2700 km), and therefore never cover the whole of Lake Tanganyika in one image. AVHRR images are large enough to cover the entire lake. Furthermore, ATSR-2 provides a return period of roughly 3 days while AVHRR acquires imagery on a daily basis.

AVHRR imagery was automatically geocoded using information in the file header. ATSR-2 images were manually geocorrected using a DEM of the area as the base image. Ground control points were located around the lake shore and RMS errors were kept below 0.55 of a pixel for this area. Both ATSR-2 and AVHRR imagery were calibrated to radiance or brightness temperature and a cosine correction was applied to correct for the effect of the solar zenith angle.

The main problem that must be faced when analysing remotely sensed data of water bodies is that of the atmosphere. Only about 20% of the radiant energy of water bodies arriving at the AVHRR sensor channel 1 (Red) is from the Earth's surface, the majority is due to atmospheric effects, and in AVHRR channel 2 (NIR) over 95% of the signal is due to the atmosphere. It is especially important in remote sensing studies of suspended sediments to obtain accurate estimates of the reflectance of the lake water since small deviations in reflectance caused by atmospheric effects will translate into erroneous identification of significant concentrations of suspended matter. Thus, atmospheric correction and cloud masking are essential. The atmospheric correction of Stumpf (1992) was applied. This technique includes processing to remove the effects of changes in the down-welling solar irradiance caused by aerosols, Rayleigh scattering and glint (Stumpf, 1992). Cloud masking was implemented in the same way as that used in the vegetation cover estimation; however, a second step that uses a ratio between NIR and Red reflectance was introduced in order to mask out the land. A simple histogram analysis of the ratio was used to identify cloud-free peaks over land (greater than unity) and water (less than unity). Since we are interested only in removing clouds from a large water body, a threshold below unity was set so that all pixels other than those representing water are masked.

Images of the lake processed in the manner outlined above will exhibit changes in reflectance due to changing suspended sediment or chlorophyll concentrations in the near surface waters. Because Lake Tanganyika is a nutrient-limited system chlorophyll is of little importance and changes in reflectance will be due to near surface sediments. Under these conditions it is possible to calibrate water reflectance to suspended sediment concentration using a comparison of water reflectance and measured suspended sediment concentrations. This final step was not employed due to the lack of field data. Instead we use an increase in reflectance of water at red wavelengths (i.e., channel 2 of AVHRR and ATSR-2) as a relative indicator of near surface sediment concentration.

8.3 Results

There was a considerable amount of rainfall throughout much of the region in April 1998 and this produced much erosion (Figure 8.2). Most of the erosion occurred in regions outside the Lake Tanganyika catchment for much of the time, only in the second decad was there

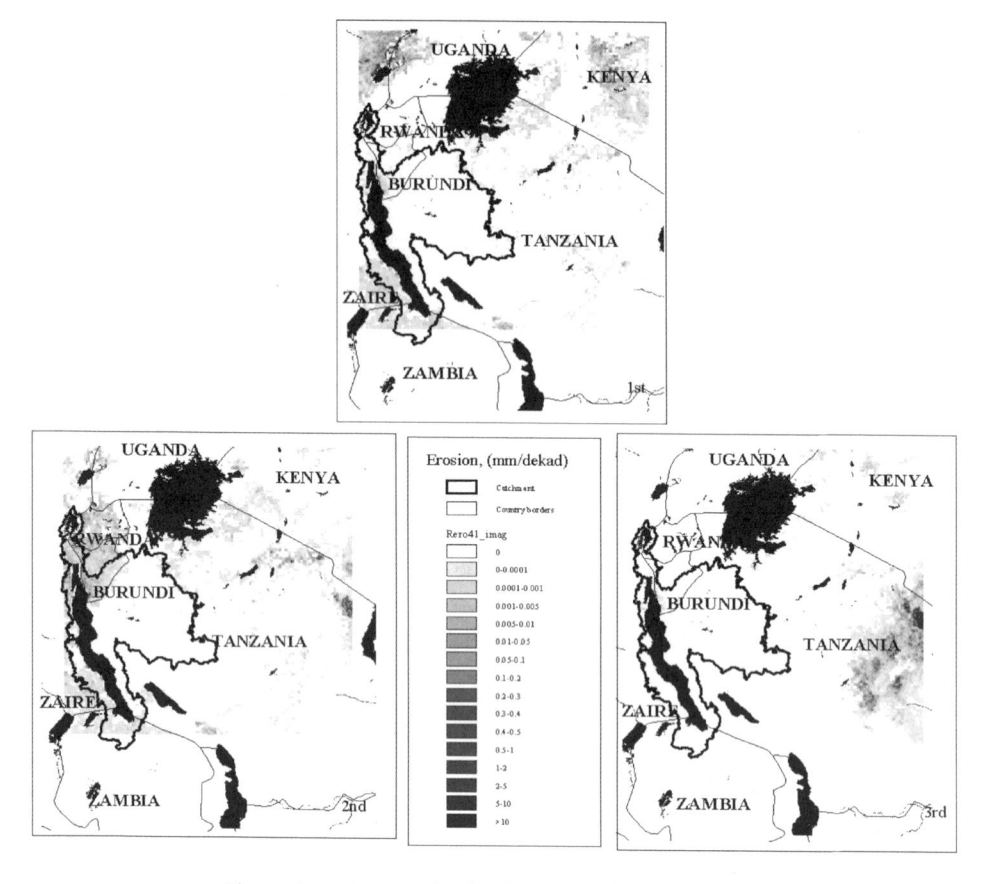

166 *Spatial Modelling of the Terrestrial Environment*

Figure 8.2 *Erosion for the three decads of April 1998*

substantial rainfall, overland flow and erosion in the catchment itself. The northern region of the Lake Tanganyika catchment centred on Rwanda and Burundi suffered severe erosion and was a likely source of lake sediments. When it is considered that slopes are high in these regions, that they tend to be densely populated, and thus vegetation cover is being reduced by human interference, it goes a long way to explaining the high erosion predicted in these regions. Thus erosion is largely restricted to the mountainous regions; however, not all mountains produce erosion, presumably because some possess high vegetation cover. Though we only have results for a single month, these erosion estimates suggest that there are a few key regions in Rwanda, Burundi and the Democratic Republic of Congo that are prone to erosion. Some of these regions have already been deforested and degraded (i.e., those in Burundi) and have the potential to produce significant sediment input to the lake. In these areas the model suggests that alleviation measures could be considered. However, before this is done, the model results need further evaluation and validation. It is interesting to note that a number of the regions of highest sediment yield are located in areas that are cloud covered throughout much of April, so vegetation cover estimates in these regions are uncertain and thus the erosion results should be treated with caution. This problem

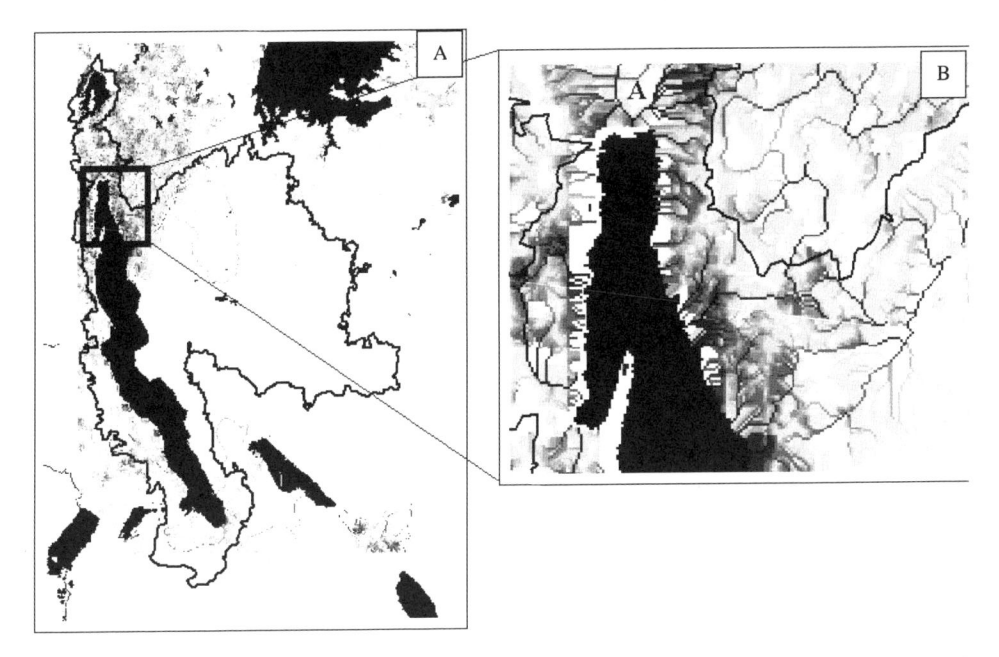

Figure 8.3 Sediment transport in the Lake Tanganyika catchment for the second decad of April 1998. Black polygons are lakes, dark grey indicates high-sediment delivery and white low-sediment delivery, (a) the entire catchment: the black line indicates the extent of the catchment, (b) the northern lake region: to the right of the letter A are the River Rusizi sub-catchments that provide the highest sediment inputs to the lake during this period

highlights a paradox when using remote sensing to estimate both rainfall and vegetation cover from AVHRR imagery. When it is actually raining, and erosion is occurring, remote estimation of rainfall is possible but that of vegetation cover is not.

Figure 8.3(a) shows the total monthly sediment transport results for the Lake Tanganyika region for the second decad of April 1998, when the catchment received up to 270 mm rainfall. At this scale the reader gets a picture of the erosion-prone areas; however, the fine detail of the sediment transport in the channel network cannot be seen. Figure 8.3(b) is zoomed in on the region of Rwanda, Burundi and eastern Zaire that shows the highest amounts of sediment transport to the lake during this decad. The figure depicts both hillslope and channel sediment transport. The dark regions are areas of high-hillslope erosion and subsequent sediment transport and the networks overlying and surrounding these blocks are areas where sediments transported from the hillslopes concentrate in river channels. Estimates of sediment yield of individual rivers during April (Plate 9) suggest that 43 rivers produced significant sediment inputs into the lake, with the majority occurring in the second decad. The Burundi sub-catchments of the River Rusizi appear to produce most of the sediment providing 17% of the total sediment yield. High erosion in this catchment can be attributed to relatively steep slopes (0.64 m/m, or 33%), high-overland flow (4 mm), and most importantly low-vegetation cover (7%). Though the model appears to work quite well when routing sediment into many parts of Lake Tanganyika, it suggests that the Malagarasi River, the catchment's largest watercourse, produces no sediment, and very little sediment reaches Lake Victoria, though this lake is known to suffer sedimentation problems. This seems to illustrate a deficiency in the sediment delivery model when it is applied to large catchments that contain a considerable amount of flood plain.

Problems with the sediment routing can, to a certain extent, be evaluated by remote sensing of the lake sediment plumes as it is possible to determine if sediment plumes are associated with rivers that the sediment delivery ratio method predicts produce no sediment. AVHRR and ATSR-2 imagery has clearly identified large near-surface sediment plumes emanating from a number of rivers in the catchment (Figures 8.4–8.6), and in the case of AVHRR even other rift valley lakes (Figure 8.4). A number of plumes are visible on the eastern side of the lake (Figures 8.4–8.6). The most extensive being the Malagarasi River plume, which is observed in all images, and consistently covers hundreds of kilometres. Patterns displayed by the Malagarasi River plumes closely match the wind-driven lake currents modelled by Huttula *et al.* (1997). The sediment yield modelling suggests that the Malagarasi River provides no sediment inputs to the lake, so there is a big discrepancy here. The study by Tiercelin and Mondegeur (1991) initially appears to supports that of the sediment routing. They suggest that Malagarasi River deposits much of its sediment in upstream swamplands and also traps sediment in its delta system. Thus, the fact that AVHRR and ATSR-2 imagery clearly identifies many large plumes is somewhat puzzling. However, Tiercelin and Mondegeur's (1991) study is based on sediment core analysis of lake sediments that indicates that the central Lake Tanganyika basin has a lower sedimentation rate than that of the north where the Ruzizi River flows into the lake. It is possible that there is not much depth of sediment in the middle basin of the lake because the Malagarasi River sediment plume is transported towards the north under prevailing near-surface currents, and dispersed over a wide area, therefore creating less observable deposition in the middle basin. Thus, it is possible that the large buoyant plumes of the Malagarasi River could cause an underestimation in the significance of this river as a sediment contributor to the lake because plume dispersal causes its sediments to be deposited over a wide area.

Only a few plumes can be seen in the northern basin where the highest sediment yields were predicted. Plumes are detected at the Ruzizi River mouth on only one occasion over the monitoring period (4 March 1998), whereas plumes can be detected from the Mugere or Ntahangwa on three dates (22 March 1998, 4 March 1998, 24 April 1998). It is interesting to note that the Ruzizi River does not have many large visible plumes, while modelling results estimated that it is responsible for a large amount of sediment input into the lake. Field measurements suggest that it transports a great deal of sediment (Sebahene *et al.*, 1999) and that the plume sinks rapidly presumably because its water is more dense than the lake. Because of this, remote sensing is unlikely to be effective at monitoring the dispersal of what appears to be the most important source of sediments into the lake.

8.4 Conclusion

The results show that regional scale erosion modelling can provide a general picture of the source areas of erosion in the Lake Tanganyika catchment, and estimates of the quantity of sediment transported into the lake. Previous spatial studies of lake sediment inputs have relied on the interpretation of satellite imagery (and topographic maps) to assess the amount of forest in a catchment, in order to determine those catchments most likely

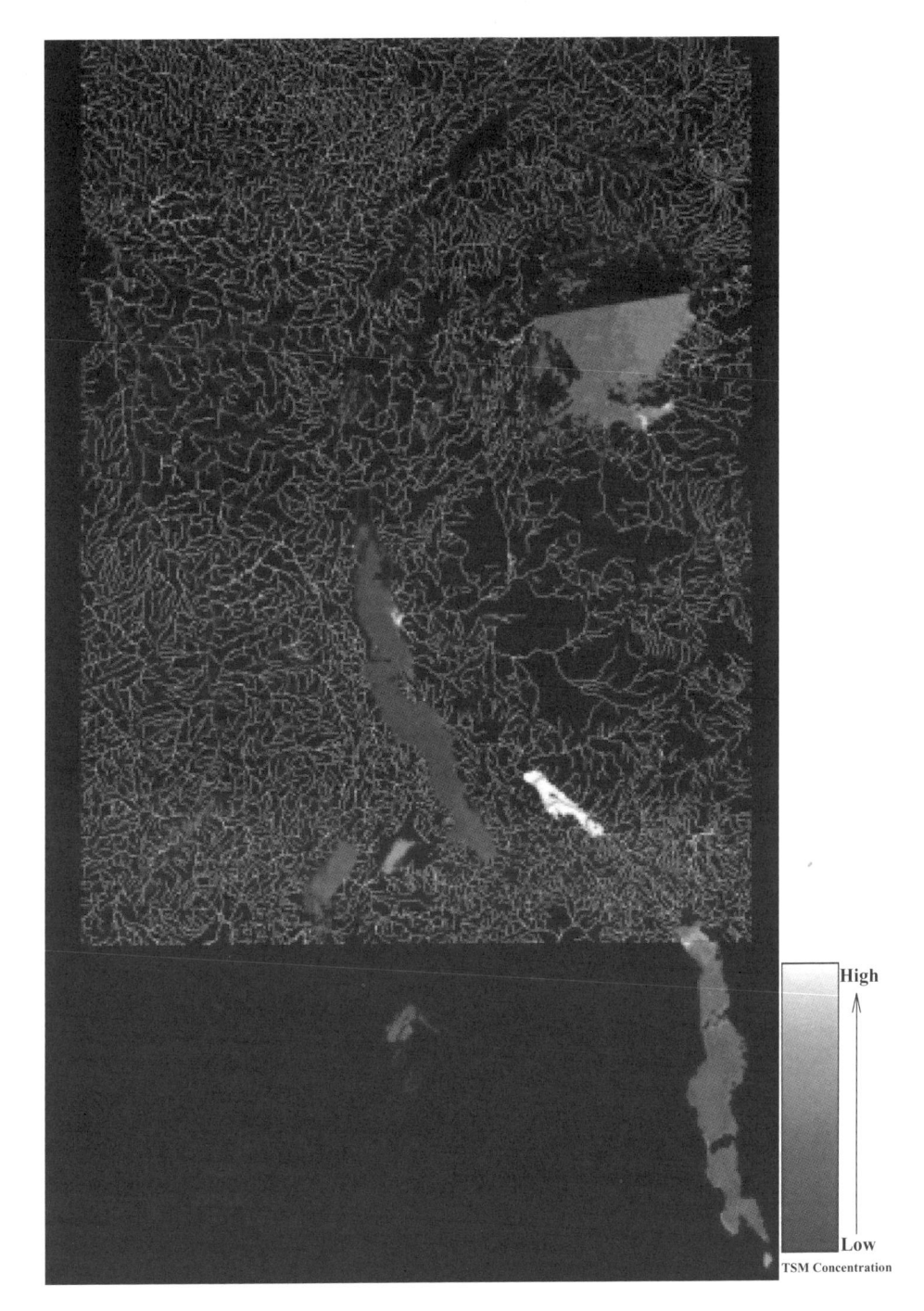

Figure 8.4 AVHRR band 2 image of Lake Tanganyika and surrounding lakes acquired on the 23 April 1998 overlaid on a map of the river network. Areas of black on the image have been cloud/land masked. It should be noted that this and the following image have been contrast stretched, in order to best display the sediment plumes around river mouths. Apparent qualitative sediment levels cannot therefore be compared between images

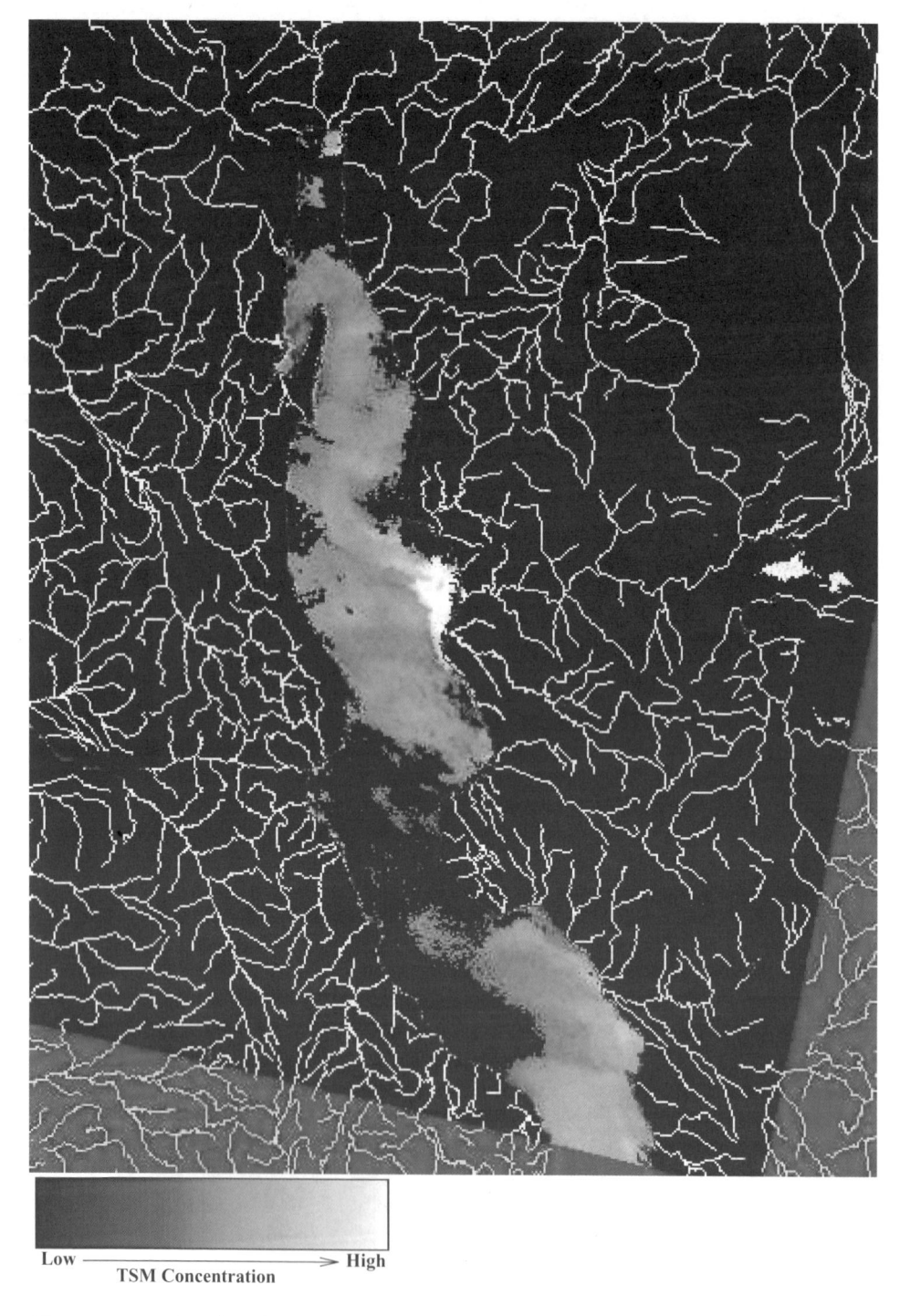

Figure 8.5 ATSR-2 band 2 image of northern and central Lake Tanganyika acquired on the 24 April 1998 overlaid on a map of the river network. Areas of black on the image have been cloud/land masked

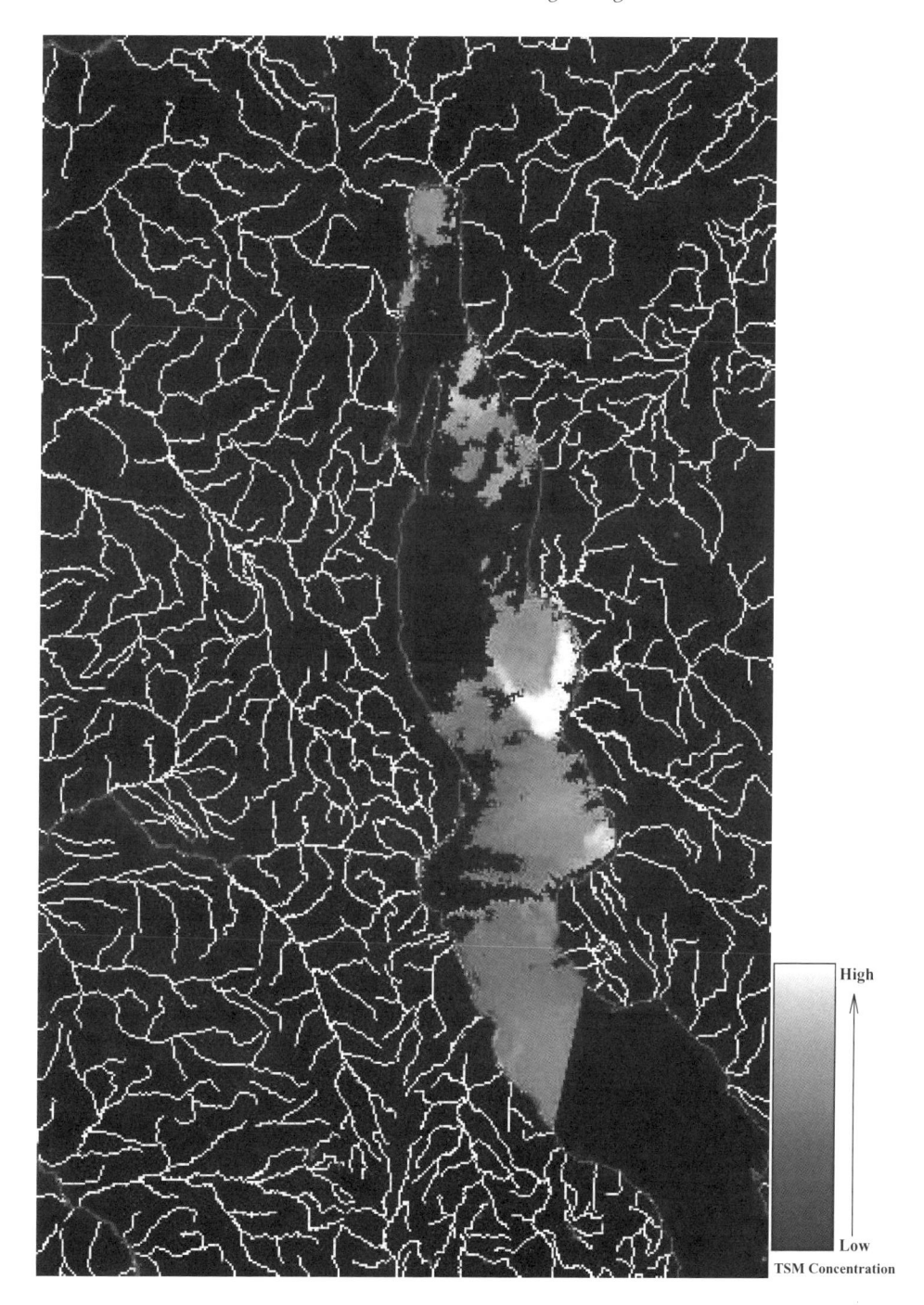

Figure 8.6 ATSR-2 band 2 image of northern and central Lake Tanganyika acquired on the 27th April 1998 overlaid on a map of the river network. Areas of black on the image have been cloud/land masked

to provide increased sedimentation due to deforestation (Cohen et al., 1993). The model presented here provides a more direct method for estimating sediment-affected areas of the lake, supplies maps of the source areas, and thus provides a significant improvement in our spatial and temporal understanding of the pollutant that provides the most immediate and significant threat to the lake ecosystem. Furthermore, if real-time monitoring were implemented on an operational basis, the model has the potential to provide the riparian nations with up-to-date information on erosion and sediment yield that could be used to target further research and co-ordinate remediation measures.

However, many of the model results should be treated with caution. The true accuracy and precision of the model, and of many of the input parameters, have not been assessed. One of the most likely sources of inaccuracy is associated with the use of NDVI to estimate vegetation cover. Vegetation cover is the most sensitive parameter of the model and is thus the most important parameter to estimate correctly. Total vegetation cover estimates are needed; however, NDVI is most sensitive to green vegetation cover and thus may underestimate cover when there is a large amount of dead, and/or senescent material on the ground or in the canopy. This will lead to an overprediction of erosion. Furthermore, NDVI estimates cannot be obtained during cloudy periods, and are depressed by partial cloud cover. All these effects will reduce predicted vegetation cover and it is thus possible that some areas of low-predicted cover may be due to these problems.

One of the weaker aspects of the model is the sediment yield estimation. The equation we have developed relies on an empirical relationship between catchment size and the sediment delivery ratio and appears to underestimate sediment yield from large catchments. This aspect of the monitoring system could be improved through efforts to model the overland flow and erosion hydraulics rather than relying on empirical relationships, and by comparison of the method with sediment yield data collected in the field. Because of these limitations no estimates of the accuracy of the model results can be provided and we would be reluctant to use solely these results to identify the areas most at risk. However, Zhang et al. (2002) applied a similar methodology at the global scale and found similar results to those obtained from long-term field measurements, so we have reason to have some confidence in the results.

Remote sensing of near surface sediments provides useful information with which to evaluate the effectiveness of the sediment routing as well as providing information on sediments dispersal within the lake. However, of the large rivers that enter the lake, only the Malagarasi River plume appears to experience near surface dispersal, the plumes of most other rivers rapidly sink and thus remote sensing is not an effective method for monitoring their dispersal.

An important issue that needs to be considered when applying a coarse resolution model is that it only allows a general assessment of the problem to be made and remediation measures such as terracing or small-scale agroforestry systems, cannot be resolved at this scale. Thus it is not possible to assess the effects of such measures using a model implemented at the scale considered here. The methodology outlined here would be most effectively used if it formed part of a more complex, multi-scale modelling package. Ideally we envisage a system whereby the coarse spatial resolution model is implemented in order to provide a general picture of the areas that are responsible for large amounts of sediment input to the lake. Erosion-prone areas are then targeted by high-resolution sensors, such as the Landsat Thematic Mapper (TM), in association with limited field measurements, in order to parameterize and validate an intermediate scale model. This model could provide a more detailed assessment, confirm coarse resolution model predictions using upscaling from ground measurements – through intermediate scale model predictions – to coarse scale estimates, and finally, to study the effects of any erosion alleviation measures that might be implemented.

Acknowledgements

We would like to thank the staff of the Lake Tanganyika Biodiversity Project for assistance with various aspects of this work. The LTBP project is funded by the United National Development Programme/Global Environment Facility.

References

Arkin, P.A. and Meisner, B.N., 1987, The relationship between large-scale convective rainfall and cold cloud over the Western Hemisphere during 1982–84, *Monthly Weather Review*, **115**, 51–74.

ASCE (American Society of Civil Engineering), 1975, *Sedimentation Engineering*, American Society of Civil Engineers, New York, Manuals and Reports on Engineering Practice, 54.

Boyce, R.C., 1975, Sediment routing with sediment-delivery ratios, *Present and Prospective Technology for Predicting Sediment Yields and Sources*, U.S. Department of Agriculture Publication, **ARS-S-40**, 61–65.

Cohen, A.S., Bills, R., Cocquyt, C.Z. and Caijon, A.G., 1993, The impact of sediment pollution on biodiversity in Lake Tanganyika, *Conservation Biology*, **7**, 667–677.

Drake, N.A., Vafeidis, A., Wainwright, J. and Zhang, X., 1995, Modelling soil erosion using remote sensing and GIS techniques, *Proceedings of the 1995 RSS Annual Symposium: Remote Sensing in Action*, 11–14 September 1995, Southampton, 217–224.

Drake, N.A., Zhang, X., Berkhout, E., Bonifacio, R., Grimes, D., Wainwright, J. and Mulligan, M., 1999, Modelling soil erosion at global and regional scales using remote sensing and GIS techniques, in P. Atkinson and N. Tate (eds), *Spatial Analysis for Remote Sensing and GIS* (Chichester: John Wiley & Sons), 241–262.

Herman, A., Kumar, V.B., Arkin, P.A. and Kousky, J.V., 1997, Objectively determined 10-day African rainfall estimates created for famine early warning systems, *International Journal of Remote Sensing*, **18**, 2147–2159.

Kothyari, U.C., Tiwari, A.K. and Singh, R., 1997, Estimation of temporal variation of sediment yield from small catchments through the kinematic method, *Journal of Hydrology*, **203**, 39–57.

Mitchel, J.K. and Bubenzer, G.D., 1980, Soil loss estimation, in M.J. Kirkby and R.P.C. Morgan (eds), *Soil Erosion* (Chichester: John Wiley & Sons), 17–62.

Patterson, G. and Mackin, J., 1998, *The State of Biodiversity in Lake Tanganyika: A Literature Review* (Chatham: Natural Resources Institute).

Rawls, W.J., Ahuja, L.R., Brakensiek, D.L. and Shirmohammadi, A., 1993, Infiltration and soil water movement, in D.R. Maidment (ed.), *Handbook of Hydrology* (New York: McGraw-Hill), 5.1–5.51.

Richards, K., 1993, Sediment delivery and the drainage network, in K. Beven and M.J. Kirkby (eds), *Channel Network Hydrology* (London: John Wiley & Sons), 221–252.

Saunders, R.W. and Kriebel, K.T., 1988, An improved method for detecting clear sky and cloudy radiances from AVHRR data, *International Journal Remote Sensing*, **9**, 123–150.

Sebahene, M., Nduwayo, M., Songore, T., Ntungumburanye, G. and Drieu, O., 1999, *Travaux Hydrologique et d'échantillonnage sédimentologique du Bassin du Lac Tanganyika (Burundi)*. Lake Tanganyika Biodiversity project Pollution Special Study Report, http://www.ltbp.org/PDD5.HTM.

Soil Conservation Service, 1972, *SCS National Engineering Handbook, Sec. 4, Hydrology*, U.S. Department of Agriculture.

Stumpf, R.P., 1992, Remote sensing of water clarity and suspended sediments in coastal waters: needs and solutions for pollution monitoring, control, and abatement, *Proceedings of First Thematic Conference on Remote Sensing for Marine and Coastal Environments (SPIE 1930)* (Ann Arbor, MI: ERIM), 293–305.

Symeonakis, E., 2001, Soil erosion modelling over sub-Saharan Africa using remote sensing and geographical information systems, unpublished Ph.D. thesis, King's College, London.

Thornes, J.B., 1985, The ecology of erosion, *Geography*, **70**, 222–234.

Thornes, J.B., 1989, Erosional equilibria under grazing, in J. Bintliff, D. Davidson and E. Grant (eds), *Conceptual Issues in Environmental Archaeology* (Edinburgh: Edinburgh University Press), 193–210.

Tiercelin, J.-J. and Mondeguer, A., 1991, The geology of the Tanganyika trough, in G.W. Coulter (ed.), *Lake Tanganyika and its Life* (London: British Museum (Natural History)/Oxford University Press), 7–48.

Vandelannoote, A., Robbertson, H., Deelstra, H., Vyumvuhore, F., Bitetera, L. and Ollevier, F., 1996, The impact of the River Ntahangwa, the most polluted Burundian affluent of Lake Tanganyika, on the water quality of the lake, *Hydrobiologica*, **328**, 161–171.

Zhang, X., 1999, Soil erosion modelling at the global scale using remote sensing and GIS, unpublished PhD thesis, King's College. London.

Zhang, X., Drake, N.A., Wainwright, J. and Mulligan, M., 1999, Comparison of slope estimates from low resolution DEMs: scaling issues and a fractal method for their solution, *Earth Surface Processes and Landforms*, **24**, 1–18.

Zhang, X., Drake, N.A. and Wainwright, J.W., 2002, Scaling land-surface parameters for global scale soil-erosion estimation, *Water Resources Research*, **38**, 1180–1189.

9

Estimation of Energy Emissions, Fireline Intensity and Biomass Consumption in Wildland Fires: A Potential Approach Using Remotely Sensed Fire Radiative Energy

Martin J. Wooster, G.L.W. Perry, B. Zhukov and D. Oertel

9.1 Introduction

Aside from modification of the Earth's surface for urban development and agricultural activity, fire is the most widespread of all terrestrial disturbance agents (Bond and van Wilgen, 1996), making it a key candidate for observation via remote sensing. Evidence from a variety of sources suggests that fire has been an important feature of the natural environment for at least 350 million years (e.g. Cope and Chaloner, 1985). Furthermore, debates continue over the role of wildfire events in climate change and species extinction at the Cretaceous/Tertiary boundary (Anders *et al.*, 1991; McLean, 1991). Without doubt, fire has influenced terrestrial ecosystems over evolutionary time, but there is now a large body of evidence suggesting that in many, and probably most, geographical regions the majority of wildland fires are now not purely 'natural' phenomena, but are ignited by humans, either deliberately or as a result of some related anthropic activity (Saarnak, 2001). Brain and Sillen (1988) suggest that the earliest evidence of the use of fire by hominids is approximately 1.5 million years BP and the rise of humans as a major fire ignition source is one reason why biomass burning is believed to have significantly increased in the last 100 000–250 000 years BP (Caldararo, 2002; Keller *et al.*, 2002).

Spatial Modelling of the Terrestrial Environment. Edited by R. Kelly, N. Drake, S. Barr.
© 2004 John Wiley & Sons, Ltd. ISBN: 0-470-84348-9.

Fires have a range of environmental effects across multiple scales in time and space. For example, fires at the local scale significantly reduce the vegetation cover, and more intense fires may completely remove all above-ground vegetation. At the appropriate time and place such processes maybe advantageous, for example, in rapidly returning nutrients to the soils after grassland senescence, or by keeping forest surface litter levels low so as to reduce the likelihood of larger, more intense fires that may kill normally fire-tolerant vegetation. Furthermore, the reproductive biology of many species in fire-prone systems is closely related to fire activity (Whelan, 1995), with, for example, heat or chemicals in the smoke emitted by the fire triggering flowering and/or seed release (Enright et al., 1997). Additionally, the nutrient-rich post-fire ash bed may be important for the establishment of seedlings in certain systems. However, at the local scale, fires can also alter the species makeup and ecology of an area and render the underlying soils more prone to erosion (Johansen, 2001). In extreme cases dramatically altered fire regimes may cause the local extinction of some species (Keith, 1996), whilst at the regional scale fire may also have significant ecological effects, resulting in altered flows of energy and matter through the landscape and the generation of spatially complex mosaic patterns (Turner et al., 1994, 1997). At these regional scales, the particulates and gases released during burning can concentrate at levels far beyond recognized safe human health limits, leading to hazardous visibility and poor air quality. A high-profile example of regional-scale fire-related pollution occurred during the 1997–1998 fires in SE Asia, when an El Niño-related drought allowed fires to become much more intense and widespread than is normally the case, and which also significantly reduced the incidence of 'cleaning' of the atmosphere by rainfall (Wooster and Strub, 2002). The widespread haze blanketed areas of SE Asia for weeks and resulted in regional economic losses of more than $1.4 billion, exposing more than 20 million people to potentially dangerous levels of air pollution (Brown, 1998; Schweithelm, 1998). Such is the magnitude of fire activity on Earth, particularly in the tropics, that pyrogenic gaseous and particulate emissions are now recognized to be highly significant at the global scale, where Earth's atmospheric chemistry, cloud and rainfall characteristics, and radiative budget are all influenced by the combustion products (Rosenfeld, 1999; Andreae and Merlet, 2001; Chou et al., 2002).

Clearly, for the above reasons, biomass burning and pyrogenic emissions must be rigorously considered when analysing the effect of wildfires on Earth's surface and atmosphere (Beniston et al., 2000). As such, the quantification of biomass burning activity is now a major focus in the global change community and the large-scale, repetitive measurements required to quantify terrestrial fire activity mean Earth Observation (EO) satellites have been identified as a key technology for the study of pyrogenic activity and the assessment of emissions (Robinson, 1991). However, whilst the emission of certain chemical species and particulates can in theory be examined directly through the use of EO (Kaufman et al., 1990; Goode et al., 2000; Kita et al., 2000; Kaufman et al., 2002), not all species are easily inventoried in this way. Instead the majority of EO-based studies of fire emissions have been limited to mapping the location of active fires or burnt areas (Scholes et al., 1996), with little information provided on the fire physical characteristics such as fire intensity, rate of spread, etc. The absence of this information has been highlighted as a potential problem in certain studies of fire ecology (Whelan, 1995; Morrison, 2002), as has the reliance on maps of burnt area to estimate the amount of biomass combusted in the fire event (Andrea and Merlet, 2001). This is because the relationship between remotely measured burnt area and the biomass actually combusted in the affected region is dependent upon highly variable factors such as the combustion efficiency and biomass density, which are very difficult to measure remotely to the required accuracy. Thus, Andrea and Merlet (2001) demonstrate that whilst average 'emission factors' for many important pyrogenic species are known to uncertainties of perhaps 20–30%, much larger uncertainties persist for regional and global fire emissions because of the difficulty inherent in estimating the amount of biomass combusted with the current EO approaches. For this reason it can be difficult to settle disagreements where significant differences exist between EO-based emissions estimates and data from non-EO sources (Barbosa et al., 1999; Andrea and Merlet, 2001).

In part to tackle this limitation, a recent development has been the suggestion that a new generation of satellite-based infra-red (IR) imagers can accurately measure the rate of emission of energy radiated by large fires, and that this may be well related to the rate of combustion of the vegetation, which is itself directly proportional to the rate of release of pollutant species (Kaufman et al., 1996; 1998a). Integrating the measurements of emitted energy over time for individual fires should then provide information on the total amounts of vegetation combusted and pollutants emitted. If correct, then this more direct remote sensing method could be a candidate for the independent EO-based emissions–estimation route called for by Andreae and Merlet (2001), which would greatly assist in helping solve the disagreements that often arise between biomass consumption estimates derived via current approaches. The purpose of this chapter is to review this new methodology, which Kaufman et al. (1996) termed fire radiative energy (FRE), to indicate the theoretical framework in which it has been derived and show how it relates to the understanding of the energy liberated in free-burning vegetation fires as modelled by fire scientists. Finally, we present some examples of FRE derivation from several new EO instruments and fire scenarios, and indicate some of the characteristics of this new EO measure that will likely influence its ability to contribute to fire science, including the estimation of biomass consumption and the related pyrogenic pollutants and the modelling of fire propagation.

9.2 Pyrogenic Energy Emissions: Combustion and the Energy Flux

9.2.1 Combustion: The Physico-Chemical Reaction

When observing an active fire, the thermal energy measured by a remote sensing instrument is a direct result of energy stored in the biomass being released as heat when the fuel combines with oxygen. This process also liberates carbon dioxide, water vapour and small amounts of other substances, the very emissions that scientists wish to quantify at global and regional scales. The chemical equation for combustion can be illustrated in the complete combustion of a simple sugar (e.g. D-glucose):

$$C_6H_{12}O_6 + 6O_2 \longrightarrow 6CO_2 + 6H_2O + 1.28 \times 10^6 kJ$$

The fuels burned in wildland fires are much more complex than the simple glucose described above. Vegetative material is made up of cellulose, hemicelluloses (together 50–75% of most dry-plant matter), lignin (15–25%), proteins, nucleic acids, amino acids and volatile extractives (Lobert and Warnatz, 1993). Furthermore, the different components of the fuel (e.g. dead leaf litter, dead wood and live foliage) contain energy stores in a variety of forms (Chandler et al., 1983; Rothermel, 1972). However, in a wild fire combustion is never complete, and therefore the actual heat produced during the combustion is but a fraction of the potential (theoretical) heat yield (Pyne, 1984) – an important point and one that we will return to. Combustion requires activation energy from an external source, which, as we have previously stated, is commonly provided as a result of human activity but may also be due to lightning, burning of underground coal deposits, volcanic activity, specific forms of vegetation decomposition or certain other phenomena (Chandler et al., 1983). The fuel is ignited when one of these processes applies enough heat for pyrolysis to occur (Albini, 1993; Whelan, 1995). Pyrolysis can be summarized as the chemical degradation of fuel through thermal decomposition, and results in the release of water vapour, carbon dioxide and other gases such as methane, methanol and hydrogen (Lobert and Warnatz, 1993). During this process the combustion reaction shifts from being endothermic to exothermic.

A spreading fire is a more complex combustion process than a simple campfire or candle flame; it is one in which the flaming front is heating and then igniting unburnt fuels, allowing the fire to propagate through the fuel bed. During the heating process, fuel moisture is initially evaporated (fuel temperature > 100° C), then the cellulose is thermally degraded and at its breakdown temperature volatized (temperature > 200° C) and finally the volatiles are ignited to form a visible flame (300–400° C). The combustion process in a spreading fire also goes through three distinct stages (Pyne 1984; Albini, 1993): (i) preheating during which the fuel ahead of the fire front is heated, dried and partially pyrolyzed; (ii) flaming combustion, which is the result of the ignition of flammable carbohydrate gases; and (iii) glowing combustion where any remaining charcoal burns as a solid, with oxidation taking place on the surface, leaving a small amount of residual ash. A consideration of the processes of pyrolysis and ignition shows that a fire can to some extent be considered as self-sustaining. Hence, the fire is able to spread away from the originally ignited region and into the surrounding landscape. In areas where conditions are optimum for this process, huge areas of hundreds of thousands of hectares maybe burnt in a single fire event.

9.2.2 Heat Generation

The spreading of fire through the heating of neighbouring unignited fuel, and the measurement of energy emitted from the fire via remote sensing, are both possible because combustion releases energy in large quantities, most obviously in the form of radiation (some of which is visible to the naked eye) but also via convective and other processes. The amount of energy liberated is of great significance because it is closely related to fire intensity, one of the most significant parameters in any wildfire. In order of importance, the factors affecting the nature of the combustion and the amount of heat released are fuel moisture, fuel chemistry and the surface area to volume ratio of the fuel particle (Chander et al., 1983). The single most important element influencing the total amount of heat released is fuel moisture content (FMC), the most variable component of a fuel's chemistry (Lobert and Warnatz, 1993; Whelan, 1995). FMC may vary from 2.5% (dead savanna grasslands) to 200% (fresh needles and leaves) of the vegetation's dry weight. High FMC has the capacity to stop a fire, or to slow down the process to a slow, intense smoldering where much of the vegetation may remain only partly combusted. Since the heat released is used to (i) raise fuel temperature to 100° C; (ii) separate bound water from the fuel; (iii) vaporize the water in the fuel; and (iv) heat the water vapour to flame temperature, it is clearly a very important characteristic. The importance of FMC is illustrated by the fact that for eucalyptus fuels, the rate of combustion of the vegetation is four times higher at a FMC of 3% than at 10% (Luke and McArthur, 1986). However, in contrast to the rather variable nature of the actual amount of heat released during combustion, the (theoretical) heat yield (kJ kg^{-1}) of most fuels, which is the amount of energy released per unit mass of matter under complete combustion, is remarkably constant at between 16 000 and 22 000 kJ kg^{-1} (Whelan, 1995); see later for a fuller discussion of this point. This heat yield parameter is, however, influenced to some degree by the presence of volatile resins and oils and inorganic minerals in the fuel (Bond and van Wilgen, 1996). Crude fats are volatile and make plants more flammable, as they are driven out of the fuel at relatively low temperatures and subsequently burn as gases. Thus, for example, eucalyptus, which contains a high percentage of oils, has a higher heat yield than savanna grasses (Whelan, 1995). Finally, the amount of heat actually released during combustion is influenced by the surface area to volume ratio of particles in the fuel bed, which for a given particle density is directly correlated with particle thickness. The combustible volatiles can only mix with oxygen after diffusing across the fuel surface boundary (Albini, 1980). For a fuel of a specific density, chemical composition and FMC, the 'residence time' (length of time a particle will support flaming) is directly proportional to its thickness. For woody fuels with a FMC in the range 4–10%, the residence time in seconds is approximately three times the fuel thickness in centimetres (Chandler et al., 1983; Lobert and Warnatz, 1993).

The amount of heat released during the burning of vegetation is also dependent upon the completeness of the combustion process, since in many cases 100% combustion is not achieved. This incomplete combustion of fuel reduces the energy output to below the potential maximum predicted by use of the theoretical heat yield parameter. Incomplete combustion describes two possible processes. First, not all components of the burning fuel may be completely combusted and so the released energy will be less than the possible maximum, the incompletely combusted material either being given off as particulate carbon compounds in the smoke pall, or remaining as charred plant material at the site. Second, not all the biomass at the site may ignite in the first place (Whelan, 1995). This is sometimes expressed as a combustion efficiency (η), which can be defined as the ratio of the actual CO_2 release to the theoretical limit of CO_2 release if all the carbon in the material were converted to CO_2 (Ward and Radeke, 1993). In fact, because the wildfire combustion very rarely approaches 100% completeness, the actual heat yield is usually only a small fraction of the potential heat yield provided by the available biomass (Pyne, 1984).

9.2.3 Fire Intensity

A common measure used to assess the amount of energy being released during the combustion process is fire intensity, possibly the most valuable quantitative descriptor of any fire (Alexander, 1982; Whelan, 1995). It is significant for two primary reasons; because the ability of living cells to tolerate heat decreases markedly at high intensities, so affecting the fire's ecological consequences; and because fires of different intensities release pollutant emissions at different rates, mainly because the rate and completeness of combustion vary with intensity. However, despite its importance, accurate quantification of wild–land fire intensity is extremely problematic (Cheney, 1990; Albini, 1976). Albini (1976, p. 75) states that "... perhaps no descriptor of wildfire behaviour is as poorly defined or as poorly communicated as are measures of fire intensity ... these measures [are] virtually unobservable but through various empirical relations they can be related indirectly to observable fire phenomena which themselves serve as indirect measures of intensity."

Fire intensity was first defined by Byram (1959) as being the effective radiative temperature of a fire front. Although Tangren (1976) and certain other physicists studying fire disagree with the use of the term fire intensity, preferring the term prefer 'fire power', we use the term intensity here since it is the most widely used in the literature. The most commonly used measure of fire intensity is Byram's (1959) fireline intensity:

$$I = HWR \qquad (1)$$

where, I is the fireline intensity (kW m^{-1}), H is the heat yield of the fuel (kJ kg^{-1}), W is the mass of fuel consumed in the *active* flaming zone per unit area (kg m^{-2}) and R is the forward rate of spread of the fire (m s^{-1}).

In quantitative terms, fireline intensity can be summarized as the heat release per unit length at the fire front. In practical terms it requires estimation of the total fuel load (typically partitioned into different particle size classes based on the time for each particle to reach its equilibrium moisture content (Pyne, 1984) and the rate of (forward) spread of the fire using model-based and field observations). It is very difficult (if not impossible) to determine with any accuracy the amount of fuel consumed in the active flaming part of the fire front under field conditions. Alexander (1982, p. 351) noted that 'the amount of fuel consumed by secondary combustion and residual burning after passage of the main fire front will increase W and consequently result in an overestimation of fire intensity.' Hence for accurate estimation of fire-line intensity, reductions in the measured value of W are necessary if significant quantities of fuel are consumed subsequent to the passage of the flaming front. Put simply, it may not be sufficient to simply compare pre-fire and post-fire fuel loads.

As mentioned earlier, the heat yield parameter (H) is remarkably constant and in many cases it is assumed static at 18 608 kJ kg^{-1} (Burgan and Rothermel, 1984). The value is actually derived by measuring oxygen consumption under complete combustion of the fuel in a cone or bomb calorimeter and published tables of H exist for many common fuel types. However, these values are usually measured under conditions (oven-dried material and complete combustion) seldom achieved in a wild-fire situation and thus adjustments may be necessary. Two reductions are generally required, one for the presence of moisture in the fuel, and another for incomplete combustion (Alexander, 1982). Reduction for latent heat absorbed when the water of reaction is vaporized is 1263 kJ kg^{-1} (Byram, 1959), with another reduction for FMC of 24 kJ kg^{-1} per FMC percentage point. This moisture-related correction to the theoretical heat yield results in the so-called low heat yield of the fuel or the low heat of combustion parameter. Unfortunately the second correction, for the (in)completeness of combustion, is extremely difficult to make due to the extreme variable and difficulty in measuring this parameter.

The length scale for determining the local fireline intensity is provided by the size of the coherent flame structure, measured normal to the fire edge (Cheney, 1990). The length of the flame at the fire front has been correlated to a fractional power of the fireline intensity by a number of authors (e.g. Byram, 1959; Nelson and Adkins, 1986; Cheney, 1990; Fuller, 1991). Fireline intensities in excess of 20 000 kW m^{-1} have been estimated for high-intensity, fast-burning fires in eucalyptus fuels (Bradstock *et al.*, 1998); and also in tropical savanna environments, where average intensities of c. 8000 kW m^{-1} have been noted (Williams et al., 1999). However, these may be best regarded as maximum potential estimates as they appear to (tacitly) assume complete combustion of the entire fuel load and are dependent on the accuracy with which the rate of fire spread is either modelled or measured. Byram (1959) believes that fireline intensities could realistically be expected to vary from 15 to 100 000 (!) kW m^{-1}; the primary uncertainties in their estimation being the rate of spread of the flaming front, followed by the fuel consumption and finally the estimation of the low heat of combustion. These huge extremes of fireline intensity described by Byram (1959) have been documented in, for example, wildfires in the Canadian forests (Alexander, 1982) and provide clear evidence as to the widely varying nature of this parameter.

Given the potential importance of accurately estimating fire intensity for a range of applications, and the inherent difficulties and uncertainties involved with estimating it in the field, there is a pressing need for remote sensing methods for calculating this parameter. In terms of the remotely sensed measurement of fire-related energy emission and fire intensity, only the fraction of the liberated energy (I) that is radiated away from the fire is available to be detected by a remote sensing device. However, if this fraction is known, because equation (1) relates the *rate* of heat release (kW m^{-1}) to the *rate* at which biomass is combusted by the fire (kg m^{-2} × m sec^{-1}), there exists the possibility of estimating the rate at which vegetation is consumed in the fire via observations of the radiant energy produced (both measures being per unit length of the fire front in this case). Furthermore, since it is expected that such biomass combustion estimates are likely to be strongly correlated to the rate at which emissions of trace gases and aerosols are produced (Kaufman et al., 1996), if sufficient radiant energy observations are available and can be integrated over the fire's lifetime, then in theory the total amount of radiative energy released can be calculated and used to estimate the total amount of biomass consumed.

Perhaps surprisingly, estimation of biomass consumption (a first-order fire effect) has not been a prime concern of researchers involved with modelling fire behaviour and dynamics (Reinhardt et al., 2001), with much more effort placed on developing physical models of fire spread and intensity (reviewed by Perry, 1998, and see Perry et al., 1999, for an example). Initial attempts at measuring fireline intensity using remote sensing were made via airborne sensors. For example, Budd et al. (1997) modelled head-fire intensity over a series of experimental bushfires in Australian eucalyptus forests using airborne IR imagery to measure the speed of fire propagation. They estimated that the head-fire intensity (averaged over 6 minutes) exceeded 1000 kW per metre of fire front (kW m^{-1}) for most of the fires when they were at their most intense, and ranged as high as 3280 kW m^{-1}. Kaufman et al. (1998a) used the Moderate Resolution Imaging Spectroradiometer (MODIS) Airborne Simulator to measure the rate of release of thermal energy from spreading cerrado fires in Brazil, and went on to successfully relate the integral of this parameter to the rate of generation of the burn scar, which was believed proportional to the rate of biomass consumption. The strong relationship between these parameters ($r^2 = 0.97$, $n = 21$) supported the idea of using such radiative energy measures to directly estimate the amount of biomass combusted and Kaufman et al. (1998a) found that the relationship between FRE and the change in burn scar size was four times stronger than that between simple fire size and the rate of change of the burn scar. This clearly illustrated the improvement provided when the actual energy emission from the fires was considered.

9.3 Remote Sensing of Fire Radiative Energy

In order to advance the understanding of the relationship between FRE and biomass consumption, Wooster (2002) conducted a small-scale experiment, observing small fires of varying size using a GER 3700 field spectroradiometer. Twelve experimental fires were conducted, each consisting of the combustion of a pile of grass of known biomass (20–1300 g) and low moisture content (< 3%). During combustion the radiative emission from the fire was measured every few seconds using the spectroradiometer to make spectral radiance observations in the range 0.35–2.5 µm. The instrument was positioned sufficiently far from the fire (7–9 m) to ensure that the burning vegetation was located completely within the spectrometer's field-of-view, and the experiment was conducted at night to remove any possibility of solar radiation contaminating the thermal signals. After each fire the post-fire materials (ash and partially burnt vegetation) were collected, weighed and used to calculate the actual mass combusted.

Figure 9.1 shows one example of a spectrum measured during the experiment. By fitting a simple two-component radiating model to the measured emission curve, the black-body 'effective emitter' characteristics that best approximate the measured spectra over the 0.35–2.5 µm range were derived and these characteristics were used to model the (unmeasured) emission curve over an extended wavelength range. Clearly the assumption of a two-component radiating blackbody is a gross simplification for approximating the conditions

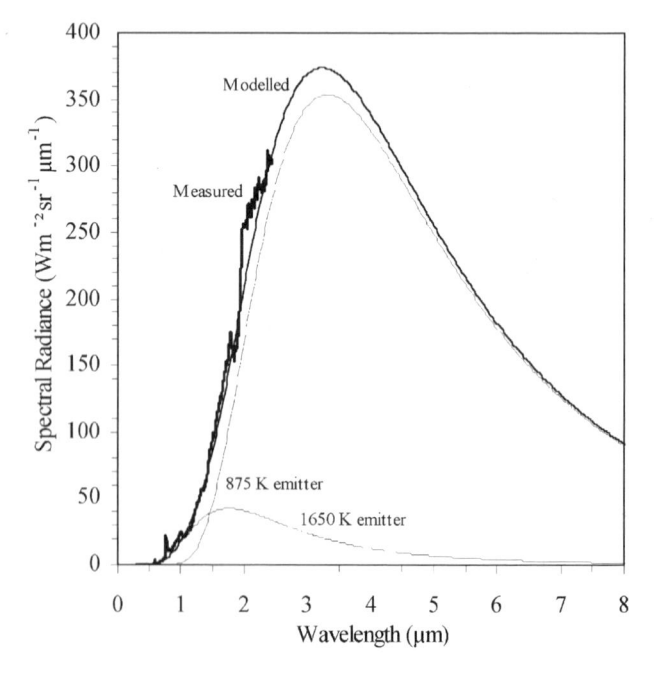

Figure 9.1 Example measured spectral radiant emission signal (0.5–25 µm) from a 1.0 kg fire 70 s after ignition, as measured with a GER 3700 field spectroradiometer. The best-fit model is shown extended to 8 µm, with the model being derived from the summed thermal radiation from two 'effective emitters' having brightness temperatures 1650 K and 875 K (also shown), with the cooler emitter being 200X larger

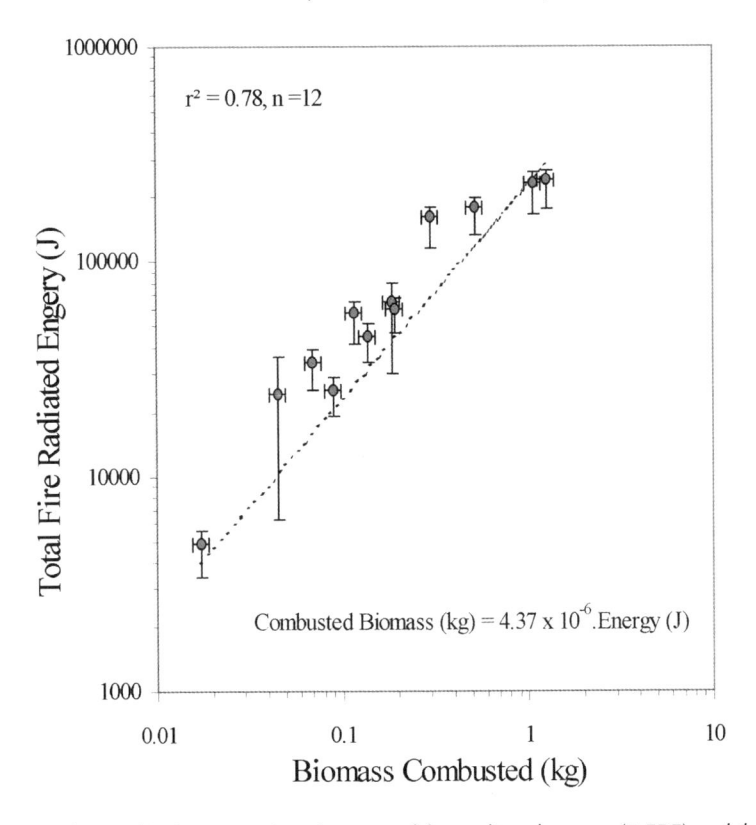

Figure 9.2 Relationship between time-integrated fire radiated energy (ΣFRE) and the amount of biomass combusted for 12 small-scale experimental fires. The best-fit linear equation passing through the origin is indicated ($r^2 = 0.78$). Error bars for biomass were determined by assuming the post-burn materials were 0 and 100% combusted respectively. Error bars for ΣFRE were calculated from the results of the model used to determine the effective emitter characteristics (see Wooster (2002) for more detail). Note that recently a correction to the calibration of the loaned GER 3700 spectroradiometer used herein was noted to be required. This correction changes the magnitude of the slope of the line shown in Figure 9.2 (and of the spectra of Figure 9.1), but the linearity of the relationship remains

at any point during a real fire. However, for this application the important points are, first, that the modelled spectra are a good fit to the measured spectra, which from Figure 9.1 appears to be the case, and second that the emission at wavelengths longer than those measured can be accurately calculated. Once the effective radiator characteristics have been obtained, the FRE (J s^{-1}) was calculated as the area under the spectral emission curve extended to longer wavelengths (until spectral radiance approaches zero). For each individual fire this procedure was carried out for each spectra recorded during the burn, and the total emitted energy release, termed the time-integrated FRE (ΣFRE, J), calculated via the integration of FRE over the fire's lifetime. Comparing ΣFRE to the biomass combusted, calculated as the difference between the pre- and post-burn masses, produced the significant relationship shown in Figure 9.2 ($r^2 = 0.78$, $n = 12$). Hence, results from this experiment support the idea that measurement of the radiative energy release can provide information on the biomass combusted, at least at the scale of these small experimental fires.

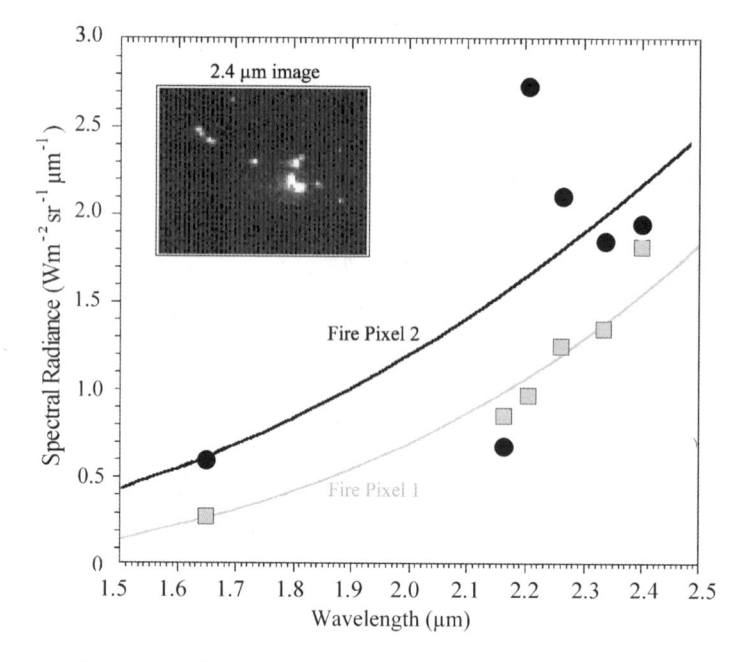

Figure 9.3 Study of a wildfire spectral emittance analysed with ASTER imagery. The fire was imaged during the night-time ASTER pass on 8 September 2000 over Zambia (16.09° S, 22.84° E). The image subset shows the 100 × 75 pixel ASTER 2.4 μm sub-image of the fire, with a ground pixel resolution of 30 m. Bright pixels correspond to areas where fire activity is causing intense thermal radiation at 2.4 μm wavelength. Similar emittance occurs in the other five ASTER SWIR bands (1.6–2.3 μm) and spectra in all six ASTER bands for two of the 'fire' pixels are shown in the graph, alongside the best-fire modelled spectra calculated via the two thermal component model used previously to analyse spectra from ground-based fires recorded with the GER 3700 instrument (see Figure 9.2). Spectra have been adjusted for atmospheric effects using MODTRAN-estimated atmospheric transmissions convolved to the ASTER bandpasses. The fit between the measured and modelled ASTER data of fire pixel 1 is reasonable, but that for fire pixel 2 is poor because the individual spectral radiances do not approximate a smoothly varying Planck-type function

However, applying this spectral-matching technique to satellite imagery poses the problem that, unlike field spectroradiometers, the majority of satellite-based IR instruments measure radiation in only a relatively few wavebands. This generally precludes use of this method to derive the effective emitter characteristics from the data shown in Figure 9.1. However, the Advanced Spaceborne Thermal Emission and Reflection Radiometer (ASTER), orbiting onboard the EOS Terra satellite, does possess sufficient shortwave IR wavebands for the application of this approach. A night-time southern Africa ASTER scene (Figure 9.3, inset) was used to test the possibility of determining fire thermal characteristics via this approach. Unfortunately, as can be seen from Figure 9.3, the individual spectral measurements from the majority of the ASTER fire pixels were not well fitted by a smoothly varying multi-thermal component Planck function of the type used to analyse the emission spectra recorded by the GER 3700. In some cases a reasonable fit was obtained, for example, in the case of Fire Pixel 1 in Figure 9.3, but it was found that the majority of the pixels more closely resemble the situation of Fire Pixel 2. One explanation for this is provided by the ASTER technical specifications (ERSDAC, 2001), which suggest the ASTER inter-band spatial registration is accurate to only 0.2 pixels. When coupled with this imprecise inter-band registration, a fire signal that varies strongly at the sub-pixel scale could easily induce the type of spectra shown in Fire Pixel 2. One solution to this problem might be to analyse ASTER pixel groups rather than each pixel individually, essentially by calculating the total spectral emittance from a fire via the sum of the spectral radiance recorded in each fire pixel at each wavelength. The resultant composite spectra could then be used to determine the effective emitter characteristics for the entire fire. Unfortunately, this approach is also limited in its application, this time by the fact that for any particular fire imaged by ASTER, a number of the pixels are saturated, particularly in the longer wavelength SWIR bands, so that the total fire thermal emittance cannot be accurately determined. We must therefore conclude that current spaceborne sensors appear not to allow the multi-thermal component spectral fitting approach to be used to determine FRE, although the hyperspectral data from the EO-1 Hyperion experimental sensor remain to be investigated.

Fortunately, an alternative method for FRE derivation exists, whereby measurement of the energy emitted by a fire in one particular, well-chosen, spectral region provides the ability to estimate the amount of energy emitted over all wavelengths. This method was first proposed by Kaufman *et al.* (1996) and was first tested with data of active fires obtained by the MODIS Airborne Simulator and the Airborne Visible/IR Imaging Spectrometer (AVIRIS). Later Kaufman *et al.* (1998a) developed empirically derived equations relating the fire pixel brightness temperature recorded in the middle IR (MIR) 3.9-µm channel of the MODIS spaceborne sensor, which also orbits on the EOS Terra (and Aqua) satellites, to the total energy emitted by all flaming and smouldering activity within that pixel. Fortunately, in comparison to the ambient temperature background, fires emit so strongly in the MIR spectral region that pixels containing fires significantly smaller than one hectare are easily identified in remotely sensed imagery of kilometre-scale spatial resolution, making FRE derivation from a wide range of fires possible via this approach. Building on the work of Kaufman *et al.* (1998a), an alternative method for the derivation of FRE from MIR radiance data has been derived *via* a more physically based approach, which allows the resultant equations to be easily adapted to sensors of differing spatial and spectral resolutions (Wooster *et al.*, 2003).

9.4 Derivation of Fire Radiative Energy from MIR Spectral Radiances

9.4.1 Theory and Testing with Polar-Orbiting Satellite Imagery

The physically based method of deriving FRE from measurements of the MIR radiance (*L*) recorded at the fire pixel is based on the fact that, over any particular temperature range of interest, the radiance emitted by a blackbody at a particular wavelength can be approximated by a simple non-linear function (Wooster and Rothery, 1997):

$$L_\lambda(T) = aT^b \qquad (2)$$

where λ is the wavelength (m), T is the temperature (K), L is the spectral radiance (W m^{-2} sr^{-1} m^{-1}) and a and b are empirically derived constants, dependent upon both wavelength and temperature range used.

At wavelengths around 4 μm in the MIR terrestrial atmospheric window (3–5 μm), the exponent b has a value equal to 4, whilst the multiplier a has a value of 3×10^{-9}. Expanding this relationship to represent a fire 'hotspot' pixel containing n subpixel thermal components of fractional area A_n and temperature T_n (which may be the different smouldering and flaming parts of the fire), provides:

$$L_{h,MIR} = a\varepsilon_{MIR} \sum_{i=1}^{n} A_n T_n^4 \qquad (3)$$

where $L_{MIR,h}$ and ε_{MIR} are the hot 'fire' pixel spectral radiance and surface spectral emissivity in the appropriate MIR spectral band.

For the same modelled fire activity within a pixel, the energy emitted over all wavelengths is given by Stefan's Law:

$$FRE_{TRUE} = A_{sampl} \cdot \varepsilon\sigma \sum_{i=1}^{n} A_n T_n^4 \qquad (4)$$

where FRE_{TRUE} is the Fire Radiative Energy (J s^{-1}) emitted over all wavelength, A_{sampl} is the ground-pixel area (m^2), ε is the surface emissivity averaged over all wavelengths, σ is the Stefan–Boltzmann constant (5.67 × 10^{-8} J s^{-1} m^{-2} K^{-4}).

Equating (3) and (4) provides the following equation relating FRE to the MIR spectral radiance recorded at the fire pixel:

$$FRE_{MIR} = \frac{A_{sampl} \cdot \sigma \cdot \varepsilon}{a \cdot \varepsilon_{MIR}} L_{MIR,h} \qquad (5)$$

Since $L_{MIR,h}$ represents the MIR radiance from the fire only, when analysing real remotely sensed data this should be calculated by subtracting the background MIR radiance $L_{MIR,bg}$ (estimated from neighbouring non-fire pixels) from that of the fire pixel ($L_{MIR,h}$). Langaas (1995) suggests that for larger wildfires (the type most likely to be analysed via moderate-resolution satellite imagery) the flames can be assumed to radiate as black or grey bodies (i.e., $\varepsilon = \varepsilon_{MIR}$), which we suggest is a reasonable assumption for our purposes, especially so since, when viewing from above, the hot surface material will form the background to any observation of the actual flames themselves. Thus, the emissivity terms are removed from the equation, and setting A_{sampl} equal to the 1 km^2 pixel area of the MODIS 3.9 μm channel provides the algorithm for estimate FRE from MIR radiances recorded at MODIS fire pixels:

$$FRE = 1.89 \times 10^7 (L_{MIR} - L_{MIR,bg}) \qquad (6)$$

where spectral radiance has units of W m^{-2} sr^{-1} μm^{-1} and FRE units of J s^{-1} or Watts.

Wooster et al. (2003) have compared the FRE estimates derived via equation (6) to those from the empirically derived equation provided by Kaufman et al. (1998a), which is used in the MODIS fire products (Kaufman et al., 1998b), and the results show excellent agreement ($r^2 = 0.99$). However, the physically based nature of equation (6) allows it to be easily adapted for use with other satellite-based sensor working in the 3–5 μm atmospheric window. The Hot Spot Recognition System (HSRS), carried by the Bi-spectral InfraRed Detection (BIRD) small experimental satellite (Briess et al., 2003), is one such sensor and Figure 9.4 shows a comparison of MODIS and HSRS MIR images for one of

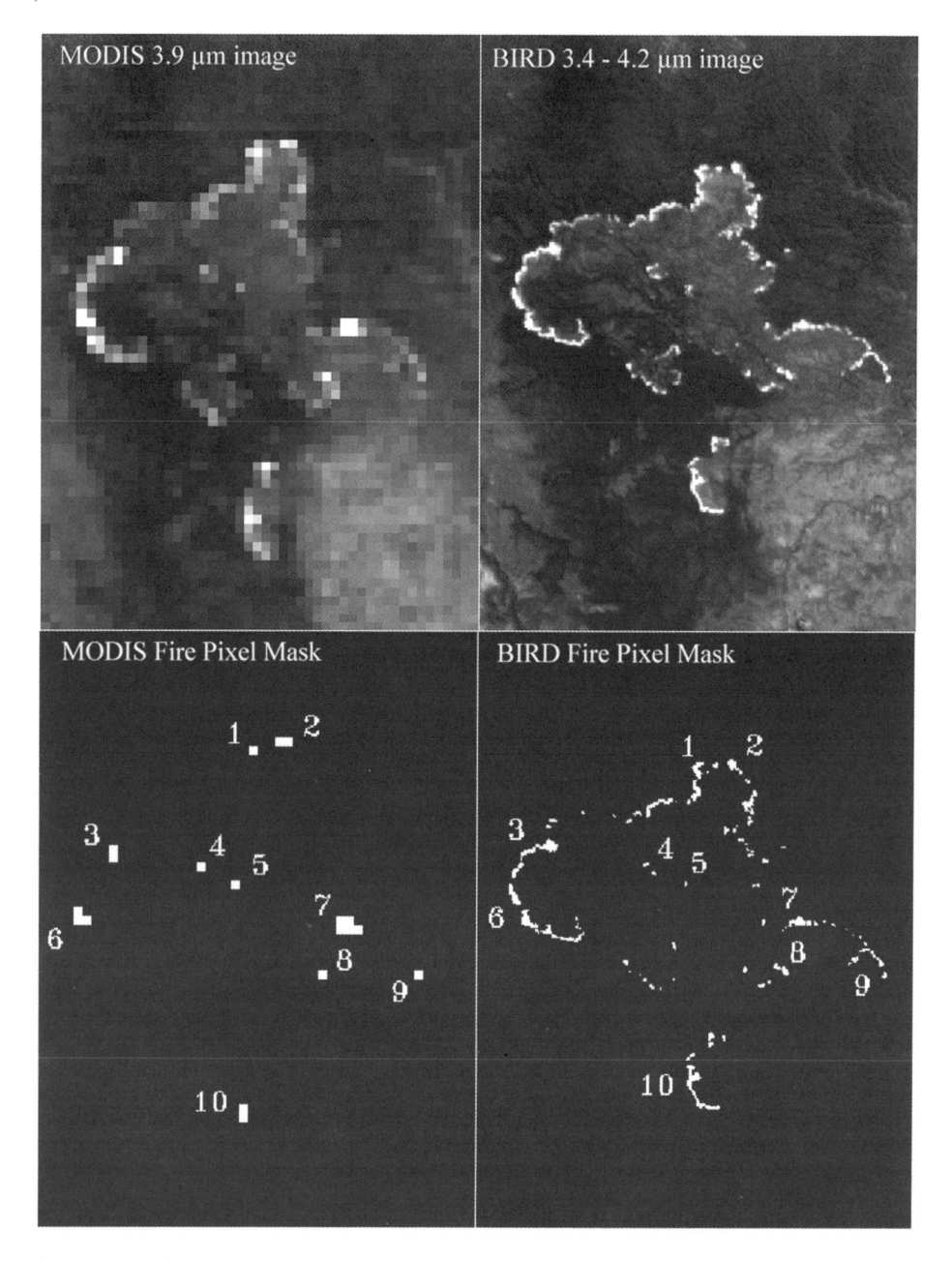

Figure 9.4 The upper row shows EOS MODIS and BIRD HSRS daytime brightness temperature subscenes, extracted from MIR imagery of the Australian wildfires occurring close to Sydney on 5 January 2002. The HSRS image was collected at 00:10 GMT, whilst the MODIS image was collected 12 min. later. The lower row shows the pixels that were detected as fires by the MODIS and BIRD automated fire pixel detection algorithms (Kaufman and Justice, 1998; Zhukov and Oertel, 2001). Ten discrete fire fronts, each composed of clusters of fire pixels are identified in both subscenes by the automated algorithms. With the 1 km spatial resolution MODIS data these 'clusters' consist of between 1 and 5 contiguous pixels, whereas with the higher spatial resolution HSRS imagery (370 m pixel size, 185 m sampling step), they consist of between 4 and 126 pixels

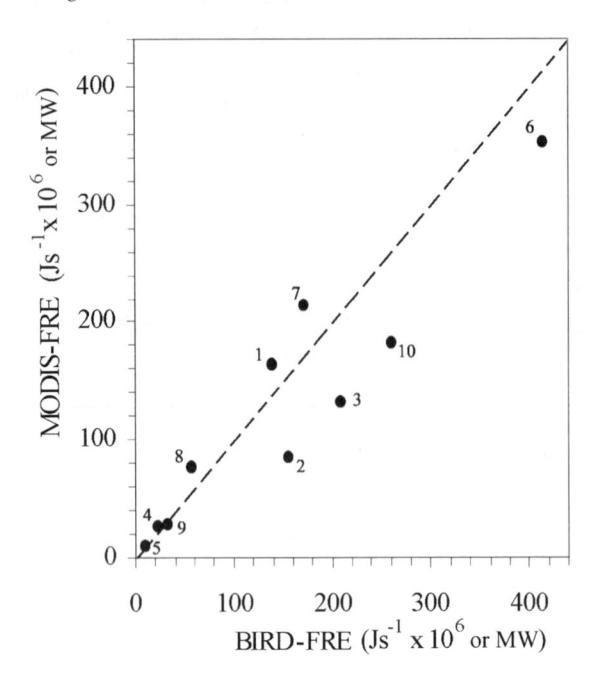

Figure 9.5 Comparison of fire radiative energies derived via equations (6) and (7) applied to the MODIS and BIRD-HSRS MIR imagery shown in Figure 9.4. Numbers correspond to those for the individual fire fronts shown in Figure 9.4 and the 1:1 line of agreement is also shown. Prior to FRE derivation the MIR spectral radiances were corrected for the atmospheric absorption using MODTRAN parameterized with radiosonde data from Sydney Airport, taken within 3.5 hr of the satellite overpass

the large wildfires that occurred close to Sydney, Australia in early 2002. Adapting equation (6) for use with BIRD-HSRS gives the following equation:

$$FRE_{MIR} = 5.93 \times 10^5 (L_{MIR} - L_{MIR,bg}) \tag{7}$$

Applying equations (6) and (7) to the corresponding fire pixels making up the fire fronts visible in the MODIS and BIRD imagery provides evidence of a strong relationship between the FRE estimates made via the two sensors (Figure 9.5). The agreement appears excellent if you consider the widely varying spatial resolution, the different spectral coverage of the sensor band passes and the slight time difference in the FRE measurements. For fires 2, 3 and 10, the underestimation of FRE by MODIS when compared to BIRD-HSRS is primarily due to the failure of the MODIS fire detection algorithm to identify all 'fire pixels' at those locations (see Figure 9.4). The MODIS algorithm was specified prior to launch and, though it works well on larger events, it is now undergoing post-launch modification to improve its sensitivity to smaller fires.

In addition to FRE determination, the higher spatial resolution of BIRD allows the fire front length and the mean radiative energy release per unit length of the fire front to be determined from the HSRS observations, the latter being a direct measure with the same units (kW m^{-1}) as fireline intensity, though in this case referring to the radiative component

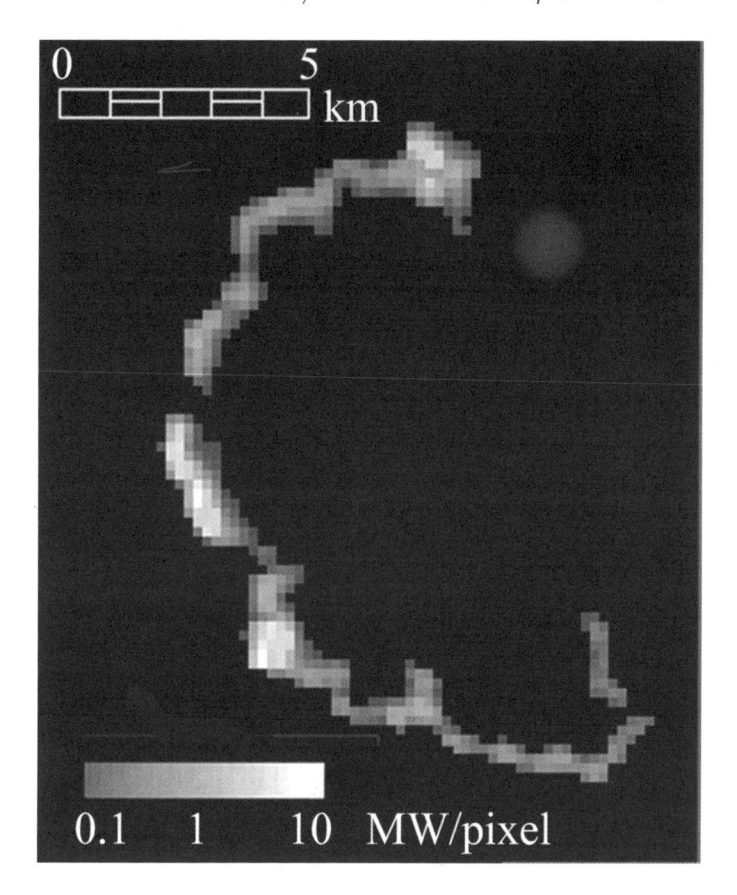

Figure 9.6 *Fire radiative energy variation for fire fronts 3 and 6 of Figure 9.4, derived from the BIRD HSRS data having a pixel resolution of 370 m and a pixel sampling step of 185 m. FRE varies by more than an order of magnitude over the cluster of pixels making up each fire front*

only. Figure 9.6 shows an example of the detailed FRE pattern associated with fronts 3 and 6 of Figure 9.4. FRE clearly varies widely along the zone of active combustion, presumably due to variations in the rate of combustion of vegetation within each pixel, and the total FRE for these fire fronts is 210 and 396 MW, respectively. The corresponding radiatively derived mean fireline intensities, calculated simply by dividing FRE by the fire front length, are 70 and 65 kW m^{-1}, though the variation in fireline intensity along the individual fire fronts mirrors that in FRE.

The remaining clusters in Figure 9.4 have radiatively derived mean fireline intensities in the range 15–75 kW m^{-1}, values all very much lower than the earlier reported 'average' fireline intensities of 8000 kW m^{-1}. However, as discussed earlier, the uncertainties inherent in the calculation of fireline intensity make direct comparison difficult. In many cases 'traditional' field-based estimations of this parameter appear to assume complete fuel combustion, with all particles in the fuel bed available to the combustion process and the degree to which the heat yield of the fuel bed is corrected for FMC is unclear (see section 9.3). Many field-based estimates are also based on head fires, with much lower intensities reported for flanking and backing fires (e.g. Roberts et al., 1988). Analysis of the raw data provided by Roberts et al. for a series of experimental grassland fires shows that the fireline intensity (average ± 1SE; KW m^{-1}) of backfires (148.1 ± 23.2; $n = 22$) was only 6% that of headfires (2306.9 ± 405.0; $n = 39$). Finally, the extent to which heat may be lost to convective and conductive transfer(s) is somewhat unclear, with Ferguson et al. (2000) quoting convection as providing 40–80% of the total heat loss process (McCarter and Broido, 1975), a range in broad agreement with the statement that only around one-third of the evolved heat may be radiated (Vines, 1981). Clearly these effects may explain some of the difference between the modelled fireline intensities, and the values obtained from remote sensing. However, there may be effects operating that are currently not accounted for, and indeed it may be that a much more complex comparison is needed between the measured and modelled parameters. Further investigations continue on this issue.

9.4.2 Application to Geostationary Satellite Imagery

Despite the current level of uncertainty in the relationship between the FRE and the total energy liberated during combustion, the strong agreement between the BIRD- and MODIS-derived FRE measurements provides confidence in the transportability of the approach between sensors. One limitation, however, is that neither of these polar-orbiting sensors can provide very high frequency observations. The experimental BIRD-HSRS sensor can at best supply only one daily FRE measurement of a fire due to its limited swath width and orbital configuration, and this can continue for perhaps four or five consecutive days before the instrument moves out of observing range. This situation is somewhat improved with the operational MODIS imager where, cloud cover permitting, the two spaceborne sensors now operating can supply four FRE observations per day of any fire large enough to be detected. More frequent FRE measurements can only be provided via a geostationary platform equipped with a suitable detector system having MIR imaging capability. The current generation of US Geostationary Operational Environmental Satellites (GOES) are suitable for this purpose, as is the new European Meteosat Second Generation satellite launched in August 2002. Figure 9.7 provides the first example of remotely sensed FRE derived from geostationary satellite data. The target was a large forest fire that occurred in Miller's Reach, Alaska (61° N, 149° W) on 3–4 June 1996 (see Hufford et al., 2000, for maps of the activity). The Miller's Reach fire is reported to have burned 1335 ha of forest on the evening of 3 June and the results in Figure 9.7 were derived by adapting equation (6) for use with the GOES imager 3.9 μm channel and applying this to the GOES data presented by Hufford et al. (1999). The values of FRE are of a similar magnitude to the largest obtained for the Australian fires by BIRD and MODIS, which appears reasonable if we recall that GOES pixels at this latitude have a spatial dimension of around 8 km, which is a large proportion of the total length of many of the fire fronts seen in Figure 9.4. Total FRE release for Fire Pixel 1 as measured by GOES is 8.1×10^{11} J, and for Fire Pixel 2 is 2.8×10^{12} J, which as a rough guide equates to $0.7 - 2.2 \times 10^5$ kg and $2.5 - 7.5 \times 10^5$ kg of vegetation combusted (in 3½ and 5 hours, respectively) if we assume Burgan and Rothermel's (1984) constant heat yield value and the assumption that between 20 and 60% of the heat is liberated in the form of radiation (see Ferguson et al., 2000). However, as already stated, the relative importance of radiation and convection may vary from fire to

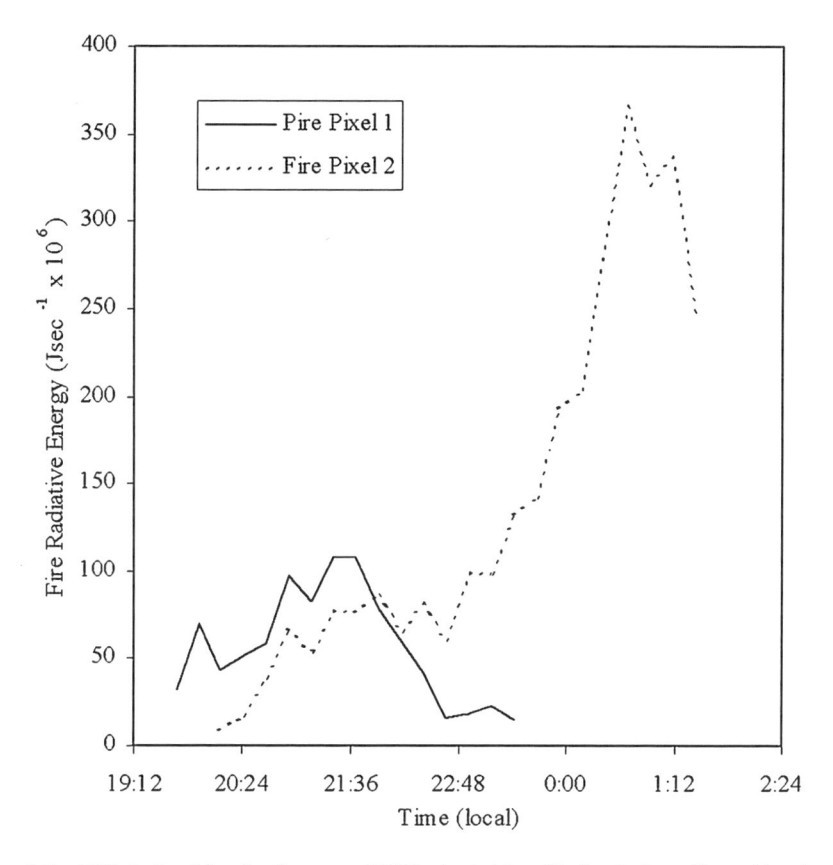

Figure 9.7 *FRE derived for the first two GOES pixels identified as being affected by the 1996 Miller's Reach Fire (Alaska), shown here between 19:40, 3 June and 01:25, 4 June. The fire reportedly decreased very rapidly in size and intensity within 1:30 hr after 01:00 on 4 June (Hufford et al., 2000)*

fire, and estimation of their exact combination is not simple and remains an important research goal (Thomas, 1971; Weber, 1991). Nevertheless, this example does indicate the feasibility of making FRE measurements from geostationary orbit and, if the relationship between FRE and biomass consumption can be accurately characterized, of relating these observations to biomass volatized and emissions produced.

9.5 Potential Use of Fire Radiative Energy in Fire Propagation Modelling

In addition to their potential use in estimating the mass of combusted vegetation and the resultant pollutant emissions, remotely sensed measures of wildfire heat output may hold significant value for modelling the spatial propagation of wildfires. For such modelling of fire dynamics, local measures of the rate of heat generation are usually more important than the total heat output rate of the fire. Thus, although the rate of heat generated per unit length of the fire front (i.e., the fireline intensity) is a parameter commonly used to describe wildfire, Anderson (1969) and Rothermel (1972) found that the heat generated per unit area (known as the combustion rate) is both less variable and more useful in the prediction of the rate of spread in surface fires. Another related measure is the reaction intensity, which also describes the heat release per unit area of ground beneath the fuel bed. The reaction intensity curve describes the consumption of the fuel as the fire burns down and through the fuel bed over some fixed point on the ground (Cheney, 1990). In some cases a third measure of intensity, the total heat output, is the most appropriate measure, being most useful in the prediction and correlation of convection plume phenomena such as spotting which can significantly increase the rate of fire spread. A final intensity parameter is convective intensity, defined as that portion of the total heat output used to lift the resultant pollutant products and entrained air above the flame zone, and calculated by subtracting the estimated conductive and radiative heat losses from the total energy release rate (Chandler et al., 1983).

To predict or estimate any of these measures of fire intensity, vegetation has had to be classified into burnt and unburnt components. Living plant matter may or may not burn in a surface fire because its FMC can lie between 80 and 200% (Pyne, 1984). By contrast, dead plant matter components have a FMC fixed by recent atmospheric activity, below a fibre saturation of ~30% (Rothermel, 1972; Albini, 1993). If the burning of dead plant matter creates a fire of sufficient intensity, then the smaller living plant components will also be burned and will contribute to the local heat release rate. In most cases, however, it is usually assumed that it is the dead fine fuels that carry most wildfires. As already discussed, however, the calculation and interpretation of fire intensity measures are often very difficult from measurements made in the field (and usually impossible for very severe events) and interpretation of these values has proved problematic (Alexander, 1982). The new remote sensing methodologies for directly measuring heat release rates we have presented provide a potential method of improving estimates of fire intensity and energy fluxes, particularly for the most difficult to measure large-scale fire events. Thus, this might not only enable improved estimation of particulate and chemical emissions, but also prove useful for the parameterization, development and improvement of a new generation of physically based fire behaviour models (see Perry, 1998).

9.6 Conclusion

Clearly, research on FRE is at a fairly early stage and remotely sensed FRE estimates and any resultant estimate of biomass consumption must be rigorously analysed for their accuracy and representativeness, a process that began with the work of Kaufman et al. (1996) on airborne remote sensing but which is now beginning to be considered at the satellite scale. The methods now available from satellite-based sensors clearly offer a new approach to determining a measure of fireline intensity, traditionally a difficult-to-measure parameter, though the radiative-based measures available from IR remote sensing still need to be understood in the context of the wider heat-loss mechanisms operating. Recently, Sazanovich and Tsvyk (2002) have commenced attempts to determine the relative partitioning between the heat loss processes of radiation, convection and conduction more precisely, initially for small experimental fires, and there is a clear need for further work in this area now methods to accurately assess the radiative component are at hand. If proved valid and of sufficient accuracy and precision, the new remotely derived estimates of heat generation, fireline intensity and vegetation combustion will be of significant value in the estimation of emissions' products and also the study of how wildfires propagate through the spatial environment. As such, research continues at a variety of spatial scales on improving the understanding of these concepts, on the relationship between them, and on the accuracy with which they can be estimated using these new EO approaches.

Acknowledgements

M. Wooster is supported by the NERC Earth Observation Science Initiative and is grateful to the NERC Equipment Pool for Field Spectroscopy for much advice and for the loan of the GER3700 field spectrometer. MODIS and ASTER data were obtained via the NASA Goddard Space Flight Center (GSFC) and EROS Data Centre Distributed Active Archive Centres (DAACs). Louis Giglio provided useful advice regarding use of the MODIS Fire Products. Radiosonde data were kindly provided by the Australian Bureau of Meteorology.

References

Albini, F.A., 1976, *Estimating Wildland Fire Behaviour and Effects*, USDA Forest Service, General Technical Report, GTR-INT 30.

Albini, F.A., 1980, *Thermochemical Properties of Flame Gases for Fine Wildland Fuels*, USDA Forest Service, General Research Paper, INT 243.

Albini, F.A., 1993, Dynamics and modelling of vegetation fires: observations, in P.J. Crutzen and J.G. Goldammer (eds), *Fire in the Environment: The Ecological, Atmospheric and Climatic Importance of Vegetation Fires* (New York: John Wiley and Sons), 39–53.

Alexander, M.E., 1982, Calculating and interpreting forest fire intensities, *Canadian Journal of Botany*, **60**, 349–357.

Anders, E., Wolbach, W.S. and Gilmour, I., 1991, Major wildland fires at the Cretaceous-Tertiary boundary, in J.S. Levine (ed.), *Global Biomass Burning* (Cambridge, MA: MIT Press), 485–492.

Anderson, H.E., 1969, *Heat Transfer and Fire Spread*, USDA Forest Research Paper, INT-69.

Andreae, M.O. and Merlet, P., 2001, Emission of trace gases and aerosols from biomass burning, *Global Biogeochemical Cycles*, **15**, 955–966.

Barbosa, P., Stroppiana, D., Gregoir, J. and Pereira, J., 1999, An assessment of vegetation fire activity in Africa (1981–1991): burned areas, burned biomass and atmospheric emissions, *Global Biogeochemical Cycles*, **13**, 933–950.

Beniston, M., Innes, J. and Verstraete, M., 2000, Biomass burning and its inter-relationships with the climate system, in *Advances in Global Change Research*, Vol. 3 (Dordrecht: Kluwer).

Bond, W.J. and van Wilgen, B.W., 1996, *Fire and Plants* (London: Chapman and Hall).

Bradstock, R.A., Gill, A.M., Kenny, B.J. and Scott, J., 1998, Bushfire risk at the urban interface estimated from historical weather records: consequences for the use of prescribed fire in the Sydney region of south-eastern Australia, *Journal of Environmental Management*, **52**, 259–271.

Brain, C.K. and Sillen, A., 1988, Evidence from the Swartkans cave for the earliest use of fire by hominids, *Nature*, **336**, 464–466.

Briess, K., Jahn, H., Lorenz, E., Oertel, D., Skrbek, W. and Zhukov, B., 2003, Fire recognition potential of the Bi-spectral InfraRed Detection (BIRD) satellite, *International Journal of Remote Sensing*, **24**, 865–872.

Brown, N., 1998, Out of control: fires and forestry in Indonesia, *Trends in Ecology and Evolution*, **13**, 41.

Budd, G.M., Brotherhood, J.R., Hendrie, A.L., Jeffery, S.E., Beasley, F.A., Costin, B.P., Zhien, W., Baker, M.M., Cheney, N.P. and Dawson M.P., 1997, Project Aquarius 4. Experimental bushfires, suppression procedures, and measurements, *International Journal of Wildland Fire*, **7**, 99–104.

Burgan, R.E. and Rothermel, R.C., 1984, *BEHAVE – Fire Behaviour Prediction and Fuel Modeling System: Fuel Subsystem*, USDA Forest Service General Technical Report, INT-238.

Byram, G.M., 1959, Combustion of forest fuels, in K.P. Davis (ed.), *Forest Fires: Control and Use* (New York: McGraw-Hill), 65–89.

Caldararo, N., 2002, Human ecological intervention and the role of forest fires in human ecology, *Science of the Total Environment*, **292**, 141–165.

Chandler, C.C., Cheney, P., Thomas, P., Trabaud, L. and Williams, D., 1983, *Fire in Forestry Volume 1: Forest Fire Behaviour and Effects* (New York: John Wiley and Sons).

Cheney, N.P., 1990, Quantifying bushfires, *Mathematical and Computer Modelling*, **13**, 9–15.

Chou, M.D., Chan, P.K. and Wang, M.H., 2002, Aerosol radiative forcing derived from SeaWiFS-retrieved aerosol optical properties, *Journal of the Atmospheric Science*, **59**, 748–757.

Cope, M.J. and Chaloner, W.G., 1985, Wildfire: an interaction of biological and physical processes, in B.H. Tiffney (eds), *Geological Factors and the Evolution of Plants* (London: Yale University Press), 257–277.

Enright, N.J., Goldblum, D., Ata, P. and Ashton, D.H., 1997, The independent effects of heat, smoke and ash on emergence of seedlings from the soil seed bank of a healthy *Eucalyptus* woodland in Grampians (Gariwerd) National Park, western Victoria, *Australian Journal of Ecology*, **22**, 81–88.

ERSDAC, 2001, *ASTER Users Guide Part 1*, Earth Remote Sensing Data Analysis Centre, Tokyo.

Ferguson, S., Sandberg, D. and Ottmar, R., 2000, Modelling the effect of landuse changes on global biomass emissions, in J.L. Innes, M. Beniston and M.M. Verstraete (eds), *Biomass Burning and its Inter-Relationship with the Climate System* (Dordrecht: Kluwer), 33–50.

Fuller, M., 1991, *Forest Fires: An Introduction to Wildland Fire Behaviour, Management, Fire Fighting and Prevention* (New York: John Wiley and Sons).

Goode, J., Yokelson, R., Ward, D., Susott, R., Babbitt, R., Davies, M. and Hao, W., 2000, Measurements of excess O_3, CO_2, CO, CH_4, C_2H_4, C_2H_2, HCN, NO, NH_3, HCOOH, CH_3COOH, HCHO, and CH_3OH in 1997 Alaskan biomass burning plumes by airborne Fourier transform infrared spectroscopy (AFTIR), *Journal of Geophysical Research*, **105**, 22147–22166.

Hufford, G.L., Kelley, H.L., Moore, R.K. and Cotterman, J.S. 2000, Detection and growth of an Alaskan forest fire using GOES-9 3.9 micron imagery, *International Journal of Wildland Fire*, **9**, 126–136.

Johansen, M.P., Hakonson, T.E. and Breshears, D.D., 2001, Post-fire runoff and erosion from rainfall simulation: contrasting forests with shrublands and grasslands, *Hydrological Processes*, **15**, 2953–2965.

Kaufman, Y.J., Kleidman, R.G. and King, M.D., 1998a, SCAR-B fires in the tropics: properties and remote sensing from EOS-MODIS, *Journal of Geophysical Research*, **103**, 31955–31968.

Kaufman, Y.J., Tanre, D. and Boucher, O., 2002, A satellite view of aerosols in the climate system, *Nature*, **419**, 215–223.

Kaufman, Y.J., Tucker, C.J. and Fung I., 1990, Remote sensing of biomass burning in the tropics, *Journal of Geophysical Research*, **95**, 9927–9939.

Kaufman, Y.J., Justice, C.O., Flynn, L.P., Kendall, J.D., Prins, E.M., Giglio, L., Ward, D.E., Menzel P. and Setzer, A.W., 1998b, Potential global fire monitoring from EOS-MODIS, *Journal of Geophysical Research*, **103**, 32215–32238.

Kaufman, Y., Remer, L., Ottmar, R., Ward, D., Rong-R, L., Kleidman, R., Fraser, R., Flynn, L., McDougal, D. and Shelton, G., 1996, Relationship between remotely sensed fire intensity and rate of emission of smoke: SCAR-C experiment, in J. Levine (ed.), *Global Biomass Burning* (Cambridge, MA: MIT Press), 685–696.

Keith, D.A., 1996, Fire-driven extinction of plant populations: a synthesis of theory and review of evidence from Australian vegetation, *Proceedings of the Linnean Society of New South Wales*, **116**, 37–78.

Keller, F., Lischke, H., Mathis, T., Mohl, A., Wick, L., Ammann, B. and Kienast, F., 2002, Effects of climate, fire, and humans on forest dynamics: forest simulations compared to the palaeological record, *Ecological Modelling*, **152**, 109–127.

Kita, K., Fujiwara, M. and Kawakami, S., 2000, Total ozone increase associated with forest fires over the Indonesian region and its relation to the El Niño-Southern oscillation, *Atmospheric Environment*, **34**, 2681–2690.

Langaas, S., 1995, A critical review of sub-resolution fire detection techniques and principles using thermal satellite data, PhD thesis, Department of Geography, University of Oslo.

Lobert, J.M. and Warnatz, J., 1993, Emissions from the combustion process in vegetation, in P.J. Crutzen and J.G. Goldammer (eds), *Fire in the Environment: The Ecological, Atmospheric and Climatic Importance of Vegetation Fires* (New York: John Wiley and Sons), 15–39.

Luke, R.H. and McArthur, A.G., 1986, *Bushfires in Australia* (Canberra: Australian Government Publishing Service).

McCarter, R.J. and Broido, A., 1975, Radiative and convective energy from wood crib fires, *Pyrodynamics*, 265–285.

McLean, D.M., 1991, Impact winter in the global K/T extinctions: no definitive evidence, in J.S. Levine (ed.), *Global Biomass Burning* (Cambridge, MA: MIT Press), 493–503.

Morrison, D.A., 2002, Effects of fire intensity on plant species composition of sandstone communities in the Sydney region, *Australian Ecology*, **27**, 433–441.

Nelson, R.M. and Adkins, C.W., 1986, Flame characteristics of wind-driven surface fires, *Canadian Journal of Forest Research*, **16**, 1293–1300.

Perry, G.L.W., 1998, Current approaches to modelling the spread of wildland fire: a review, *Progress in Physical Geography*, **22**, 222–247.

Perry, G.L.W., Sparrow, A.D. and Owens, I.F., 1999, A GIS-supported model for the prediction of the spatial structure of wildland fire, Cass Basin, New Zealand, *Journal of Applied Ecology*, **36**, 502–518.

Pyne, S.J., 1984, *Introduction to Wildland Fire: Fire Management in the United States* (New York: John Wiley and Sons).

Reinhardt, E.D., Keane, R.E. and Brown, J.K., 2001, Modeling fire effects, *International Journal of Wildland Fire*, **10**, 373–380.

Roberts, F.H., Britton, C.M., Wester, D.B. and Clark, R.G., 1988, Fire effects on tobosagrass and weeping lovegrass, *Journal of Range Management*, **41**, 407–409.

Robinson, J., 1991, Fire from space: global fire evaluation using infrared remote sensing, *International Journal of Remote Sensing*, **12**, 3–24.

Rosenfeld, D., 1999, TRMM observed: first direct evidence of smoke from forest fires inhibiting rainfall, *Geophysical Research Letters*, **26**, 3105–3108.

Rothermel, R.C., 1972, *A Mathematical Model for Predicting Fire Spread in Wildland Fuels*, USDA Forest Service, General Technical Report, GTR-INT 115.

Saarnak, C.F., 2001, A shift from natural to human-driven fire regime: implications for trace-gas emissions, *Holocene*, **11**, 373–375.

Sazanovich, V.M. and Tsvyk, R.Sh., 2002, Experimental study of the convective plume above burning forest material, *Atmospheric and Oceanic Optics*, **15**, 336–343.

Scholes, R.J., Kendall, J. and Justice C.O., 1996, The quantity of biomass burned in Southern Africa, *Journal of Geophysical Research*, **101**, 23667–23676.

Schweithelm, J., 1998, *The Fire This Time: An Overview of Indonesia's Forest Fires in 1997/98*, World Wide Fund for Nature Discussion Paper, WWF Indonesia Programme, Jakarta, Indonesia.

Tangren, C.D., 1976, The trouble with fire intensity, *Fire Technology*, **12**, 261–265.

Thomas, P.H., 1971, Rate of spread of some wind-driven fires, *Forestry*, **44**, 155–175.

Turner, M.G., Dale, V.H. and Everham, E.H., 1997, Fires, hurricanes and volcanoes: comparing large-scale disturbances, *BioScience*, **47**, 758–768.

Turner, M.G., Hargrove, W.W., Gardner, R.H. and Romme, W.H., 1994, Effects of fire on landscape heterogeneity in Yellowstone National Park, Wyoming, *Journal of Vegetation Science*, **5**, 731–742.

Vines, R.G., 1981, Physics and chemistry of rural fires, in A.M. Gill, R.H. Groves and I.R. Noble (eds), *Fire and the Australian Biota* (Canberra: Australian National Academy of Sciences), 129–149.

Ward, D.E. and Radeke, L.F., 1993, Emissions measurements from vegetation fires: a comparative review of methods and results, in P.J. Crutzen and J.G. Goldammer (eds), *Fire in the Environment:*

The Ecological, Atmospheric and Climatic Importance of Vegetation Fires (New York: John Wiley and Sons), 54–77.

Weber, R.O., 1991, Modeling fire spread through fuel beds, *Progress in Energy and Combustion Science*, **17**, 67–82.

Whelan, R.J., 1995, *The Ecology of Fire* (Cambridge: Cambridge University Press).

Williams, R.J., Cook, G.D., Gill, A.M. and Moore, P.H.R., 1999, Fire regime, fire intensity and tree survival in a tropical savanna in northern Australia, *Australian Journal of Ecology*, **24**, 50–59.

Wooster, M.J., 2002, Small-scale experimental testing of fire radiative energy for quantifying mass combusted in natural vegetation fires, *Geophysical Research Letters*, **29**, 21.

Wooster, M.J. and Strub, N., 2002, Study of the 1997 Borneo fires: quantitative analysis using Global Area Coverage (GAC) satellite data, *Global Biogeochemical Cycles*, 16, 1–12.

Wooster, M.J. and Rothery, D.A., 1997, Thermal monitoring of Lascar Volcano, Chile, using infrared data from the Along Track Scanning Radiometer: a 1992–1995 time-series, *Bulletin of Volcanology*. **58**, 566–579.

Wooster, M.J., Zhukov, B., and Oertel, D., 2003, Fire radiative energy for quantitative study of biomass burning: derivation from the BIRD experimental satellite and comparison to MODIS fire products, *Remote Sensing of Environment*, **86**, 83–107.

PART III
SPATIAL MODELLING OF URBAN SYSTEM DYNAMICS

Editorial: Spatial Modelling of Urban System Dynamics

Stuart L. Barr

With the world's urban population expected to double between the years 2000 and 2020 (Harrison and Pearce, 2000), many urban areas will become increasingly dynamic environments, experiencing growth that will result in changes in their social and economic structure, spatial pattern of land use and their physical appearance. A consequence of such growth will be that an increasing amount of the Earth's natural and financial resources will have to be directed towards urban areas, in order to supply the increased demand for energy, food and water, as well as to provide the required housing stock and related education and health services. Concerns over how to effectively manage such resources, in order to minimize the negative impacts of urban growth, have increased the awareness that sustainable urban planning policies need to be adopted that allow resources to be spatially targeted and managed effectively.

Computational mathematical models of urban systems provide one potential mechanism by which sounder theories of sustainable urban development may be developed as they allow the likely effects of population growth on the physical, functional and socio-economic structure of urban areas to be estimated (Wilson, 2000). Such models, whether coarse-scale models of urban growth (Batty and Longley, 1994) or small-area level models of urban activity and function, require a range of detailed geographically referenced information for their initialization, calibration and validation. In many cases, particularly in developed nations, such information is provided in the form of national censuses, which often provide the most complete, consistent and objective information at a single point in time on the socio-economic structure of urban areas (Wilson, 2000). A number of concerns have been raised, however, in relation to the suitability of certain national censuses for studying and modelling urban systems; these include, the long time period between censuses (often

---

Spatial Modelling of the Terrestrial Environment. Edited by R. Kelly, N. Drake, S. Barr.
© 2004 John Wiley & Sons, Ltd. ISBN: 0-470-84348-9.

10 years), the course level of spatial aggregation often employed, their inability to describe directly the general physical form and land use function of urban areas, and their poor coverage or lack of availability in many developing nations (Openshaw, 1995).

Potentially, recent developments in sensor technology within field Earth observation allow a number of the above-mentioned concerns to be addressed. For example, very high spatial resolution (2–4 m) optical satellite images (i.e., IKONOS and QuickBird) and high spatial-density airborne LiDAR (Light Detection and Ranging) data, potentially allow accurate, consistent and timely information on the physical form and land use organization of urban areas to be obtained at scales of between 1:10 000 and 1:25 000 (Ridley *et al.*, 1997). However, if the full potential of Earth-observed images for studying and modelling urban systems is to be realized, sensor development needs to be matched not only by improved approaches to the inference of policy-relevant information, but also by an improved integration of such information into the current computational methodologies employed in urban system modelling (Donnay *et al.*, 2001).

In this section on urban systems two studies are presented that address the above observation. In the chapter by Barr and Barnsley (Chapter 10) the spatial topological structure of urban land cover information, such as one may typically try to derive directly from very high spatial resolution remotely sensed images, is studied in order to ascertain whether it provides a relatively simple means by which to infer land use. They show for the urban area under investigation, that while the spatial topology of land cover allows the broad land use categories present to be distinguished (e.g. residential versus industrial), it does not allow a more detailed land use typology to be characterized (e.g. different types of residential development). The chapter by Devereux *et al.* (Chapter 11) demonstrates how Earth-observed information and UK census data can be integrated in order to spatially model sustainable traffic emission scenarios for the county of Cambridgeshire, UK. They demonstrate that land cover information derived from a Landsat-TM image can be used to derive traffic emission coefficients for the parameterization of a spatial interaction model of traffic emissions. Moreover, the utility of LiDAR data for visualizing the relationship between modelled traffic emissions and urban density is also demonstrated.

References

Batty, M. and Longley, P.A., 1994, *Fractal Cities: A Geometry of Form and Function* (London: Academic Press).

Donnay, J.P., Barnsley, M.J. and Longley, P.A., 2001, Remote sensing and urban analysis, in J.P. Donnay, M.J. Barnsley, and P.A. Longley (eds), *Remote Sensing and Urban Analysis* (London: Taylor and Francis), pp. 3–18.

Harrison, P., and Pearce, F., 2000, *AAAS Atlas of Population and Environment* (Berkeley, California: University of California Press).

Openshaw, S., 1995, The future of the census, in S. Openshaw (ed.), *Census Users Handbook* (Cambridge: Geoinformation International) pp. 389–411.

Ridley, H., Atkinson, P. M., Aplin, P., Muller, J.-P., and Dowman, I., 1997, Evaluating the potential of forthcoming commercial U.S. high-resolution satellite sensor imagery at the Ordnance Survey, *Photogrammetric Engineering and Remote Sensing*, **63**, 997–1005.

Wilson, A.G., 2000, *Complex Spatial Systems: The Modelling Foundations of Urban and Regional Planning* (London: Prentice Hall).

10

Characterizing Land Use in Urban Systems via Built-Form Connectivity Models

Stuart Barr and Mike Barnsley

10.1 Introduction

It has been argued that information on urban land use cannot be *directly* inferred from automated analyses of multispectral, remotely sensed images, when performed at the level of individual pixels (Sadler *et al.*, 1991; Barr and Barnsley, 1995, 1999). While the spectral response of the latter may be functionally related to *land cover* (i.e., the physical materials, such as grass, tarmac, concrete and water, at the Earth surface), *land use* is an altogether more complex construct – one that is defined principally in terms of socio-economic activity and that is typically expressed spatially across multi-pixel regions of the urban scene (Barnsley and Barr, 2001; Barnsley *et al.*, 2001). Human photo-interpreters, on the other hand, are generally able to recognize and classify a range of different urban land-use categories in airborne- and satellite-sensor images. One might posit that they do so, whether consciously or subconsciously, by identifying key land-cover objects/entities within the urban scene (e.g. buildings, roads and various types of open space), by considering the size and shape of these objects, and by examining the spatial and semantic relations between them. It seems likely, therefore, that automated analyses of urban land use in remotely sensed images can best be achieved using a similar multi-stage approach: in other words, by inferring land use from an analysis of previously identified land-cover types (Wharton, 1982a; Moller-Jenson, 1990; Barnsley and Barr, 1992, Johnsson, 1994, 1996, 1999;).

Spatial Modelling of the Terrestrial Environment. Edited by R. Kelly, N. Drake, S. Barr.
© 2004 John Wiley & Sons, Ltd. ISBN: 0-470-84348-9.

The multi-stage approach can be expressed more formally as a composition of two functions:

$$I \circ S = \{(i \mapsto s) : i \in I, s \in S, \exists f \in F([i \mapsto f] \wedge [f \mapsto s])\} \quad (1)$$

where I is the set of multi-spectral responses in an image ($i_{m,n} = p_{m,n}(\lambda_1, \lambda_2, \ldots, \lambda_w)$, $i \in I$), F is a set of first-order (i.e., land cover) themes ($F = \{f_1, f_2, \ldots, f_x\}$), each of which exhibits a distinct spectral response and S is a set of second-order (i.e., land use) themes ($S = \{s_1, s_2, \ldots, s_x\}$). Equation (1) states that there is a unique and unambiguous mapping from the image domain, I, for each geometric entity, $i \in I$, to a theme in the second-order thematic domain, S, that is itself a composition of two mappings: (i) from the image domain to the first-order thematic domain ($\exists f \in F(i \mapsto f)$) and (ii) from the first-order to the second-order thematic domains ($\exists f \in F(f \mapsto s)$) (Barr and Barnsley, 1999).[1]

The approach formalized in equation (1) has been employed in a number of studies seeking to derive information on urban land use from Landsat Thematic Mapper (TM) and SPOT-HRV XS images. These have typically made use of kernel-based (i.e., moving window) re-classification techniques to perform the second-stage mapping, that is, to infer land use from an analysis of either the frequency (Wharton, 1982a, 1982b; Treitz et al., 1992; Fung and Chan, 1994; Steinnocher, 1996; Steinnocher and Kressler, 1999) or the spatial pattern (Barnsley and Barr, 1992, 1996) of land-cover labels within an $N \times N$ pixel window. The assignment of the most appropriate land-use label to the central pixel in the window is variously achieved by using frequency-histogram clustering algorithms (Wharton, 1982a, 1982b), frequency-proportion rules (Fung and Chan, 1994; Steinnocher, 1996; Steinnocher and Kressler, 1999), statistical measures of frequency proportion (Treitz et al., 1992) or statistical measures of spatial pattern (Barnsley and Barr, 1992, 1999). Although each of these has produced encouraging results for specific case studies, with consistently higher levels of accuracy than traditional per-pixel classification techniques (Kontoes, 1999; Kontoes et al., 2000), the general utility of this approach has been questioned on a number of grounds. These include: (i) the smoothing effect of the moving window on region boundaries; (ii) the difficulty of selecting a single window size appropriate to the identification of all land-use categories; and (iii) issues relating to spatial scale and scale dependence (Forster, 1985; Toll, 1985; Barnsley and Barr, 1996; Donnay and Unwin, 2001; Donnay et al., 2001; Weber, 2001).

In terms of the last point, spatial reclassification techniques were originally designed for application to images with a spatial resolution in the order of tens of metres (e.g. Landsat TM and SPOT-HRV). In these data, the pixels are approximately of the same size as many of the key spatial entities/objects in urban scenes (e.g. houses), so that their spatial disposition can sensibly be explored within a reasonably small $N \times N$ pixel window. Recently, however, new optical satellite sensors have been launched that acquire images at a spatial resolution of 1 − 4 m (McDonald, 1995; Fritz, 1996; Corbley, 1996; Aplin et al., 1997; Ridley et al., 1997; Barnsley, 1999). While these clearly have considerable potential for urban land-use mapping at spatial scales between 1:25 000 and 1:10 000 (Corbley, 1996; Ridley et al., 1997; Donnay et al., 2001), they are not well suited to the use of

---

[1] Barr and Barnsley (1999) show formally that, for equation (1) to be valid, the sets F and S must be defined in terms of the cross. product of the possible members of the first-order and second-order themes (denoted by C and U, respectively) and the underlying set of geometrical entities ($R = \{r_1, r_2, r_3, \ldots, r_n\}$).

spatial reclassification procedures of the type outlined above. This is because the pixels are typically much smaller than the principal objects within the urban scene, such that a large-moving window is required to capture the relevant spatial patterns of buildings, roads and open spaces. Apart from increasing the required computation time, this has the unwanted effect of significantly smoothing/blurring the boundaries between land-use parcels.

An alternative approach – one that is arguably much better suited to the new generation of high spatial resolution (<5 m) satellite-sensor images – is to perform the second-stage mapping in equation (1) by using structural pattern recognition techniques. These have been widely employed in automated analyses of very high spatial resolution ($\ll 1$ m), digitized aerial photography and airborne scanner images (Nagao and Matsuyama, 1980; McKeown, 1987; Nicolin and Gabler, 1987; McKeown *et al.*, 1989; Mehldau and Schowengerdt, 1990). They have also been adapted to explore the spatial and semantic relations between discrete land-cover parcels (regions) in satellite-sensor images (Barr and Barnsley, 1999; Wilkinson, 1999; Barnsley *et al.*, 2001). The overall approach adopted in this latter set of studies is represented schematically in Figure 10.1. Assuming that task 1 in Figure 10.1 (i.e., the first mapping in equation (1)) can be successfully achieved using per-pixel, multi-spectral classification techniques, the challenge is to develop a method for inferring urban land use from an analysis of the morphological characteristics and spatial organization of discrete land-cover parcels (i.e., tasks 2–4 in Figure 10.1): in other words, to infer *function from form* (Batty and Longley, 1994).

Space syntax theory (Hillier and Hanson, 1984) presents a conceptual framework that could potentially be employed for this purpose. It has been used to describe and partition urban structures based on unweighted topological measures of the urban road network, expressed in graph-theoretic terms (e.g. road-network axial maps). Analysis of the road network on its own, however, is unlikely to facilitate unambiguous inference of urban land use. This requires further information on, among other things, the spatial disposition of individual buildings within distinct topological components of the road network. A set of techniques, known as *built-form connectivity models*, has been developed that incorporates both of these elements (Kruger, 1979a, 1979b).

The remainder of this chapter explores the utility of built-form connectivity models as a means of inferring urban land use from an initial set of land-cover parcels. In doing so, we use land-cover information derived from fine spatial scale digital map data (cf. a remotely sensed image); the intention is to evaluate the suitability of built-form models to perform the second mapping in equation (1), given a land-cover map free from thematic or geometric error. Of course, land-cover classifications derived from remotely sensed images typically contain a number of such errors. These will undoubtedly make the inference of land use less straightforward and may even alter the nature of the process from a semi-deterministic to a probabilistic one. That, however, is the subject of a separate study, for which the work presented here may be considered to be the 'ideal' or reference case.

10.2 Built-Form Connectivity Models

10.2.1 Kruger's Original Model and Implementation

Built-form connectivity models attempt to represent the spatial organization of urban systems by analysing the distribution of buildings within the urban fabric, as well as their

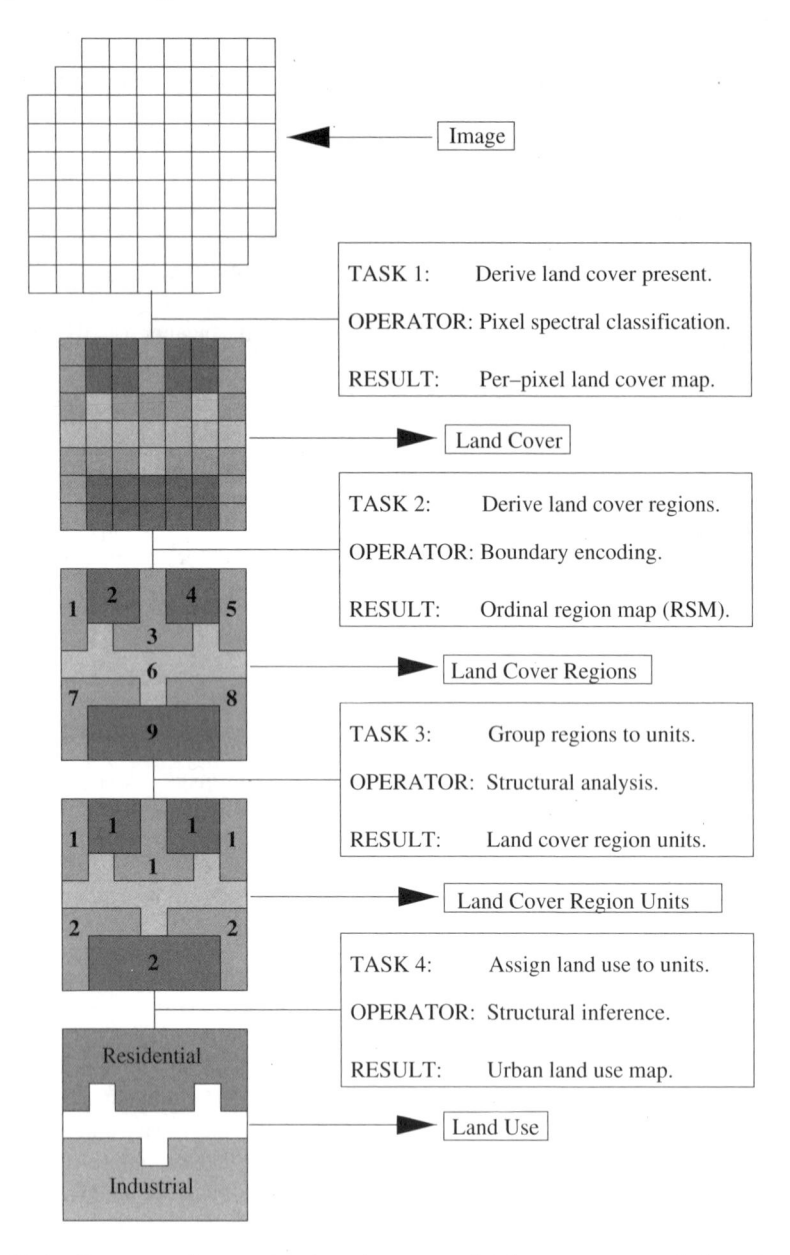

204 Spatial Modelling of the Terrestrial Environment

Figure 10.1 *Key stages in a region-based, structural inference system designed to derive information on urban land use from very high spatial resolution remotely sensed images*

spatial relationships with other urban features, most notably the road network. Thus, they describe the way in which buildings are 'connected and packed over an area of land' (Crowther and Echenique, 1972; Kruger, 1979a). Studies making use of these models have tended to focus on an analysis of the spatial, topological organization of built forms,

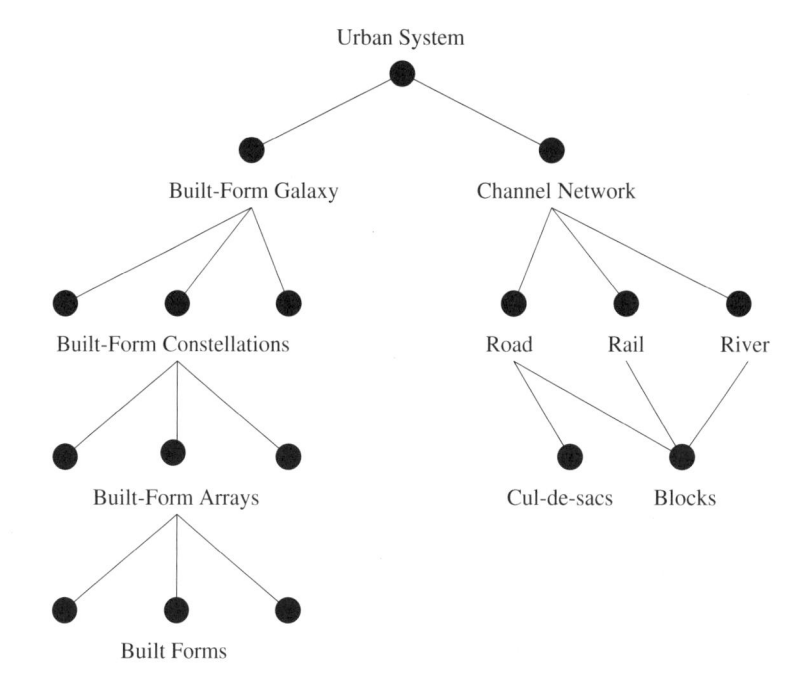

Figure 10.2 Kruger's (1979a) conceptual hierarchy of built-form in an urban system. Reproduced by permission of Pion Limited

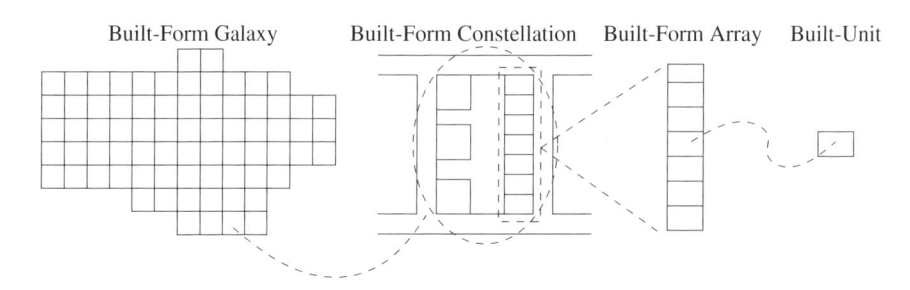

Figure 10.3 Diagrammatic representation of the different levels in Kruger's (1979a) built-form hierarchy

rather than an explicit evaluation of their potential to map such patterns onto different categories of urban land use. Kruger (1979a, b), for example, models urban systems as structured trees stratified into Galaxies, Constellations, Arrays and Units on the basis of their channel-network components (i.e., Roads, Railways and Rivers; Figure 10.2). In these studies, Kruger distinguishes different elements of the urban system using two topological properties, namely containment (Urban-System→Galaxy→Constellation) and adjacency (Constellation→Built-Form Array↔Built-Forms; Figure 10.3). Thus, constellations are defined as collections of buildings that are wholly contained within a closed loop of either the road, river or rail network. Within each constellation, the spatial organization of buildings is characterized by a number of distinct built-form arrays. These are, in turn,

Table 10.1 Kruger's (1979a) graph-theoretic measures of built-form connectivity

Measure	Equation	Description
Cyclomatic	$E - V + C$	Number of independent cycles in a graph
α	$\frac{E_1 - V_1 + C_1}{V_1 - 2C_1}$	Ratio between the observed and maximum number of cycles
β	$\frac{E_n}{V_n}$	Average number of partitions per built-form
δ	$\frac{E_n}{C_n}$	Number of partitions per array of built-forms
ρ	$\frac{V_n}{C_n}$	Average number of built-forms per array
γ	$\frac{E_n}{3V_n - 6}$	Ratio between observed number of lines and maximum possible number
η	$\frac{E_n + V_n + C_n}{E_n}$	Proportional to the sum of the number of vertices and components
θ	$\frac{E_n + V_n + C_n}{V_n}$	Proportional to the sum of the number of edges and components
ψ	$\frac{E_n + V_n + C_n}{C_n}$	Proportional to the sum of the number of edges and vertices
ξ (perimeter)	$\frac{E_2}{V_1}$	Average number of walls per built-form
π (shape)	$\frac{E_2}{E_1}$	Ratio of external walls to the number of partitions
ζ (compactness)	$\frac{E_2}{C_1}$	Ratio of the number of external walls to the number of built-form arrays

Note: V represents the number of vertices, E is the number of edges and C is the number of disjoint independent connected components. V_n, E_n and C_n are the corresponding features for the nth graph universe. Reproduced by permission of Pion Limited.

defined by the number and nature of adjacency events that exist between built-form units (Figure 10.3). Kruger (1979a) proposes 12 graph-theoretic measures (Table 10.1) that can be used to quantify the built-form connectivity of an urban system, based on an analysis of vertices (V), edges (E) and independent connected components (C). These represent built-form connections in the urban system as a whole and in individual constellations, as well as the interrelationships between built-form arrays: each of these three levels is referred to as a universe (Kruger, 1979a). The first measure in Table 10.1 is a global measure of spatial structure, while α through to ψ are measures applied to individual universes, and ξ through to ζ are ratio measures used to compare universes at different levels in the urban hierarchy.

Kruger (1979a) uses the measures listed in Table 10.1 to investigate the spatial organization of one real and two simulated urban systems. Each of the simulated systems has the same number of built-form units, although their spatial organization differs markedly – in one, the units are physically connected (Figure 10.4, left); in the other, they are dispersed (Figure 10.4, right). Most of the 12 graph-theoretic measures in Table 10.1 produce substantially different values for these two simulated urban systems (Table 10.2), which suggests that they might provide a suitable means of characterizing the spatial organization of these and other urban systems. For the real urban system (Reading, UK), however, only five of the measures (δ, γ, η, ξ and ζ) are shown to be statistically independent (Kruger, 1979b). Significantly, though, the ξ and ζ measures identify discrete constellations (i.e., areas of land defined and delineated in terms of the road network) representing different categories of urban land use (Kruger, 1979b).

Table 10.2 Values of Kruger's (1979b) graph-theoretic measures for the simulated built-form arrays in Figure 10.4

Graph measure	Built-form array No One	Two
V_1	6	6
E_1	9	0
C_1	1	6
E_2	10	24
Cyclomatic	4	0
α	1	0
β	1.5	0
δ	9	0
ρ	6	1
γ	1	0
η	1.78	∞
θ	2.67	2
ψ	16	2
ξ (perimeter)	1.67	4
π (shape)	1.11	∞
ζ	10	4

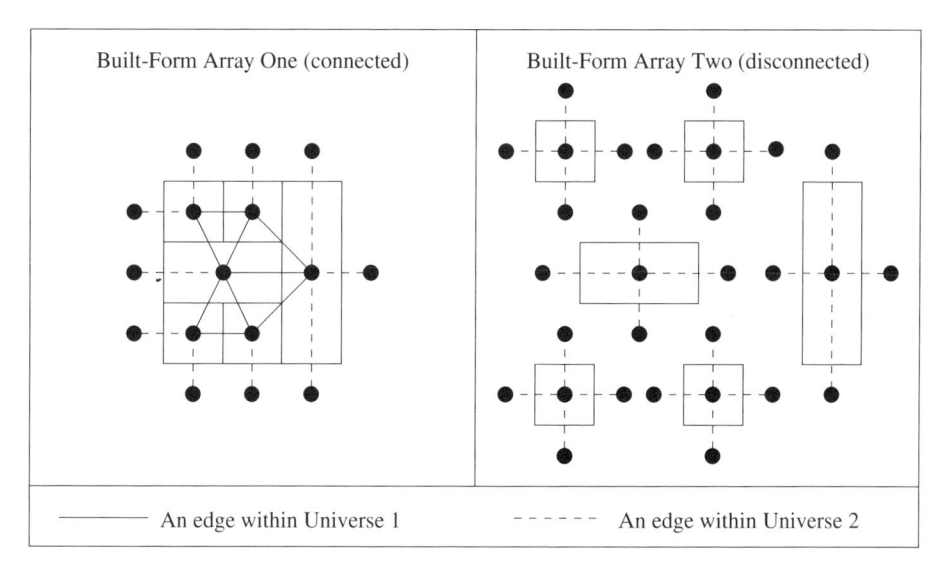

Figure 10.4 Two simulated built-form arrays, each with the same number of built-form units but with different spatial organizations. The figure shows relationships at two different levels (universes) of built-form connectivity

10.2.2 A Region-Based, Graph-Topological Implementation

Although Kruger's (1979a, b) studies provide the conceptual basis for the work presented in this chapter, a number of modifications have been made to the original built-form connectivity model, taking into account the type of information that might reasonably be derived from the new generation of satellite-sensor images. For example, the land-cover parcels produced by multi-spectral image classification do not, in general, permit individual dwellings in a row of terraced houses to be distinguished as separate entities: in other words, several dwellings will typically be represented as a single parcel. Thus, as a general rule, we are unable to characterize the topological interconnections between built-form units in a built-form array. For this reason, we limit the built-form connectivity models used here to a subset of the complete hierarchy considered by Kruger (1979a), i.e., Constellation → Built-Form Units (Regions), as opposed to Constellation → Built-Form Arrays ↔ Built-Form Units.

On the other hand, the use of a parcel- or region-based representation of land cover allows us to analyse built-form morphology (i.e., size and shape) – a feature not included in Kruger's original studies – which is likely to assist in the identification of different urban land-use types (Barr and Barnsley, 1999; Barnsley and Barr, 2001). Moreover, the advent of high-level, dynamic programming languages, which were less widely available at the time of Kruger's studies, means that we are able to implement an automated, graph-theoretic system to analyse built-form connectivity.

Representation of Region Morphology and Spatial Structure. In this study, a region-based structural analysis system, Structural Analysis and Mapping System (SAMS), is used to analyse the built-form constellations present in a raster land-cover dataset (Barr and Barnsley, 1997). SAMS operates by deriving structural information about the regions (i.e., land-cover parcels) present in these data. The boundaries of each region are identified using a simple contour-tracing algorithm (Gonzalez and Wintz, 1987; Gonzalez and Woods, 1993), represented by using Freeman chain codes (Freeman, 1975), and stored in a Region Search Map (RSM). The RSM can be processed to derive further information about the structural characteristics of the observed scene, including various morphological properties (e.g. region area, perimeter and various measures of shape) and spatial relations (e.g. region adjacency, containment, distance and direction). This information is used to populate a graph-theoretic data model, known as XRAG (eXtended Relational Attribute Graph), which is defined by the heptuple:

$$XRAG = \{N, E, EP, I, L, G, C\} \tag{2}$$

where N is the set of nodes (i.e., regions), such that $N \neq \emptyset$, E is the set of spatial relations between $n \in N$ (e.g. adjacency and containment), EP is the set of properties associated with the relations in E (e.g. distance and direction), I is the set of properties relating to $n \in N$ (e.g. area and perimeter), L is the set of labels (interpretations) assigned to $n \in N$ (e.g. grass, tree, urban, non-urban, *etc.*), G is the set of groups 'binding' $\forall l \in L$ to the context of a scene interpretation (e.g. the label tarmac is bound to the group land cover, while the label residential is bound to the group land use) and C is the set stating the confidence to which $l \in L \to n \in N$ (e.g. 'the probability that region n belongs to land-cover class l is 0.9').

Each region is represented by a node, $n \in N$, in the XRAG model. A relationship between two regions n_x and n_y for a given relation, $r_i \in E$, is represented by an edge (i.e., $(n_x, n_y) \in r_i$). Thus, the node set N, in combination with the relation r_i, is equivalent to a standard relational graph, $\{N, r_i\} \equiv G$ (Barr and Barnsley, 1997). Non-relational properties of the regions (e.g. their area and perimeter) are represented as attributes of the nodes, while relational properties (e.g. distance and cardinal direction (orientation)) between any two regions are represented as attributes of the edges in the set *EP*.

Measuring Constellation→Built-Form Spatial Structure. The overall objective of this study is to quantitatively compare the spatial and morphological structure of different urban constellations. In this context, it is hypothesized that constellations of the same land-use category will tend to exhibit similar structural properties, while those of different land-use categories will tend to exhibit dissimilar properties. As the built-form connectivity model used here is limited to Constellation → Built-Form Unit relationships, we are unable to employ the set of 12 structural measures originally proposed by Kruger (1979a) to evaluate the degree of structural similarity between constellations. Instead, this is analysed using SAMS/XRAG, which permits a range of simple descriptive statistics (e.g. the mean and standard deviation (SD) of built-form unit area), measures of built-form density (e.g. the number of built-form units per hectare) and spatial statistics (e.g. Moran's I, a measure of spatial autocorrelation) to be computed.

More specifically, we use a distance-weighted version of Moran's I for point samples, where the graph nodes (i.e., the geographical centroids of the corresponding land-cover parcels) provide the point data. For the unique, pairwise combinations p of a set of n observations on variable x, Moran's I is given by:

$$I = \frac{n \sum_p w_{ij}(x_i - \bar{x})(x_j - \bar{x})}{\left(\sum_p w_{ij}\right) \sum (x - \bar{x})^2} \quad (3)$$

where $\bar{x}$ is the mean of the observations on x and w_{ij} is a weighting factor applied to the values of x for points i and j (Ebdon, 1985; Fotheringham et al., 2000). Here, w_{ij} is most commonly expressed as a reciprocal of the Euclidean distance (d) between points i and j (e.g. $\frac{1}{d_{ij}}, \frac{1}{d_{ij}^2}, \ldots, \frac{1}{d_{ij}^n}$).

Moran's I has bounds of $-1.0 \leq I \leq 1.0$. A value of I close to -1.0 indicates that there is no spatial autocorrelation in terms of the values of variable x, while a value close to 1.0 indicates that there is strong spatial autocorrelation. Values of I tending to 0.0 are indicative of a random spatial distribution in the values of x. Assuming that values of x are randomly distributed spatially, the expected value of I, E_I, is given by:

$$E_I = \frac{1}{n-1} \quad (4)$$

and its SD by:

$$\sigma(E_I) = \sqrt{\frac{n[(n^2 + 3 - 3n)A + 3B^2 - nC] - k[(n^2 - n)A + 6B^2 - 2nC]}{(n-1)(n-2)(n-3)B^2}} \quad (5)$$

where $A = \sum_p w_{ij}^2$, $B = \sum_p w_{ij}$ and $C = \sum_i (\sum_j w_{ij})^2$ and k is the kurtosis of x (i.e., $\frac{\sum (x-\bar{x})^4}{n\sigma^4}$). Finally, the standard normal deviate (z_I) of I is given by:

$$z_I = \frac{I - E_i}{\sigma_I} \quad (6)$$

If the calculated value of z_I falls within the critical values of z_I for a particular significance

210   Spatial Modelling of the Terrestrial Environment

level, the null hypothesis cannot be rejected and the spatial distribution of x in terms of n is considered to be random. If $I < 0.0$ and z_I is significant, on the other hand, it may be concluded that there is no significant spatial autocorrelation of n in terms of x (i.e., x exhibits significant spatial dispersion). Finally, if $I > 0.0$ and z_I is significant, n exhibits significant spatial autocorrelation in terms of x (i.e., x exhibits significant spatial clustering, such that similar values of x are likely to be found close to one another). In the analysis that follows, a standard $\frac{1}{d}$ weighting and a significance level of 0.001 are used.

10.3   Study Area and Data Pre-Processing

The town of Orpington in the London Borough of Bromley is used to evaluate the built-form connectivity approach outlined above. Orpington contains numerous different types of urban land use within a relatively small area. These include several different residential districts, ranging in age from turn-of-the-century terraced houses (1900s), through semi-detached developments built during the 1920s/1930s and 3–4-storey blocks of flats constructed in the 1960s/1970s, to modern townhouses (1980s) and detached 3–4 bedroom residences (1990s). There are also several major schools in the area, as well as a large hospital complex. The existence of a 1:25 000 scale land-use map of the town and its environs, generated for a previous study (Barnsley and Barr, 1996), allows an objective evaluation of the analyses performed here.

The primary dataset employed in this study is a 1 m spatial resolution land-cover map that has been generated from Ordnance Survey (OS) 1:1250 scale Land-Line.93+ digital map data (Figure 10.5). This covers an area of 2 km × 2 km. The original Land-Line.93+ data were provided by the OS as sixteen 500 m × 500 m tiles in standard National Transfer Format (NTF). Each tile contains vector data for up to 32 feature codes. These are processed using ARC/INFO to select the feature codes of interest from each tile, to mosaic the resultant data together into a single 2 km × 2 km coverage, and generate a topologically structured vector dataset. Each of the polygons in the resultant coverage is assigned a land-cover label drawn from the set ROAD, BUILT, TREE (representing areas of continuous, extensive tree cover), OPEN SPACE (representing all other extensive areas of non-wooded open space) and WATER (representing all open water bodies, such as lakes and rivers). This coverage is then transferred to the Grid module of ARC/INFO and resampled to a spatial resolution of 1 m to generate a raster land-cover map, before being exported from ARC/INFO and converted to the image format used by SAMS.

Based on the data shown in Figure 10.5, an XRAG is compiled for the spatial, topological relation containment and the morphological properties area, compactness ($\frac{perimeter^2}{4\pi area}$) and geographical centroid. Containment is selected as it is the key spatial relation used by Kruger (1979a) to recognize built-form constellations. The three morphological properties, outlined above, provide a means by which to evaluate region (building) size and shape.

10.4   Recognition of Built-Form Constellations

To identify the built-form constellations present in Figure 10.5, we must first analyse the containment relation for the corresponding node set in the XRAG. Figure 10.6 shows a

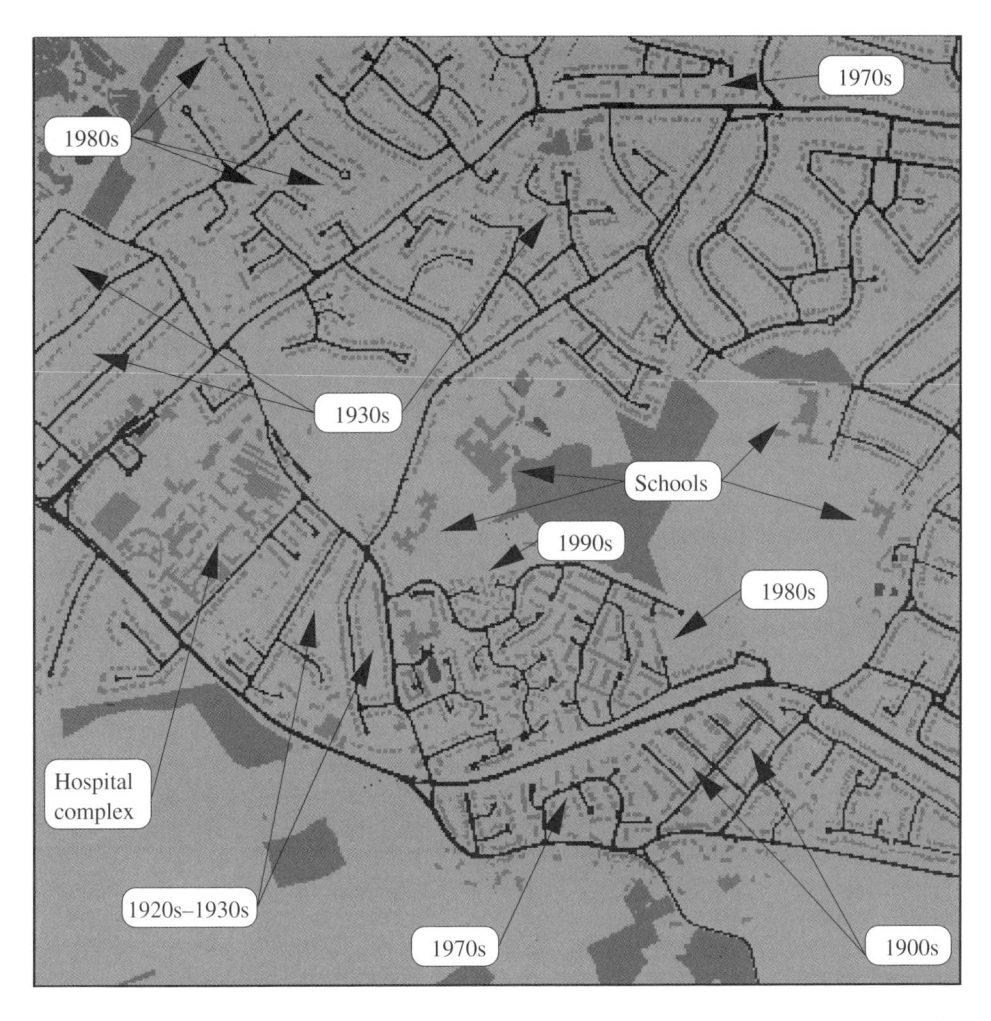

Figure 10.5 2 km × 2 km raster (1 m) land-cover dataset centred on Locksbottom in the town of Orpington, London Borough of Bromley, annotated with examples of the principal land-use categories in the study area. The data were produced from selected vector features derived from Ordnance Survey 1:1,250 Land-Line.93+ digital map data © Crown Copyright Ordnance Survey

visualization of the containment graph for the data presented in Figure 10.5, generated using the Vgraph software incorporated in SAMS. It is clear from this that there is a pronounced structure to the containment graph, such that distinct, localized subgraphs[2] can be visually distinguished. These correspond to Kruger's built-form constellations (Kruger, 1979a).

---

[2] Strictly speaking, a graph $\bar{G} = (\bar{N}, \bar{R})$ is a subgraph of the graph $G = (N, R)$ iff $\bar{N} \subseteq N$ and $\bar{R} \subseteq R$ (Piff, 1992). In the context of the current discussion, however, a subgraph is defined more loosely as a collection of nodes and edges that form a distinct, localized, spatial pattern within a larger graph.

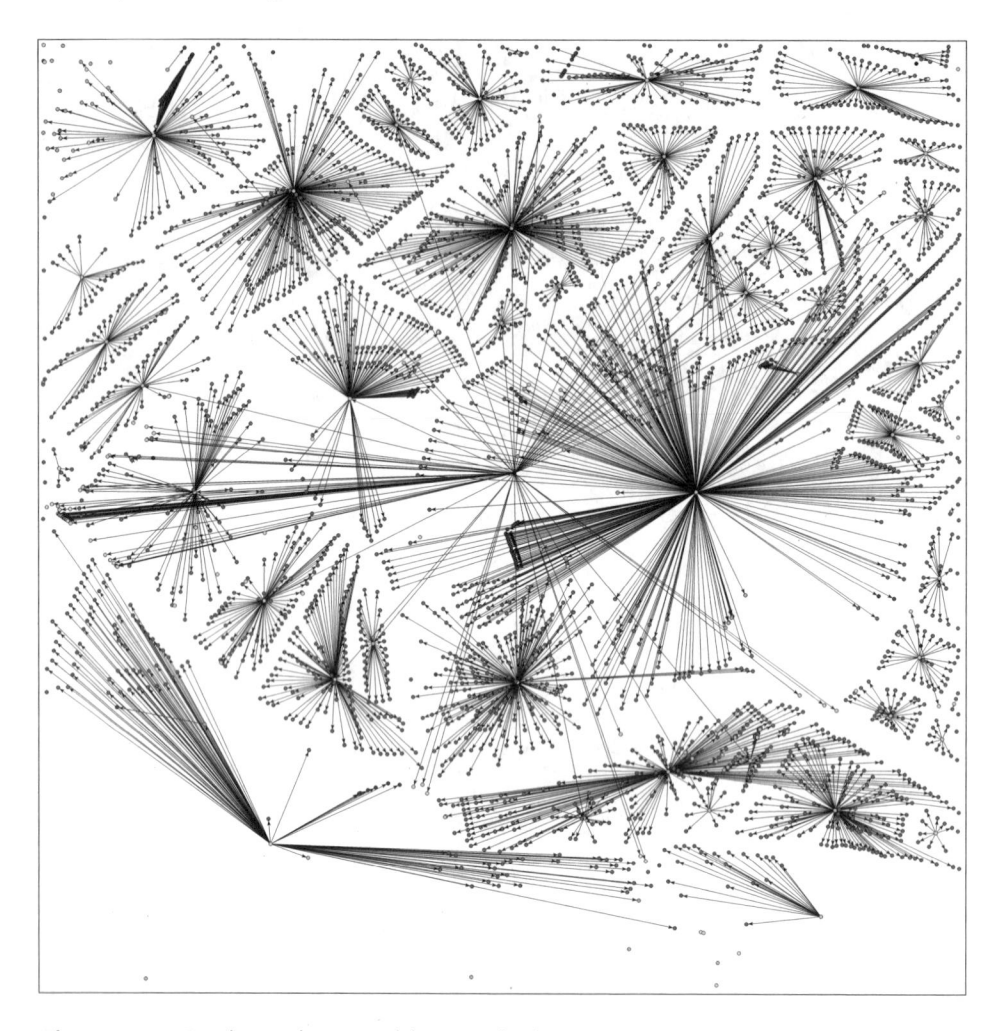

212 *Spatial Modelling of the Terrestrial Environment*

Figure 10.6 *Graph visualization of the spatial relation containment for the land-cover nodes (regions) shown in Figure 10.5*

Tables 10.3–10.5 present summary information on the containment relation for the scene (graph) as a whole, as well as on a cover-type by cover-type basis. In Table 10.3, $\delta(v)$ denotes node degree (i.e., the number of edges incident on a node), while $\overline{\delta(v)}$ denotes the mean node degree (Piff, 1992). In Tables 10.3 and 10.5, the double-column format for the number of nodes and $\overline{\delta(v)}$ refers to $N \in XRAG$ and $N \in r_{containment}$, respectively.

Tables 10.3–10.5 suggest that containment relationships are dominated by the ROAD, OPEN SPACE and BUILT nodes. More specifically, two points suggest that there is a clear containment hierarchy (ROAD→ OPEN SPACE→ BUILT)[3] within Figure 10.5:

---

[3] Hereafter we use the symbol → to denote the topological relation containment, such that ROAD→OPEN SPACE denotes an area of open space wholly contained within (i.e., surrounded by) some part of the road network.

Table 10.3 Summary statistics on the spatial relation containment for the land-cover regions identified in Figure 10.5

Land-cover type	No of nodes	No of edges		$\overline{\delta(v)}$
ALL	3061	64	2791	0.91	43.61
ROAD	6	3	85	14.17	28.33
BUILT	2617	3	3	0.0012	1.00
OPEN SPACE	254	52	2690	10.59	51.73
TREE	60	4	11	0.18	2.75
WATER	124	2	2	0.02	1.00

Note: The table summarizes the number of nodes, number of edges and mean node degree (mean number of edges per node).

Table 10.4 Percentage of containment edges between pairs of land-cover regions in Figure 10.5

Land-cover type	ROAD	BUILT	OPEN SPACE	TREE	WATER
ROAD	—	0.07	2.97	0.00	0.00
BUILT	0.00	—	0.11	0.00	0.00
OPEN SPACE	0.00	91.15	—	1.65	3.58
TREE	0.00	0.00	0.04	—	0.36
WATER	0.00	0.00	0.07	0.00	—

Note: The table reads row-wise; thus, 91.15% of the containment edges represent BUILT parcels contained within areas of OPEN SPACE.

Table 10.5 Mean node degree $(\overline{\delta(v)})$ for the land-cover regions in Figure 10.5

Land-cover type	ROAD		BUILT		OPEN SPACE		TREE		WATER	
ROAD	—	—	0.33	2.00	13.83	27.67	0.00	0.00	0.00	0.00
BUILT	0.00	0.00	—	—	0.001	1.00	0.00	0.00	0.00	0.00
OPEN SPACE	0.00	0.00	10.02	48.92	—	—	0.18	9.20	0.39	20.00
TREE	0.00	0.00	0.00	0.00	0.02	1.00	—	—	0.17	3.33
WATER	0.00	0.00	0.00	0.00	0.02	1.00	0.00	0.00	—	—

Note: The first column for each cover type refers to $N \in XRAG$, while the second column refers to $N \in r_{containment}$. The table reads row-wise and is not symmetrical. For instance, OPEN SPACE regions contain on average 10 BUILT regions (buildings), while only one in a thousand (0.001) buildings contain an open space.

- The ROAD nodes typically contain a large number of other nodes (Table 10.5) and have a $\overline{\delta(v)}$ of 28.3 for $N \in r_{containment}$ (Table 10.3), albeit only 3 (i.e., half) ROAD nodes in the scene contribute to this value. Moreover, virtually all of the nodes contained by ROAD correspond to the OPEN SPACE cover type ($\overline{\delta(v)} = 27.7$ for ROAD→OPEN SPACE and $N \in r_{containment}$; Table 10.5); and,
- The OPEN SPACE nodes contain the largest number of other nodes, both in absolute terms (91.2% of the total; Table 10.4) and in terms of $\overline{\delta(v)}$ for $N \in r_{containment}$ (51.7; Table 10.3). Nearly all of the nodes contained by OPEN SPACE are of the BUILT land-cover type ($\overline{\delta(v)} = 48.9$ for OPEN SPACE→BUILT and $N \in r_{containment}$; Table 10.5).

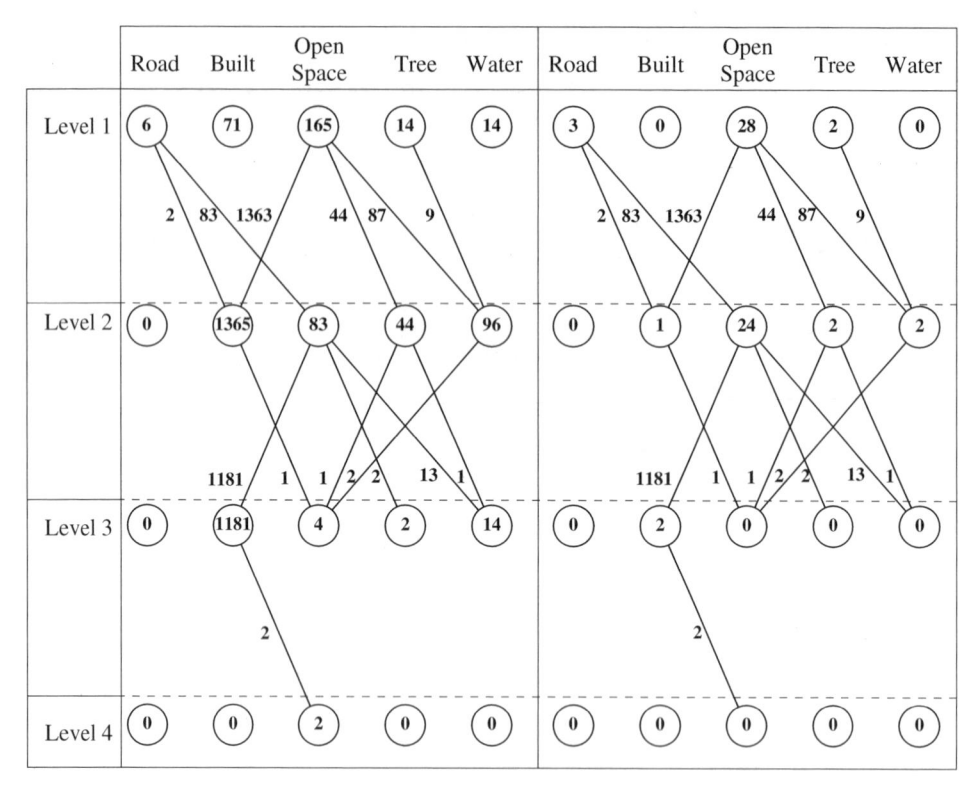

214 *Spatial Modelling of the Terrestrial Environment*

Figure 10.7 Two graph visualizations of the containment hierarchy for the land-cover nodes (regions) in Figure 10.5. The left-hand panel expresses the containment hierarchy in terms of $N \in XRAG$ and the right-hand panel in terms of $N \in r_{containment}$. See text for details

It may be possible, therefore, to exploit this containment hierarchy to perform an analysis of built form as a function of land use.

A depth-first graph-searching algorithm (Sedgewick, 1990), implemented in SAMS, is used to examine the containment hierarchy for the graph (study area) as a whole. Figure 10.7 presents a visualization of this hierarchy in terms of both $N \in XRAG$ (left-hand panel) and $N \in r_{containment}$ (right-hand panel). Level 1 in the hierarchy indicates those nodes (parcels) of each land-cover type that are *not* contained within any other node (parcel) in the study area. Level 2 nodes are wholly contained within a Level 1 node, Level 3 nodes are wholly contained within a Level 2 node, and so on. The values in the circles indicate the number of nodes of each land-cover type at the given level. Thus, for example, there is a total of 165 OPEN SPACE nodes at Level 1 (left-hand panel; $N \in XRAG$), of which 28 contain nodes from Level 2 in the hierarchy (right-hand panel; $N \in r_{containment}$). Figure 10.7 also shows that these 28 nodes contain 1363 of the total 1365 Level 2 BUILT nodes (i.e., OPEN SPACE $\xrightarrow{1363}$ BUILT).

Figure 10.7 indicates that the Orpington scene is only partly characterized by the hypothesized ROAD→OPEN SPACE→BUILT containment hierarchy (i.e., ROAD$\xrightarrow{83}$OPEN SPACE$\xrightarrow{1181}$BUILT at Levels 1 to 3). There is also a strong containment hierarchy in terms of OPEN SPACE and BUILT at Levels 1 and 2 (i.e., OPEN SPACE$\xrightarrow{1363}$BUILT), accounting for

65% and 52.3% of the total OPEN SPACE and BUILT nodes, respectively. This suggests that the land-cover containment hierarchy for the Orpington scene is, in fact, characterized by two distinct patterns, rather than just the one postulated earlier.

It is also important to note that the overwhelming majority (97.6%) of OPEN SPACE nodes occur at Levels 1 and 2 (165 and 83, respectively) in the containment hierarchy, although the actual number of these that contain further nodes is quite small – 28 and 24 at Levels 1 and 2, respectively. In other words, 79.0% of the total number of OPEN SPACE nodes at these two levels do not contain any other nodes. Closer inspection of Figures 10.5 and 10.6 reveals that many of this latter set of OPEN SPACE nodes constitute either: (i) small, grassy roundabouts within the road network; (ii) small, isolated parcels of grass in wooded areas; or (iii) small, isolated parcels of grass between buildings and the road network.

On the basis of the analysis outlined above, it is possible to define formally the containment relations needed to identify distinct built-form constellations. Two hierarchical containment patterns are of particular interest: the first is ROAD→OPEN SPACE→BUILT (i.e., buildings wholly contained within areas of open space that are themselves wholly contained within sections of the road network); the second is OPEN SPACE→BUILT (i.e., buildings wholly contained within areas of open space, where the latter are not wholly contained within the road network). Critically, identification of constellations based on these relations must avoid the inclusion of OPEN SPACE nodes that do not contain any BUILT nodes (i.e., OPEN SPACE↛BUILT).

The relations that encapsulate these hierarchical containment patterns are defined more formally as follows. First, the relation for all OPEN SPACE nodes that contain BUILT nodes is given by:

$$r_{\text{Open Space} \to \text{Built}} = r(x, y) \in r_{\text{containment}} \\ \wedge l(x) \mapsto \text{Open Space} \\ \wedge l(y) \mapsto \text{Built} \\ \wedge ((l(x) \wedge l(y)) \in g_{\text{land cover}}), \forall x, y \in N \quad (7)$$

where $r(x, y)$ is an edge in the containment relation and $l(x)$ and $l(y)$ represent the labels assigned to nodes x and y from the LAND-COVER group. Second, the relation for all ROAD nodes that contain OPEN SPACE nodes is given by:

$$r_{\text{Road} \to \text{Open Space}} = r(x, y) \in r_{\text{containment}} \\ \wedge l(x) \mapsto \text{Road} \\ \wedge l(y) \mapsto \text{Open Space} \\ \wedge ((l(x) \wedge l(y)) \in g_{\text{land cover}}), \forall x, y \in N. \quad (8)$$

It should be clear that, while $r_{\text{Open Space} \to \text{Built}}$ and $r_{\text{Road} \to \text{Open Space}}$ form new relations, they are also sub-graphs of $r_{\text{containment}}$ defined in terms of the label set, L, in XRAG (i.e., $r(x, y) \in r_{\text{Open Space} \to \text{Built}} \wedge r(x, y) \in r_{\text{containment}}$ and $r(x, y) \in r_{\text{Road} \to \text{Open Space}} \wedge r(x, y) \in r_{\text{containment}}$).

Having derived $r_{\text{Open Space} \to \text{Built}}$ and $r_{\text{Road} \to \text{Open Space}}$, it is possible to define a graph relation for built-form constellations characterized in terms of ROAD→OPEN SPACE→BUILT containment relationships:

$$G(N, r_{\text{Road} \to \text{Open Space} \to \text{Built}}) = (y, z) : r_{\text{Road} \to \text{Open Space} \to \text{Built}} \mid \\ r(x, y) \in r_{\text{Road} \to \text{Open Space}} \\ \wedge r(y, z) \in r_{\text{Open Space} \to \text{Built}}, \forall x, y, z \in N \quad (9)$$

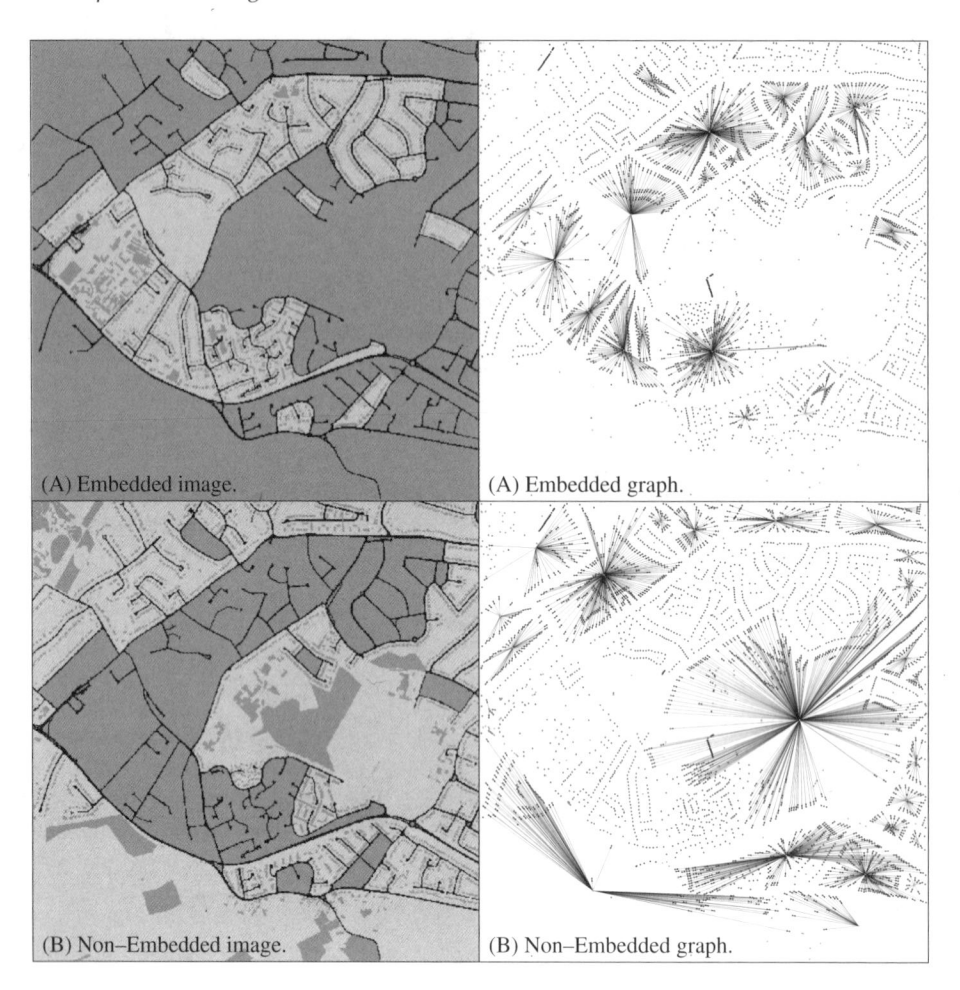

216   Spatial Modelling of the Terrestrial Environment

(A) Embedded image.   (A) Embedded graph.

(B) Non–Embedded image.   (B) Non–Embedded graph.

Figure 10.8 Images (a) and (c) and their corresponding containment graphs (b) and (d) showing separate built-form constellations identified in Figure 10.5 in terms of whether they fall within ($r_{\text{Road}\rightarrow \text{Open}\rightarrow \text{Built}}$; (a) and (b) or outside ($r_{\text{Road}\not\rightarrow \text{Open}\rightarrow \text{Built}}$; (c) and (d) the road network

and a further graph relation for built-form constellations characterized by Road$\not\rightarrow$OPEN SPACE$\rightarrow$BUILT containment relationships (i.e., only OPEN SPACE$\rightarrow$BUILT):

$$G(N, r_{\text{Road}\not\rightarrow \text{Open}\rightarrow \text{Built}}) = r_{\text{Open}\rightarrow \text{Built}} \backslash r_{\text{Road}\rightarrow \text{Open}\rightarrow \text{Built}}$$
$$= r(x, y) \in r_{\text{Open}\rightarrow \text{Built}} \qquad (10)$$
$$\wedge r(x, y) \notin r_{\text{Road}\rightarrow \text{Open}\rightarrow \text{Built}}, \forall x, y \in N.$$

By applying these relationships to the XRAG of the Orpington study area, we are able to identify 52 separate built-form constellations: 24 for $G(N, r_{\text{Road}\rightarrow \text{Open}\rightarrow \text{Built}})$ (Figure 10.8a and b) and 28 for $G(N, r_{\text{Road}\not\rightarrow \text{Open}\rightarrow \text{Built}})$ (Figure 10.8c and d). In graph-theoretic terms, each of these constellations consists of a disjoint, independent connected component, encoded in XRAG using the external relations given in equations (9) and (10), and the union of these (i.e., $G(N, r_{\text{Road} \to \text{Open} \to \text{Built}}) \cup G(N, r_{\text{Road} \not\to \text{Open} \to \text{Built}})$).

10.5 Analysis of Built-Form Constellation Structure

Having analytically derived a number of discrete built-form constellations from the data presented in Figure 10.5, we now examine whether the structural patterns that they exhibit – in terms of their constituent built-form units (i.e., BUILT regions) – can be mapped consistently and unambiguously onto specific categories of urban land use. We do so by evaluating: (i) the packing and density of built-form units; (ii) the spatial and statistical patterns of built-form area; and (iii) the spatial and statistical patterns of built-form compactness in each of the constellations.

10.5.1 Built-Form Unit Packing and Density

Figures 10.9(a) and (b) presents information on the packing and density of built-form units in each constellation. In this context, packing is defined as the number of BUILT nodes (regions) per hectare, while density is defined as 100 times the total area of built-form units in a constellation divided by the total area of that constellation (i.e., $\frac{100 \times \Sigma \, built \, area}{constellation \, area}$).

In terms of spatial packing, Figure 10.9a shows that there is a broadly inverse, monotonic relationship between constellation size (area) and the number of BUILT nodes per hectare. In other words, large constellations tend to have a proportionally smaller number of built-form units per hectare, while small constellations have a proportionally large number. Setting aside for the moment those constellations that fall outside the road network, and which are therefore characterized by ROAD $\not\to$ OPEN SPACE $\to$ BUILT containment relations, it is tempting to conclude that the observed variations in built-form packing are indicative of different types of urban land use and, more specifically, different categories of residential land. Thus, for example, constellations with a proportionally large number of built-form units per hectare might be taken to represent compact areas of high-density housing (small housing units built in close proximity to one another). While this may sometimes be the case, we also find that constellations representing areas as diverse as traditional 1920/1930s houses and modern (1990s) residential developments have similar values of built-form packing. Likewise, the hospital complex has a built-form packing value similar to that of the 1980s' residential constellation immediately to the north of it. Built-form packing cannot, therefore, be used on its own as an indicator of urban land-use type.

Noting that Figure 10.9b is scaled linearly, relative to the maximum areal proportion of a constellation in this scene that is occupied by buildings (i.e., 38%, for the hospital complex), it is evident that most of the constellations have low built-form densities. This is an interesting observation in its own right and suggests a relatively low utilization of the available space by buildings in each of the constellations. Indeed, few of the constellations that are known to represent areas of residential land exhibit a built-form density greater than 20%. Moreover, there is little variability around this figure between different types and age of housing. Given the range of residential types present in the scene, this is quite remarkable and suggests that built-form density is also of limited value, on its own, in terms of distinguishing different categories of urban land use. Even where information on

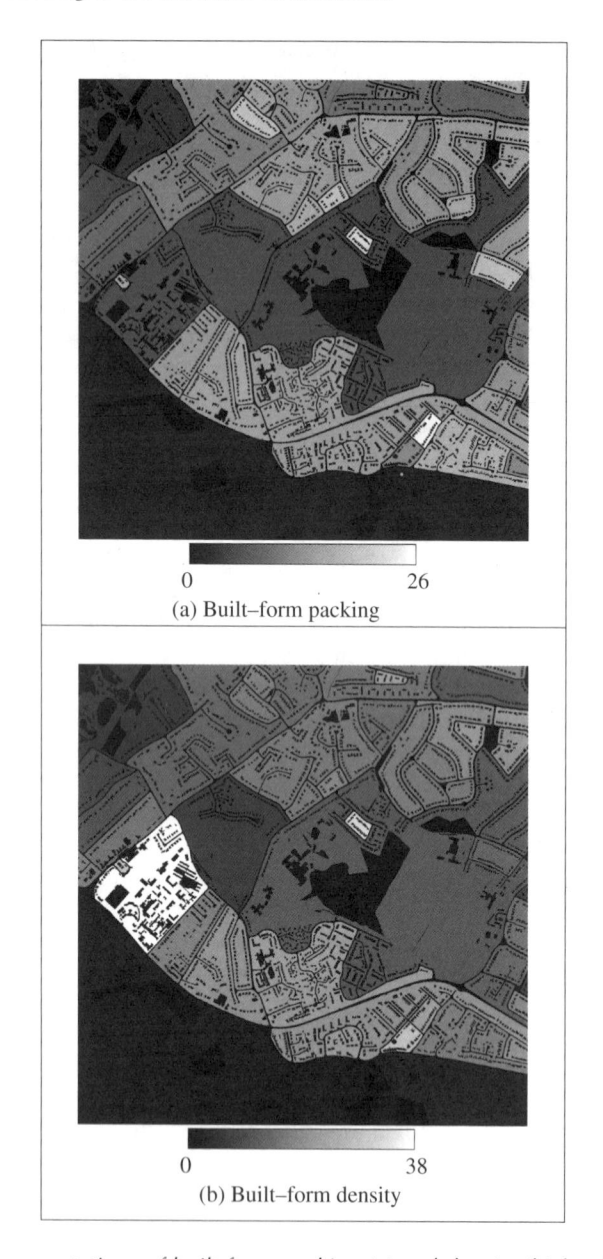

218 Spatial Modelling of the Terrestrial Environment

Figure 10.9 *Representations of built-form packing (a) and density (b) for the constellations identified from Figure 10.5. Built-form packing is computed from the number of B*ᴜɪʟᴛ *nodes (regions) per hectare, while built-form density is given by* $\frac{100 \times \Sigma \text{ built area}}{\text{constellation area}}$

the packing (Figure 10.9a) and density (Figure 10.9b) of built-form units is combined, it may only be possible to distinguish reasonably unambiguously the hospital complex from the other constellations within the scene. A finer typological stratification of residential type on the basis of these structural properties alone is likely to be highly problematic.

10.5.2 Built-Form Unit Area

Figures 10.10a and b show three clear patterns in the statistical distribution of the area of each of the built-form units located within the constellations identified in Figure 10.5. First, the built-form units in the constellation covering the hospital complex have a large mean area (608.3 m^2), while those in the remaining constellations (mostly residential districts) tend to have a much smaller mean area ($\approx$135.0 m^2). The spatially extensive constellation in the centre of the scene is an exception to this general rule, having a mean area of built-form units that lies somewhere between the two. We note, however, that this constellation contains several different land-use types (see Figure 10.5). As a result, it also has a relatively large SD in terms of the area of its constituent built-form units (Figure 10.10b). More generally, Figure 10.10b suggests that most of the constellations exhibit very limited variation in terms of the area of their constituent built-form units, although, once again, the hospital complex is the most notable exception to this general rule.

Figure 10.10c indicates that very few of the constellations in this scene exhibit either strong positive or strong negative spatial autocorrelation (i.e., spatial clustering or spatial dispersion, respectively) in terms of built-form unit area. Indeed, the values of Moran's I typically lie between 0.28 and -0.36, with an overall mean and SD of 0.03 and 0.05, respectively. Thus, the majority of constellations exhibit a random spatial distribution in terms of the areas of their constituent built-form units, with only 12 exhibiting significant spatial autocorrelation (i.e., spatial clustering; Figure 10.10d). This is a somewhat surprising result, since one might have expected to find similar-sized buildings located close to one another in, for example, a residential development of a given age. This does not, however, appear to be the case. There are two possible reasons for this: the first is that Moran's I is an unsuitable measure in this context, although we have no direct evidence either way to evaluate this; the second is that some of the constellations contain a mixture of two or more land-use categories evaluated in this study, which alters the statistics for the constellation as a whole. The latter is a more fundamental issue, since the implicit assumption is that each constellation represents a single, discrete land-use type. Referring back to Figure 10.5, there is some evidence to indicate that this may be the cause of the problem. Unfortunately, as a consequence, there does not appear to be a clear and unambiguous relationship between the spatial autocorrelation of built-form unit area within individual constellations and land use at the level of categorization employed in this study; although this does not imply that there is no relationship at another level of categorization (e.g. residential versus commercial/industrial).

10.5.3 Built-Form Unit Compactness

The compactness of built-form units within individual constellations is illustrated in Figure 10.11. These follow a similar patterns to those of built-form unit area (Figure 10.10). Thus, in constellations where the built-form units have a large mean area, they also exhibit a large mean value for compactness: in other words, the built-form units tend to have a more complex, irregular morphology. Once again, the hospital complex has both the highest mean and SD in terms of the compactness of its constituent built-form units, while the residential constellations generally exhibit much lower mean and SD values.

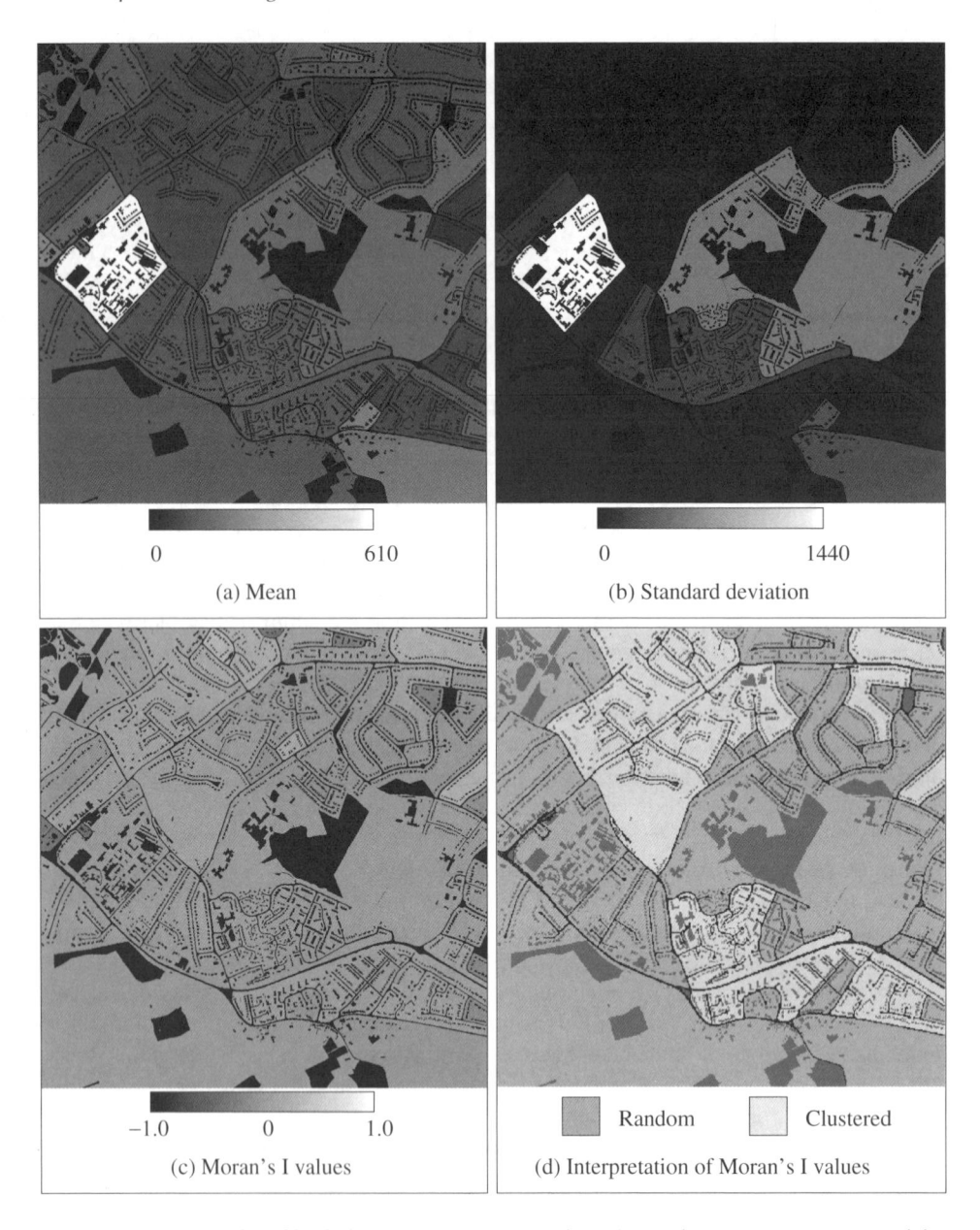

220 *Spatial Modelling of the Terrestrial Environment*

Figure 10.10 Results of built-form constellation analysis for Built area (a) mean area of the Built nodes in each constellation, (b) standard deviation of the area of Built nodes in each constellation, (c) Moran's I value for Built node area in each constellation, and (d) interpretation of Moran's I value for Built node area in each constellation

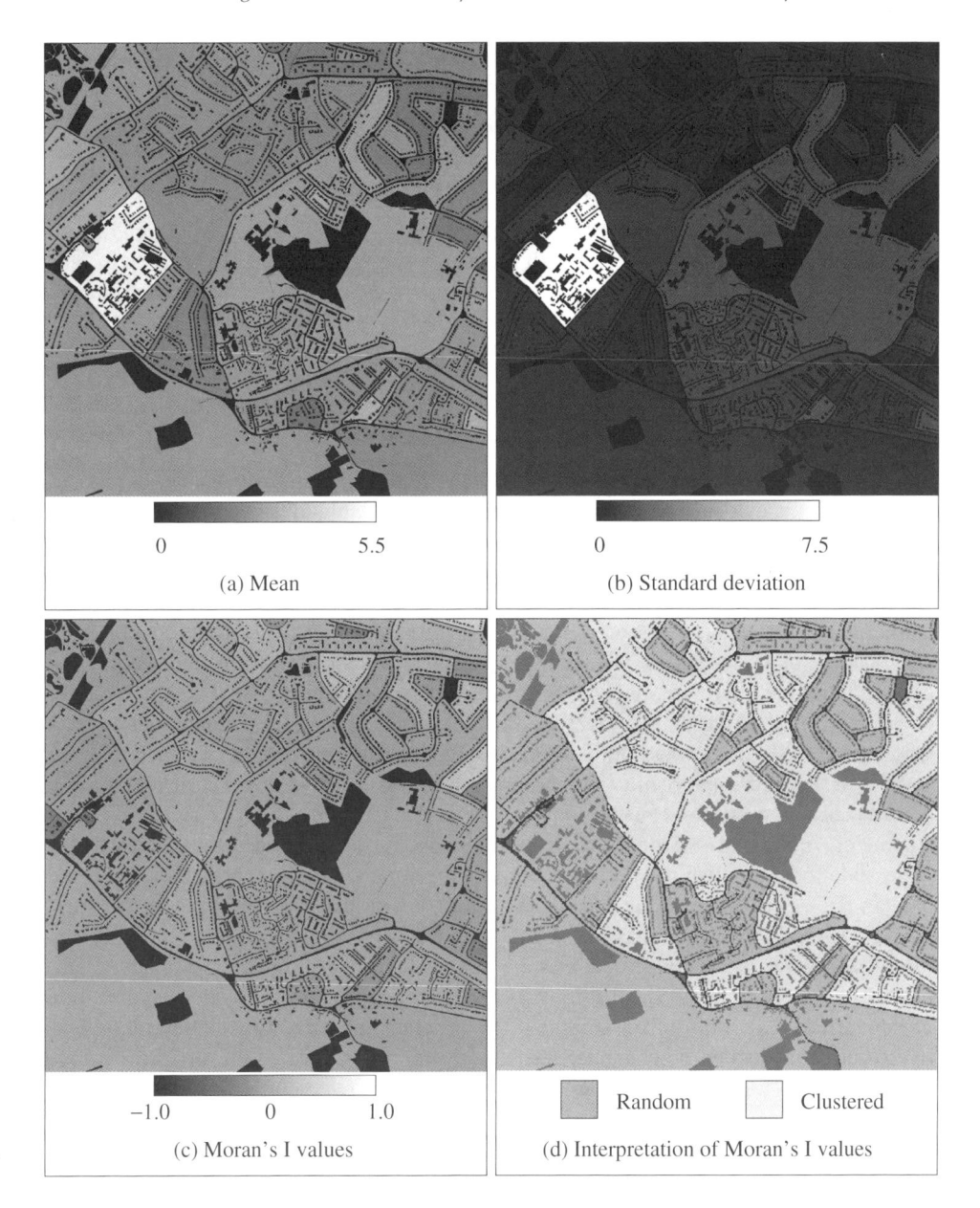

Figure 10.11 Results of built-form constellation analysis for Built compactness, (a) mean compactness of the Built nodes in each constellation, (b) standard deviation of the compactness of Built nodes in each constellation, (c) Moran's I values for Built node compactness in each constellation, and (d) interpretation of Moran's I values for Built node compactness in each constellation

Figure 10.11c shows the values of Moran's I for built-form compactness within each of the constellations. These display a similar overall spatial pattern to the corresponding values for built-form area (Figure 10.10c). The absolute range of values for Moran's I is also similar (-0.37 to 0.34), with an overall mean and SD of 0.03 and 0.05, respectively, such that most of the constellations exhibit a random spatial distribution of built-form compactness values (Figure 10.11d). Only 12 of the constellations were found to exhibit significant positive spatial autocorrelation (i.e., spatial clustering), of which eight also exhibit spatial clustering in terms of built-form area (Figure 10.10d). The reasons for this are thought to be the same as those outlined in the previous subsection, raised in the context of built-form unit area.

10.6 Conclusion

In terms of the original objective of this study, the results are largely negative. In particular, there is little evidence to suggest that a clear and unambiguous relationship exists between the morphological properties and spatial disposition of built-form units within urban constellations and the dominant land use that they represent. A number of reasons were suggested to explain why this might be the case. First, and most importantly, constellations defined and delineated in terms of closed loops of the road network frequently contain more than one type of urban land use. The statistics calculated for the built-form units within such constellations are, therefore, likely to be complex composites of their constituent land-use categories. In other words, they may be thought of as defining a mixed structural class whose properties overlap with those of the discrete land-use categories of which they are composed. Second, and arising from the first point, it is possible that a clear relationship nevertheless exists at some broader categorization of urban land use (e.g. residential versus commercial/industrial). Evidence to support this assertion is provided by the fact that it is possible to distinguish the hospital complex from the remaining constellations on the basis of built-form unit packing and density. Third, but probably least likely, the structural measures used in this study may not be the optimum ones to distinguish the types of land use present in the scene concerned. Despite this, the built-form connectivity analysis employed in this study has helped to provide a deeper insight into the spatial organization of built structures within urban areas and has provided a means by which the morphology of urban areas can be characterized at very fine spatial scales. In this context, one could envisage extending the spatial and structural measures employed in this investigation across the wider urban fabric to explore how they vary as function of constellation contiguity.

Acknowledgements

The authors would like to acknowledge the support of the U.K. Natural Environment Research Council through the provision of research grant number GR3/10186. The Ordnance Survey digital map data are reproduced with kind permission of the Controller of Her Majesty's Stationery Office, Crown Copyright.

References

Aplin, P., Atkinson, P.M. and Curran, P.J., 1997, Fine spatial resolution satellite sensors for the next decade, *International Journal of Remote Sensing*, **49**, 545–560.

Barnsley, M.J., 1999, Digital remotely-sensed data and their characteristics, in P.A. Longley, M.F. Goodchild, D.J. Maguire and D.W. Rhind (eds), *Geographical Information Systems: Principles and Technical Issues*, Vol. 1 (Chichester: John Wiley and Sons), 451–466.

Barnsley, M.J. and Barr, S.L., 1992, Developing kernel-based spatial re-classification techniques for improved land-use monitoring using high spatial resolution images, in *Proceedings of the XXIX Conference of the International Society for Photogrammetry and Remote Sensing, International Archives of Photogrammetry and Remote Sensing: Commission 7*, Washington, DC, 646–654.

Barnsley, M.J. and Barr, S.L., 1996, Inferring urban land use from satellite sensor image using kernel-based spatial reclassification, *Photogrammetric Engineering and Remote Sensing*, **62**, 949–958.

Barnsley, M.J. and Barr, S.L., 2001, Monitoring urban land use by Earth observation, *Surveys in Geophysics*, **21**, 269–289.

Barnsley, M.J., Moller-Jensen, L. and Barr, S.L., 2001, Inferring urban land use by spatial and structural pattern recognition, in J.P. Donnay, M.J. Barnsley and P.A. Longley (eds), *Remote Sensing and Urban Analysis* (London: Taylor and Francis), 115–144.

Barr, S.L. and Barnsley, M.J., 1995, A spatial modeling system to process, analyse and interpret multi-class thematic maps derived from satellite sensor images, in P. Fisher (ed.), *Innovations in GIS 2* (London: Taylor and Francis), 53–65.

Barr, S.L. and Barnsley, M.J., 1997, A region-based, graph-theoretic data model for the inference of second-order thematic information from remotely-sensed images, *International Journal of Geographical Information Science*, **11**, 555–576.

Barr, S.L. and Barnsley, M.J., 1999, A syntactic pattern-recognition paradigm for the derivation of second-order thematic information from remotely-sensed images, in P.M. Atkinson and N.J. Tate (eds), *Advances in Remote Sensing and GIS Analysis* (Chichester: John Wiley & Sons), 167–184.

Batty, M. and Longley, P.A., 1994, *Fractal Cities: A Geometry of Form and Function* (London: Academic Press).

Corbley, K.P., 1996, One-meter satellites: practical applications by spatial data users – part three, *Geographic Information Systems*, **6**, 39–43.

Crowther, D. and Echenique, M., 1972, Development of a model of urban spatial structure, in L. Martin and L. March (eds), *Urban Space and Structures* (Cambridge: Cambridge University Press), 175–218.

Donnay, J.P., Barnsley, M.J. and Longley, P.A., 2001, Remote sensing and urban analysis, in J.P. Donnay, M.J. Barnsley and P.A. Longley (eds), *Remote Sensing and Urban Analysis* (London: Taylor and Francis), 3–18.

Donnay, J.P. and Unwin, D., 2001, Modelling geographical distributions in urban areas, in J.P. Donnay, M.J. Barnsley and P.A. Longley (eds), *Remote Sensing and Urban Analysis* (London: Taylor and Francis), 205–224.

Ebdon, D., 1985, *Statistics in Geography* (Oxford: Blackwell).

Forster, B.C., 1985, An examination of some problems and solutions in monitoring urban areas from satellite platforms, *International Journal of Remote Sensing*, **6**, 139–151.

Fotheringham, A.S., Brunsdon, C. and Charlton, M., 2000, *Quantitative Geography: Perspectives on Spatial Data Analysis* (London: Sage).

Freeman, J., 1975, The modeling of spatial relations, *Computer Graphics and Image Processing*, **4**, 156–171.

Fritz, L.W., 1996, The era of commercial earth observation satellites, *Photogrammetric Engineering and Remote Sensing*, **62**, 39–45.

Fung, T. and Chan, K.C., 1994, Spatial composition of spectral classes: a structural approach for image analysis of heterogeneous land-use and land-cover, *Photogrammetric Engineering and Remote Sensing*, **60**, 173–180.

Gonzalez, R.C. and Wintz, P., 1987, *Digital Image Processing* (Wokingham: Addison-Wesley).
Gonzalez, R.C. and Woods, R.E., 1993, *Digital Image Processing* (Wokingham: Addison-Wesley).
Hillier, B. and Hanson, J., 1984, *The Social Logic of Space* (Cambridge: Cambridge University Press).
Johnsson, K., 1994, Segment-based land-use classification from spot satellite data, *Photogrammetric Engineering and Remote Sensing*, **60**, 47–53.
Kontoes, C., 1999, Image analysis techniques for urban land use classification: the use of kernel based approaches to process very high resolution satellite imagery, in I. Kanellopoulos, G.G. Wilkinson and T. Moons (eds), *Machine Vision and Advanced Image Processing in Remote Sensing* (London: Springer Verlag), 121–133.
Kontoes, C., Raptis, V., Lautner, M. and Oberstadler, R., 2000, The potential of kernel classification techniques for land use mapping in urban areas using 5m-spatial resolution IRS-1C imagery, *International Journal of Remote Sensing*, **21**, 3145–3151.
Kruger, M.J.T., 1979a, An approach to built-form connectivity at an urban scale: system description and its representation, *Environment and Planning B*, **6**, 67–88.
Kruger, M.J.T., 1979b, An approach to built-form connectivity at an urban scale: variations of connectivity and adjacency measures amongst zones and other related topics, *Environment and Planning B*, **6**, 305–320.
McDonald, R.A., 1995, Opening the cold war sky to the public: declassifying satellite reconnaissance imagery, *Photogrammetric Engineering and Remote Sensing*, **61**, 385–390.
McKeown, D.M., 1987, The role of artificial intelligence in the integration of remotely sensed data with geographic information systems, *IEEE Transactions on Geoscience and Remote Sensing*, **GE-25**, 330–348.
McKeown, D.M., Harvey, W.A. and Wixson, L., 1989, Automated knowledge acquisition for aerial image interpretation, *Computer Vision, Graphics and Image Processing*, **46**, 37–81.
Mehldau, G. and Schowengerdt, R., 1990, A C-extension for rule-based image classification systems, *Photogrammetric Engineering and Remote Sensing*, **56**, 887–892.
Moller-Jenson, L., 1990, Knowledge-based classification of an urban area using texture and context information in Landsat-TM imagery, *Photogrammetric Engineering and Remote Sensing*, **56**, 899–904.
Nagao, M. and Matsuyama, T., 1980, *A Structural Analysis of Complex Aerial Photographs* (London: Plenum Press).
Nicolin, B. and Gabler, R., 1987, A knowledge-based system for the analysis of aerial images, *IEEE Transactions on Geoscience and Remote Sensing*, **GE-25**, 317–329.
Piff, M., 1992, *Discrete Mathematics: An Introduction for Software Engineers* (Cambridge: Cambridge University Press).
Ridley, H., Atkinson, P.M., Aplin, P., Muller, J.-P. and Dowman, I., 1997, Evaluating the potential of forthcoming commercial U.S. high-resolution satellite sensor imagery at the Ordnance Survey, *Photogrammetric Engineering and Remote Sensing*, **63**, 997–1005.
Sadler, G.J., Barnsley, M.J. and Barr, S.L., 1991, Information extraction from remotely sensed images for urban land-use analysis, in *Proceedings of the Second European Conference on Geographical Information Systems (EGIS'91)*, Brussels, 955–964.
Sedgewick, R., 1990, *Algorithms in C* (New York: Addison-Wesley).
Steinnocher, K., 1996, Integration of spectral and spatial classification methods for building a land-use model of Austria, in *Proceedings of the XXXI Conference of the International Society for Photogrammetry and Remote Sensing (ISPRs'96), International Archives of Photogrammetry and Remote Sensing: Commission 7*, Vienna, 379–383.
Steinnocher, K. and Kressler, F., 1999, Application of spectral mixture analysis for monitoring urban development, in *Proceedings of the Earth Observation: From Data to Information, Proceedings of the 25th Annual Conference of the Remote Sensing Society*, Nottingham, 97–104.
Toll, D.L., 1985, Landsat-4 Thematic Mapper scene characteristics of a suburban and rural area. *Photogrammetric Engineering and Remote Sensing*, **51**, 1471–1482.
Treitz, P.M., Howarth, P.J. and Gong, P., 1992, Application of satellite and GIS technologies for land cover and land use mapping at the rural-urban fringe. *Photogrammetric Engineering and Remote Sensing*, **58**, 439–448.

Weber, C., 2001, Urban agglomeration delimitation using remote sensing data, in J.P. Donnay, M.J. Barnsley and P.A. Longley (eds), *Remote Sensing and Urban Analysis* (London: Taylor and Francis), 145–159.

Wharton, S.W., 1982a, A context-based land use classification algorithm for high resolution remotely sensed data, *Journal of Applied Photographic Engineering*, **8**, 46–50.

Wharton, S.W., 1982b, A contextual classification method for recognising land use patterns in high resolution remotely-sensed data, *Pattern Recognition*, **15**, 317–324.

Wilkinson, G.G., 1999, Recent developments in remote sensing technology and the importance of computer vision analysis techniques, in I. Kanellopoulos, G.G. Wilkinson and T. Moons (eds), *Machine Vision and Advanced Image Processing in Remote Sensing* (London: Springer), 5–11.

11

Modelling the Impact of Traffic Emissions on the Urban Environment: A New Approach Using Remotely Sensed Data

Bernard J. Devereux, L.S. Devereux and C. Lindsay

11.1 Introduction

For most of the last decade there has been significant concern over the impact of exhaust emissions from road traffic on the environment. Substantial increases in road transport (CEC, 1992) throughout the developed world have established this source of pollution as a major factor in promoting global climate change (IPCC, 1990). Furthermore, there is a growing body of evidence that constant exposure to vehicular emissions in the form of carbon monoxide (CO), nitrogen oxides (No$_x$), sulphur dioxide (SO$_2$), hydrocarbons (HC), particulates (PM$_x$), lead (Pb) and benzene (C$_6$H$_6$) has a significant harmful effect from a community health perspective (Schwartz, 1989; Collins *et al.*, 1995). DETR (1998) suggest that an increase in daily PM10 concentrations of 10 µg/m^3 would result in one additional increase to hospitals every other day for lung-related illnesses. As a consequence, significant energy is being devoted to tackling the problem of road traffic emissions.

European countries have implemented strategies for air quality monitoring within a framework established by the European Union with the objective of meeting clearly defined emissions standards for all the major pollutants. In the United Kingdom, Part IV of the Environment Act of 1995 (see DETR, 2000) made it the responsibility of local authorities to review air quality and to implement a local 'Air Quality Strategy' to ensure that the required standards are being met within the statutory time frames set down for each pollutant.

Spatial Modelling of the Terrestrial Environment. Edited by R. Kelly, N. Drake, S. Barr.
© 2004 John Wiley & Sons, Ltd. ISBN: 0-470-84348-9.

A National Atmospheric Emissions Inventory (NAIE) has been established along with a network of monitoring stations which provide a basis for national mapping (see Bush *et al.*, 2001) and monitoring. Whilst direct measurement of emissions is an expensive process, there has been a proliferation of models which take into account line and area sources of emissions and these provide a basis for local evaluation and testing (see Hassea, 2000).

As a result of this increasing awareness and understanding, the emissions problem is becoming less acute. However, despite the major progress which is being made with the implementation of standards and cleaner, more efficient forms of engine technology, road traffic continues to increase in volume and continued progress is needed if widely supported objectives of long-term sustainable development are going to be achieved (Mazmanian and Kraft, 1999). It is particularly important that sustainable planning strategies which tend to reduce the level of emissions can be identified. It is also critical that there is a clear understanding of the impact of emissions on both the population at large and the landscape ecology of the local environment.

This chapter presents a case study illustrating how a powerful methodology for testing sustainable planning policy can be created within the framework of a GIS (SHIRE 2000) containing a wide range of remotely sensed data linked to socioeconomic sources derived from the UK Census and ground-based surveys. The modelling system is used to examine planning strategy within the County of Cambridgeshire, UK, between 1991 and 2006 with a view to evaluating the impact of different planning strategies on exposure to road traffic emissions. The results demonstrate the value of LANDSAT TM satellite imagery and associated land cover information derived by automated classification techniques as a basis for evaluating and visualizing the impact of emissions on settlement patterns. Furthermore, they raise important questions about whether supposedly sustainable planning policy will always lead to sustainable outcomes.

11.2 Integrating Models within the Framework of the SHIRE 2000 GIS

In order to measure the impact of road traffic emissions on the population at large it is necessary to have accurate maps of long-run, annual concentration over wide areas and to be able to overlay and compare these maps with settlement and population distributions. Given the relatively low density of the national network of monitoring stations, the high cost of ground-based monitoring and the high levels of spatial variability that emissions exhibit, it is extremely difficult to produce and validate detailed concentration surfaces. Furthermore, population is also extremely variable in its spatial pattern and the highest level of resolution available from major sources such as the 10 yearly population census (Enumeration Districts) is too low to enable a reliable linkage between population and concentration to be established.

The approach adopted here to resolve these problems makes two key assumptions:

- First, detailed maps of settlement boundaries derived from an automated, maximum likelihood classification of widely available satellite imagery can provide a surrogate variable that provides a basis for an index of population exposure.
- Second, current methods of modelling emissions concentrations are sufficiently accurate to provide a reliable indication of spatial patterns in concentration. In particular, the reliance of such models on estimates of traffic speed, volume and mode by road link generated by transport modelling systems leads to robust results.

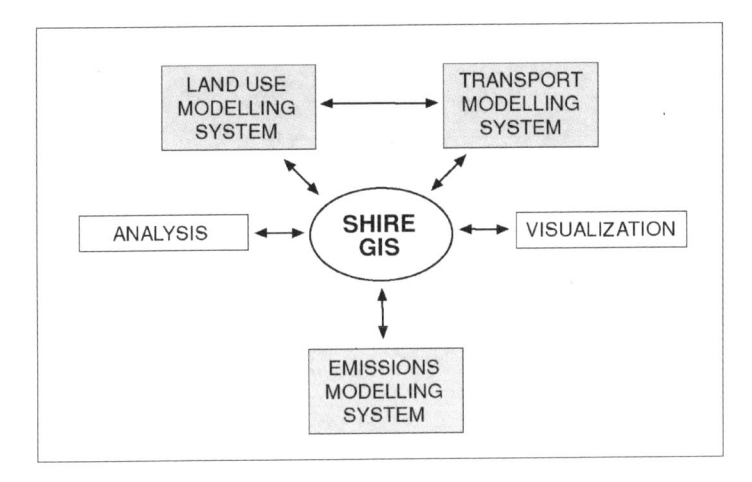

Figure 11.1 The integration of project modelling components using the SHIRE 2000 GIS

Given these assumptions a framework for modelling emissions impact has been created and is shown in Figure 11.1. The framework combines three distinct models dealing with land use, transport and emissions.

The GIS acts as a powerful integrating tool for maintaining and managing all of the model input and output data. It also provides a basis for analysis and visualization of results. Table 11.1 shows that there are three broad categories of primary information maintained within SHIRE 2000 in the form of image products, cartography/census geography and socio-economic data. There is also a substantial volume of ancillary datasets and intermediate, analytical results that would be almost impossible to organize effectively outside a reliable database management system.

The land use and transport models provide an effective basis for predicting the movement of vehicles by mode on the County transportation network. They function as a closely coupled module that is capable of representing the feedback relationships between changes in land use and changes in the transportation network through time. They have been developed over a period of some 25 years (see Echenique, 1994) and have a proven track record in large-scale modelling of complex, urban systems. In addition to modelling of the actual situation or 'reference case' that is a product of past planning policy, they also enable prediction of the outcomes for a range of 'policy cases' which are the results of alternative, proposed policy packages.

The emissions model was developed within the framework of the ArcView GIS using the methodology set out in DMRB (1999). For the purposes of this study it has the benefits of being widely recognized as a robust screening model with relatively modest data input requirements. It takes predictions of vehicle numbers by mode and average speed for each transport network link from the land use/transport modules and it converts these into maps of emissions concentration on a 25-m raster grid covering the county-wide study area.

Having provided a broad overview of the infrastructure created for emissions impact modelling, a more detailed description of the land use, transport and emissions models will now be provided.

Table 11.1 Major components of the SHIRE 2000 database

Data type	Coverage	Resolution (metres)
Imagery		
LANDSAT TM (1989; 1999; 1999)	County	30
Spot panchromatic mosaic	County	10
ERS1 RADAR (2 PRI; 2 COMPLEX)	County	12.5
Interferometric DEM	County	25
Land cover map of Great Britain	County	30
LC 2000 CLASSIFICATION	County	30
Air photo mosaic	Cambridge	2
Air photo-derived DEM	Cambridge	2
LIDAR DEM	West Cambridge	2

Data type	Coverage
Cartography	
Ward boundaries	Cambridgeshire
Enumeration district boundaries	Cambridgeshire
Mentor land use model boundaries	Cambridgeshire
Meplan transport model boundaries	Cambridgeshire
Meplan multi-modal transport network	Cambridgeshire
Bartholomews vector map data	Cambridgeshire
OS 1:2500 Digital map data	West Cambridge

Data source	Areal units	DATE
Socioeconomic data		
Census	Wards and EDs	1991
Population	Wards and EDs	1991
Dwellings	WARDS and ED's	1991
Mid-term census estimates	Wards	1999
DETR floorspace stats	Districts	1983 onwards
Traffic flows	Links	Various

11.3 The Integrated Land Use and Transport Modelling Framework

The integrated land use and transport modelling framework consists of a large suite of modules designed to represent various aspects of the land use and transport systems. It also enables representation of the feedback effects between the two systems (Echenique, 1994). The brief description given here relies heavily on the account given by Williams (1994).

The land use system is grounded in the interregional, input–output framework established by Leontif (1951) and described by a number of workers including Isard (1960). It provides a mechanism for linking demand Y_i^m, for goods and services (production factors m) in each zone i of a regional economy to production, X, via a table of production coefficients, a_i^{mn}. The model defines all the sectoral and regional linkages within and beyond the local economy. It also enables representation of the economic multiplier effect arising from changes in production caused by changes in demand:

$$Y_i^m = Y_i^{m0} + \sum_n \left[a_i^{mn} \left(X^{n0} + X_i^n \right) \right] \tag{1}$$

where for each zone i:

Y_i^m = total demand for factor m
Y_i^{m0} = final demand for factor m
a_i^{mn} = technical coefficient defining the units of m required to produce one unit of n
X^{n0} = exogenous production of factor n
X_i^n = endogenous production of factor n.

Implicit within the table of technical coefficients a_i^{mn} is the demand for movement of goods and labour between zones of the economic system. The modelling framework enables this demand to be converted into a spatial pattern of trade using the gravity model formulation defined in equation (2). Terms allowing for the effects of economic concentration and a calibrated coefficient ensuring realistic journey lengths extend the basic interaction model:

$$T_{ij} = Y_j \frac{S_i \exp[-\lambda(c_i + d_{ij} - w_i)]}{\sum_i S_i \exp[-\lambda(c_i + d_{ij} - w_i)]} \tag{2}$$

where for each factor m:

T_{ij} = trade between origin zone i and destination zone j
Y_j = total demand in zone j
S_i = size of zone i
c_i = cost of production in zone i
d_{ij} = cost of transport between zone i and j
w_i = zonal specialization/concentration factor
λ = calibrated parameter controlling factor journey lengths.

Additional elements of the modelling framework enable the calculation of household demand for each zone using a utility maximizing model and the demand for labour, floorspace and other variable cost factors on the production side. These demand elements within the system can be compared with the supply elements defined by equation (2) to give a rent for each zone and the modelling system then uses the rent values to find an equilibrium situation in which zonal supply equals demand. The search for an equilibrium solution is based on an iterative procedure described by Williams (1979).

The pattern of trade between zones provides a basis for driving the transport model side of the system. A node/link-based representation of the transportation network is used to represent each of the main transport modes. Each link is characterized by a range of attributes including origin and destination, length, cost of use, travel time/speed, traffic volume and capacity. Again, the transport model has an iterative structure, which begins with the computation of shortest routes through the network between all origin and destination zones. A modal split procedure is used to allocate flows of goods and people from the land use model to each of the travel types available. A discrete choice, log linear regression framework is then used to allocate flows to modes followed by an assignment procedure which converts the flows to vehicles and allocates them to routes based on the shortest paths. This assignment defines an initial demand for use of the transport network which can be compared with a measure of supply based on travel time, cost and congestion. A further iterative procedure is used to adjust the pattern of flows on the transport network until an equilibrium pattern is reached in which utility is maximized.

As the model iterates through time, feedback effects between the land use and transport elements are incorporated. Land use adjustments take place according to the pattern of rents and transport adjustments are controlled by journey costs and accessibility. The explicit treatment of land use and transport interaction is a particularly powerful feature of the modelling framework.

Given this structure the land use and transport model offers two important features for monitoring the environmental impact of traffic emissions:

- First, it can generate estimates of traffic flow by vehicle type and speed for each link in the regional transportation network. These variables represent primary data for calculating emissions.
- Second, it offers an immensely flexible and powerful tool for predicting the effects of planning policy packages. Changes in transport policy such as new network links or improvements to capacity can be evaluated by modifying the network structure and attributes in the model. Policy changes affecting land use (e.g. permission for new settlements, industrial installations or controls on growth) can similarly be evaluated by adjusting the structure of the input-output matrix.

The addition of an emissions impact framework to the land use and transport model thus provides a basis for evaluating the air quality outcomes of different planning policy packages.

11.4 The Emissions Model

The increasing importance of meeting air quality standards has led to a rapid growth in the number of models available for estimating emissions concentration and dispersion. Around 20 different modelling systems are identified by DETR (2000), giving rise to the need for central government guidance on the basic principles, properties and performance of these models. The important differentiating features appear to be level of detail, extent of data, computational demands, complexity and, inevitably, cost. Validation of model results is a major issue and as yet, there seems to be no guarantee that more detailed, complex models will give better results in all situations. The conventional wisdom for air quality modelling is thus to use a multi-stage approach in which low-cost, relatively simple models are used for screening. Where application of these models reveals potential air quality problems, more detailed, sophisticated models are used to make accurate assessments.

Given the policy-based nature of the work conducted here it was felt appropriate to employ a modified version of the emissions screening methodology currently published in the Design Manual for Roads and Bridges (DMRB, 1999). This approach was developed by the UK Transport and Road Research Laboratory and has been widely used since the early 1980s for assessment of vehicle emissions in relation to air quality standards. It has the major advantages of being straightforward to implement, computationally efficient and being robust in a wide range of situations. Furthermore, the underlying assumptions that limit the performance of the model have been clearly evaluated by Hickman *et al.* (2000).

The DMRB methodology enables evaluation of the emissions impact for specific road alteration/upgrade schemes and makes provision for calculation of concentration estimates at selected receptor locations in the vicinity of the scheme which might be important in terms of their environmental sensitivity or population impact. It also provides a basis for estimating the net contribution of a transport infrastructure scheme to regional and global totals. It involves:

- Separation of peak hour traffic flows into light duty and heavy-duty components for the relevant link(s) in the transport network.
- Calculating a 'relative emission rate' for each type of traffic which takes into account the national composition of the vehicle fleet and reflects changes in engine and fuel technology which are tending to reduce emissions levels through time. The basis for this calculation is a graphical relationship between relative emission rate and year, which is in turn based on a broad analysis of average fleet composition, vehicle type and age.
- Calculating a speed correction factor to account for the variation of emissions relative to the average speed on the link. This is based on an empirical understanding of average vehicle performance at different speeds.
- Calculating an emissions concentration for the receptor point based on the relative emissions rate, the speed correction factor for each vehicle type and the distance of the receptor location from the source. This calculation uses a distance decay relationship specific to each pollutant derived from a Gaussian dispersion model. It assumes a constant wind speed of 2 m/s with wind directions being evenly distributed around the points of the compass.
- Conversion of the peak period emissions values to annual totals based on an empirical relationship.

The basic model described above is designed to be used for a small number of links and receptors. If necessary, it enables emissions estimates to be made using 'pencil and paper' methods in conjunction with readings taken from graphs of the underlying relationships employed.

In this study wide area estimates of emissions concentrations for a regional transport network were required. The procedure was thus modified to treat every cell in a 25-m resolution grid positioned over the Cambridgeshire study area as a receptor point. Taking CO as an example, regression relationships were established for the light (equation (3)) and heavy duty (equation (4)) speed correction factors:

$$LC_1 = 14.6 - 0.4718S + 5.29E - 03S^2 - 1.9E - 05S^3 + \varepsilon$$
$$R^2 = 0.98 \qquad (3)$$
$$HC_1 = 4.3 - 0.102S + 7.47E - 04S^2 - 6.33E - 07S^3 + \varepsilon$$
$$R^2 = 0.99 \qquad (4)$$

Where LC_1 is the light duty vehicle correction factor, HC_1 is the heavy duty vehicle correction factor and S is the average speed on the link.

These were then used in conjunction with the appropriate relative emission rate (E) to compute emissions totals for each link in the transport network. Finally, the relationship between distance (D) and concentration (C) was modelled (equation (5)) and used to estimate concentrations for each cell in the grid on the basis of the distance to the closest point in the transport network:

$$C = 0.551 - 1.14E - 2D + 8.15E - 5D^2 - 1.93E - 7D^3 + \varepsilon$$
$$R^2 = 0.93 \tag{5}$$

As the analysis was concerned with just road traffic emissions, no attempt was made to deal with background emissions levels which are normally treated as a constant at this stage in an air quality analysis. All analyses and calculations were carried out within the ArcView GIS version 3.2, which allowed for display and analysis of results.

11.5 Modelling Emissions Impact in Cambridgeshire

Evaluation of the relationship between planning policy decisions and emissions outcomes for Cambridgeshire involved implementing and running the land use/transport model for both reference and policy cases. The *reference case* aimed to reproduce the actual behaviour of land use and transport during the study period and provided a basis for model validation. The *policy case* then enabled a package of alternative 'what if' policies to be tested. The transportation patterns associated with each case were used for mapping emissions concentration, which was in turn overlaid onto the pattern of settlement to provide an index of impact. Each stage in this process will now be described in turn.

11.5.1 Building the Cambridgeshire Model

Work on the construction of a model for Cambridgeshire was carried out between 1998 and 2000. The study area, key towns and road links are shown in Figure 11.2, whilst an overview of the model structure is provided in Figure 11.3. The work depended heavily on the extension of an existing model of the City of Cambridge to cover the 108 wards within the Districts of Cambridge, South Cambridgeshire, East Cambridgeshire and Huntingdonshire (see Figure 11.3). For the purposes of analysis, wards were aggregated into 67 internal zones and further 10 zones were used to represent the external study area including the surrounding administrative districts, London and the rest of the country. The model was calibrated largely with 1991 Census and other survey data and operated through time in 5-year steps from a starting forecast year of 1996 through to 2006.

Calibration data sources included:

- employment by ward from 1991 census Special Work Place Statistics;
- households and employed residents by ward from 1991 Census Local Base Statistics;
- commercial floorspace by District, interpolated from Department of the Environment time series statistics.

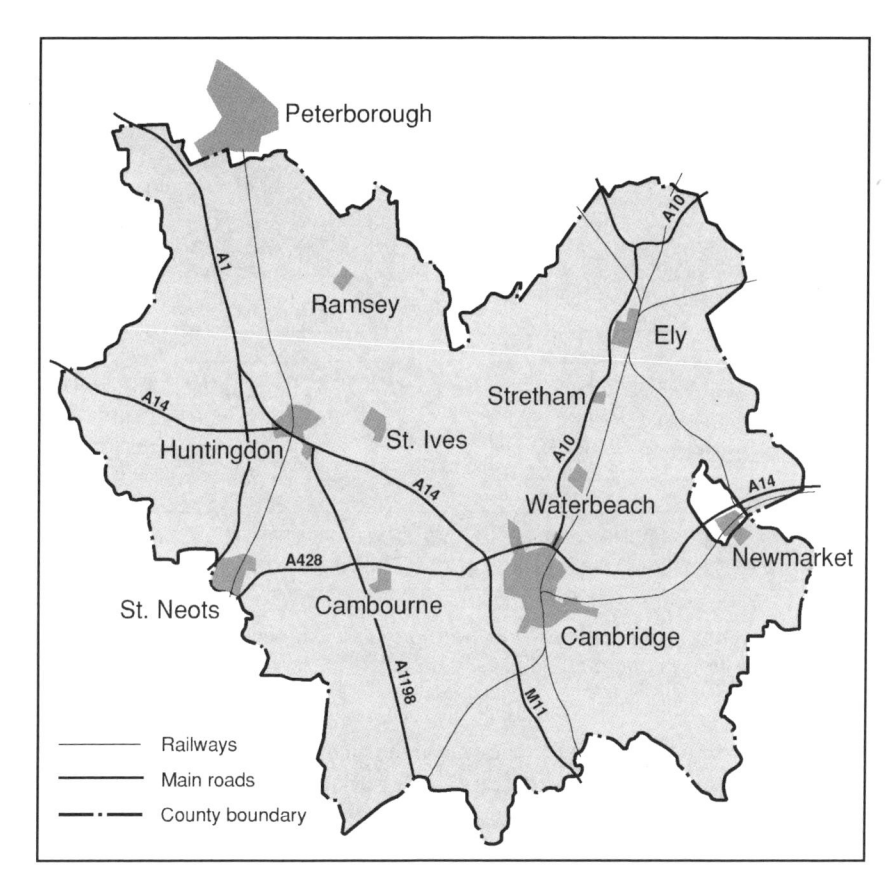

Figure 11.2 The Cambridgeshire study area showing key transportation links and settlements referred to in the policy testing and analysis

Data for the model-starting year included:

- change in dwellings by ward from 1991;
- change in jobs from 1991 by employment sector;
- change in non-residential floorspace from 1991;
- change in non-employed households (retired and inactive) from 1991.

The land use model was used to produce future year forecasts for a large number of socioeconomic variables including population by socioeconomic group, car ownership, employment by economic sector, land use and property prices. The forecast pattern of activities from the allocation model (equation (2)) was then used as the basis for estimating inter-zonal travel patterns and the pattern of trip origins and destinations.

The Cambridgeshire transport model was calibrated with local traffic survey data for the peak hour and base transport network data. The four administrative districts within the study area were divided into 176 transport zones and a further 10 zones were defined

LAND USE MODEL ⇄ FEEDBACK ⇄ TRANSPORT MODEL

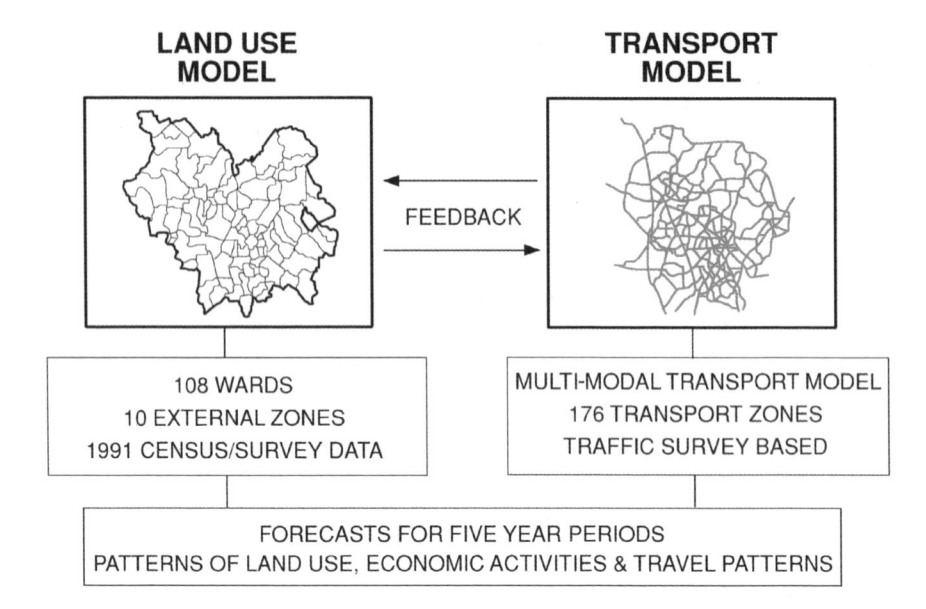

| 108 WARDS
10 EXTERNAL ZONES
1991 CENSUS/SURVEY DATA | MULTI-MODAL TRANSPORT MODEL
176 TRANSPORT ZONES
TRAFFIC SURVEY BASED |

FORECASTS FOR FIVE YEAR PERIODS
PATTERNS OF LAND USE, ECONOMIC ACTIVITIES & TRAVEL PATTERNS

Figure 11.3 Structure of the land use and transport model showing key spatial components, feedback relationships and outputs

to represent the external area. Each zone was given a centroid location, with at least one access link to the actual road network. Five modes of transport were modelled (car, bus, rail, walk and cycle). Ten link types were used to represent different types of road and rail infrastructure and each was given distance, cost, time and capacity characteristics (motorway, dual carriageway, single carriageway (rural), non-urban minor, urban minor, access, single carriageway (urban), rail, bus only, rail/road access).

The model was validated using screen line traffic count data provided by Cambridgeshire County Council supported by other locally available, observed data sources where possible. In addition, the transport model's modal split results were compared to published National Travel Survey data for journeys by mode and purpose. Details of the validation can be found in Devereux *et al.* (2001), who demonstrate that all of the validation tests generate results within acceptable limits.

11.5.2 Formulating and Testing a Sustainable Policy Package

During the last three decades Cambridgeshire has become one of the fastest growing areas in the United Kingdom and inevitably, rapid growth has brought with it a range of planning problems. The growth of high technology industry in the region has resulted in an economic boom and an inflow of workers in the high socioeconomic groups. High levels of income coupled with a relatively limited supply of housing have led to rapid increases in house prices and substantial pressure for the construction of new housing. As a consequence of constraints on development in the historic centre of Cambridge, a policy of allocating settlement in the villages of the county has been used to try and alleviate these problems. The outcome has been a large increase in the levels of commuting and pressure on the local road network. This, coupled with major growth in traffic using the strategic trunk route network, has resulted in high levels of congestion on the region's roads. The A14 Corridor, despite recent major upgrades, is now one of the most congested roads in the country.

Consequently, the county faces a number of environmental problems that need a solution. First, the section of the A14 Corridor between Huntingdon and Cambridge (see Figure 11.2) experiences particularly high levels of traffic and congestion and as a consequence is a significant source of emissions. Commuting from the villages in the corridor to Cambridge has similarly brought about a growth of emissions from the adjacent minor roads. The enormous growth of commuting to Cambridge and to a lesser extent, Huntingdon, has also led to peak-hour congestion in both of these towns and inevitably this has brought about an associated increase in traffic emissions. Despite this general pattern of growth, the last County structure plan also had to deal with problems of rural stagnation and contained policies to try and promote economic development in the rural, fenland areas of the County between Cambridge and Ely.

For the purposes of the modelling exercise, a reference case was thus designed to represent the actual situation observed in the area over the last decade. Model inputs were based on local authority development assumptions, which aimed to preserve the Cambridge Green Belt and allocate most land for housing in the market towns and larger villages. Some housing development was permitted within Cambridge and a new village at Cambourne was allowed some 15 km to the west of the City. Industrial development was confined to existing sites and office development was concentrated in Cambridge, the peripheral science parks and the main market towns. Retail development was mainly focused on existing town centre sites. The number of jobs by employment sector and the total number of households were constrained to Cambridgeshire County Council figures. No large-scale transport improvements were modelled because there had been few significant changes during the study period.

The alternative 'policy case' scenario was designed with the benefit of hindsight and aimed to examine what might have happened if a very different planning approach based on sustainable, albeit draconian policy decisions had been taken in the late 1980s and early 1990s. In general, the overall amount of development was kept to the same levels as in the reference case scenario for comparability. However, the aim of the alternative scenario was to focus most new development along the main rail corridor which runs from the north east of the study area, through Littleport, Ely, Cambridge and on to London adjacent to the A10 trunk route (see Figure 11.2). A new town was developed adjacent to the railway at Waterbeach with both housing and employment uses. In addition, existing business and residential sites throughout the corridor were expanded. Accompanying transport investment was assumed to be focused on improved rail and feeder bus facilities within the A10 corridor. Specific transport improvements included new railway stations on the existing line at Stretham and at the Science Park, near the A14 North of Cambridge. Access and waiting times at all stations in the study area were improved to reflect more frequent services and some rail journey times were also improved for villages close to Cambridge along the A10 to both the north and the south.

It is important to emphasize that the policy case package was not one that could have been achieved in practice for a number of reasons. First, it ignored the inertia of previous policy and effectively began with a 'clean sheet'. Second, it is doubtful whether the package

238   Spatial Modelling of the Terrestrial Environment

tested would have been politically[1] acceptable during the time periods modelled. Finally, it is also doubtful whether it would have been achievable in practice, given economic and social trends within the County. Nevertheless, had the policies tested been achievable, they would all have met commonly accepted, sustainability standards of today.

11.5.3   Emissions Impact of the Policy Scenarios

Emissions were calculated for both the reference and policy scenarios using the methodology described in previous sections. As anticipated, high levels of traffic emissions were found in the A14 corridor to the west of Cambridge, in Cambridge itself and in the vicinity of Huntingdon. In a number of 'hot spots' levels approached or exceeded current air quality standards. Plate 10 shows the difference in CO emissions between the scenarios superimposed on a LANDSAT image background. Areas shaded in dark blue represent a large decrease in emissions levels whilst those in pale blue represent a small decrease. Similarly, pale pink represents a small increase whilst dark pink represents a major increase. Areas of settlement are picked out in yellow.

A number of patterns are immediately apparent. Firstly, despite the very strong policy measures, the A14 immediately to the west of Cambridge still shows an increase in emissions levels. However, villages in the A14 Corridor show clear improvements and substantial improvements are evident on the Huntingdon ring road, in the suburbs to the south west of Cambridge and on many of the routes into the town. Conversely, substantial increases are evident in the A10 corridor immediately to the North of Cambridge and around the rural settlements between Waterbeach and Ely. The main feature of the pattern of change is that there has been a clear reduction of emissions in the expanded dormitory villages to the west of Cambridge and a sharp increase in the rural areas to the north. Overall, the net level of emissions is lower.

Given this pattern of change, an attempt was next made to evaluate the impact of the policy packages on settlement exposure. Plate 11 shows how the SHIRE 2000 GIS can be used for visualization of the relationship between traffic emissions concentration and the pattern of settlement. Building outlines have been taken from Ordnance Survey Landline data and used as a basis for creating 3D models of built form by attaching height attributes to them from airborne LiDAR imagery. The buildings data are superimposed on an ortho-photo image of Cambridge viewed from the South West. Emissions concentrations are depicted as an opaque cloud overlaid on the photography. Areas of low concentration are transparent with no colour and as the concentration level increases, the overlay becomes increasingly red and opaque. Visualization of the data in this way is extremely useful for investigation of concentration 'hot spots' and assessment of their local impact on buildings and population. However, a more quantitative approach is needed for strategic impact assessment.

As growth in both the reference and policy cases had been largely restricted to existing development envelopes, the most straightforward way of achieving this was to compute an index based on concentration by pixel and settlement area. The index calculated the proportion of settlement pixels in each ward with non-zero emissions values weighted

---

[1] Although Cambridgeshire County Council played an important role in the work described here, no suggestion is made or inferred that the policies tested relate in any way to their actual policy objectives, views or intentions.

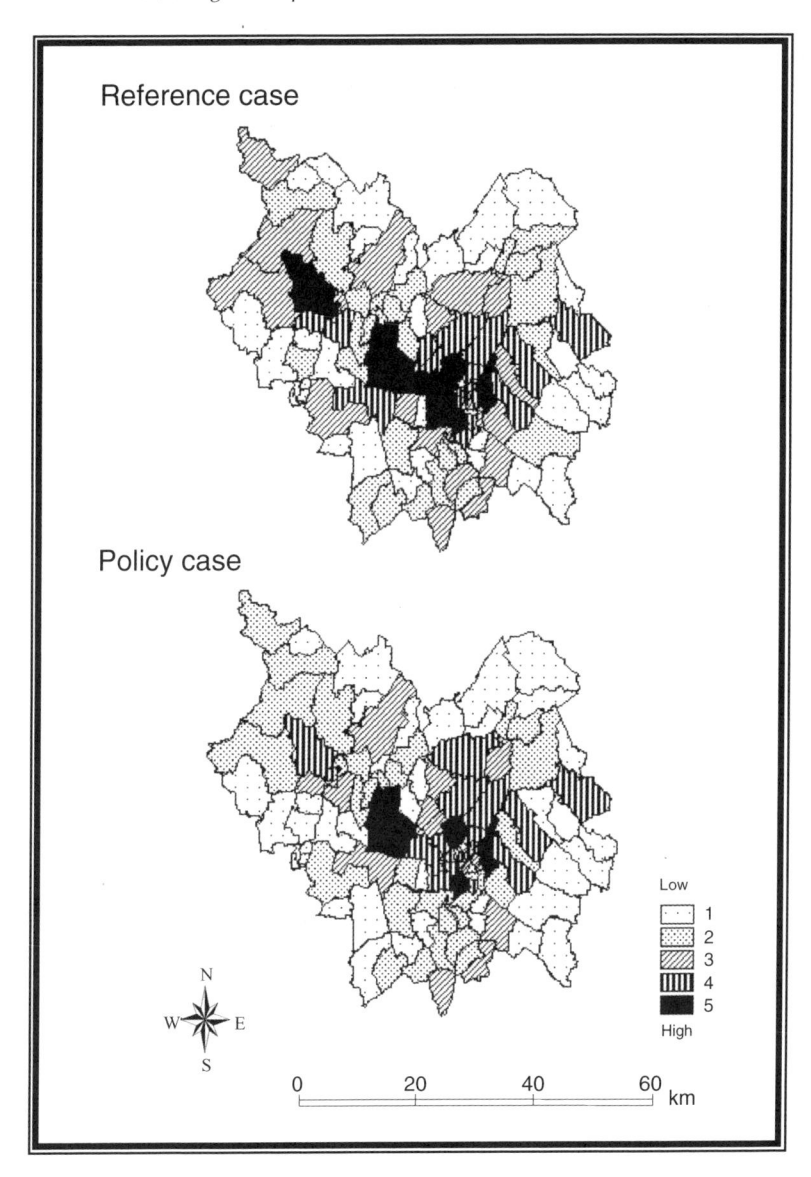

Figure 11.4 Index of emissions' impact on settlement for the reference case (top) and the policy case (bottom)

by the concentration. Resulting values were grouped into five classes ranging from low (class 1) to high (class 5). Whilst a very large number of possible indices might suggest themselves for this purpose, it was felt that a simple but robust measure was in keeping with the underlying emissions model which was designed for screening rather than detailed, accurate analysis.

Figure 11.4 shows the index by ward for both the reference case and the policy case. In the reference case, high levels of exposure are evident along the A14 corridor to the west of Cambridge and in the vicinity of Huntingdon. Wards with low values along the corridor reflect relatively free-flowing stretches of the A14 with low-settlement density. The dormitory villages west of Cambridge also have fairly high levels, as do the immediate suburbs of the City in the West and East. The policy case reveals that the alternative policy package has achieved noticeable reductions in the A14 corridor, especially in the vicinity of Huntingdon and in suburban West Cambridge. Values also tend to drop in the city centre. Significantly, however, there is a very clear increase of exposure in the A10 corridor in the north of the county, especially in the vicinity of Ely. This is despite the low density, rural population pattern in this area and the wide expanses of open countryside.

11.6 Conclusion

To date, remotely sensed data from satellites has seen remarkably little take-up by the planning community despite the fact that the complexities of data processing associated with its use have largely been removed and the costs of data purchase are now extremely low. Land use and transport models have become well established and their use as a basis for policy testing is proved. Emissions modelling exercises, however, are usually carried out by different sectors of the community with little reference to wide area environmental impacts of underlying policy. This research has demonstrated how GIS can provide a powerful basis for integrating these disparate areas of activity, thereby providing a powerful tool for modelling the environmental impacts of different planning policy packages. The Cambridgeshire case study has illustrated the power of the approach in the context of emissions impacts on settlement, but the same techniques could be used in a wide variety of planning and environmental assessment situations.

The results of the case study raise a number of interesting questions. In reality, the policies tested are so extreme that it would be almost impossible to implement them within a single plan review period. Despite this, the models suggest that they only achieve a partial solution to the problems of traffic, congestion and emissions in the study area. This suggests that development has a high level of momentum, which makes effective implementation of policy change extremely difficult over relatively short time horizons. Furthermore, although the policies tested have strong sustainability credentials in terms of encouraging reductions in car usage in favour of public transport, the predicted outcomes are not so clearly sustainable. The index of emissions shows a shift of impact away from areas that are heavily populated to more rural areas where development is much less dense. Outcomes measured in terms of the impact on public health might well be very positive but different impact measures based on environmental quality would clearly show a substantial, net deterioration. Equally, it seems clear that from an ecological point of view, the same policy may prove sustainable in the context of one species and quite the opposite in the context of another.

Perhaps the ultimate conclusion of this study is that integrated GIS that bring together complex models go far beyond offering approaches to environmental analysis. They can provide the basis for testing and learning about the likely outcomes of policy measures before they are implemented and before their undesirable effects become too difficult to reverse. Unless it can be clearly demonstrated that planning policies really do lead to sustainable outcomes, the whole concept of sustainability will not only lose credibility but will also turn out to be counterproductive in terms of its own objectives.

11.7 Acknowledgements

The research described in this chapter was conducted as part of 'Resolution E—Remote Sensing of Land Use and Transport Interaction for the Environment' and the authors would like to express their gratitude to the British National Space Centre for their financial support. Thanks are also made to Cambridgeshire County Council for their assistance in Resolution E and to the many individuals who contributed to the project. The input of Robin Fuller from the Centre for Ecology and Hydrology is particularly acknowledged together with the support of Jackie Hinton from BNSC and Professor Mike Barnsley from the University of Swansea. The Environment Agency provided access to the airborne LiDAR data and their assistance is gratefully acknowledged.

References

Bush, T., Stedman, J. and Murrells, T., 2001, *Projecting and Mapping Carbon Monoxide Concentrations in Support of the Air Quality Strategy Review*, DEFRA, The Scottish Executive, The National Assembly for Wales and The Department of the Environment in Northern Ireland.

CEC, 1992, *Towards Sustainability: A Programme of Action on the European Environment* (Brussels: Commission of the European Communities).

Collins, S., Smallbone, K. and Briggs, D., 1995, A GIS approach to modelling small area variation in air pollution within a complex urban environment, in P. Fisher (ed.), *Innovation in GIS 2: Selected Papers from the Second National Conference on GIS Research UK* (London:). Taylor and Francis.

Devereux, B.J., Devereux, L.S., Fuller, R. and Amable, G., 2001, *Resolution E*, Final BNSC Project Report, University Cambridge Unit for Landscape Modelling.

DETR, 1998, *The Effects of Carbon Monoxide on Human Health*, Expert Panel on Air Quality Standards, Department of Environment Transport and the Regions.

DETR, 2000a, *Review and Assessment: Monitoring Air Quality*, LAQM. TG1(00), Scottish Executive, National Assembly for Wales and Department of the Environment, Transport and the Regions.

DETR, 2000b, *Review and Assessment: Selection and Use of Dispersion Models,* LAQM.TG3(00), Scottish Executive, National Assembly for Wales and Department of the Environment, Transport and the Regions.

DMRB, 1999, *Design Manual for Roads and Bridges*, DETR, The Scottish Office and The Welsh Office, V11.

Echenique, M., 1994, Urban and regional studies at the Martin Centre: its origins, its present, its future, *Environment and Planning B*, **21**, 517–533.

Hassea, R., 2000, A GIS approach to modelling vehicle emissions, unpublished Master of Philosophy dissertation, University Cambridge, Unit for Landscape Modelling.

Hickman, A., McCrae, I. and Price, S., 2000, Air pollution impact assessment of roads – a revised UK approach, *TRL Journal of Research*, **3**, 2, 31–37.

IPCC, 1990, *Climate Change: The IPCC Scientific Assessment*, WMO/UNEP Intergovernmental Panel on Climate Change (Cambridge: Cambridge University Press).

Isard, W., 1960, *Methods of Regional Analysis* (Cambridge, MA: MIT Press).

Leontif, W., 1951, Input-output economics, *Scientific American*, **185**, 15–21.

Mazmanian, D. and Kraft, M., 1999, *Towards Sustainable Communities: Transitions and Transformations in Environmental Policy* (Cambridge. MA: MIT Press).

Schwartz, J., 1989, Lung function and chronic exposure to air pollution: a cross-sectional analysis of NHANES II, *Environmental Research*, **50**, 309–21.

Williams, I., 1979, An approach to solving spatial allocation models with constraints, *Environment and Planning A*, **11**, 3–22.

Williams, I., 1994, A model of London and the South East, *Environment and Planning B*, **21**, 535–553.

PART IV
CURRENT CHALLENGES AND FUTURE DIRECTIONS

12

Land, Water and Energy Data Assimilation

David L. Toll and Paul R. Houser

12.1 Introduction

12.1.1 Land–Water–Energy Systems

The land surface stores (temperature, soil moisture, snow, etc.) and modulates global energy and water fluxes that pass between the surface and the atmosphere. Hydrological cycle fluxes move energy as water. Energy used to evaporate water may be released hundreds of kilometres away during condensation, producing clouds and precipitation. Rainfall-runoff processes, weather and climate dynamics, and ecosystem changes all are highly dependent on land surface water and energy budgets.

As people alter the land surface, concern grows about the ensuing consequences for weather and climate, water supplies, crop production, biogeochemical cycles and ecological balances at various time scales. Therefore, it is crucial that any natural or human-induced water cycle changes in the land and atmosphere be assessed, monitored and predicted. For example, Gornitz *et al.* (1997) report that nearly 1% of the total, global annual stream flow is reduced by human activities such as irrigated agriculture contributing to a sea level lowering of 0.8 ± 0.4 mm per year. This offsets the predicted 1–2 mm/year sea level rise attributed to global warming. Accurately assessing land surface hydrology and energy flux spatial and temporal variation is essential for understanding and predicting biospheric and climatic responses. Data assimilation is a key means of improving our knowledge of these processes by optimally constraining model predictions with observational information.

Spatial Modelling of the Terrestrial Environment. Edited by R. Kelly, N. Drake, S. Barr.
© 2004 John Wiley & Sons, Ltd. ISBN: 0-470-84348-9.

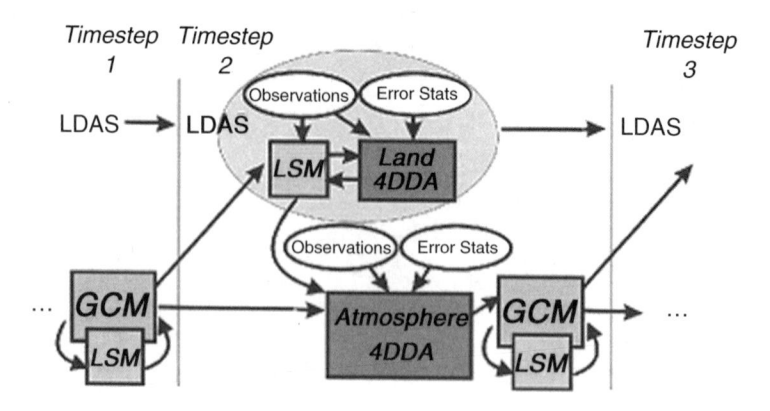

Figure 12.1 *Interaction of the Land Data Assimilation Scheme (LDAS) with an operational Numerical Weather Prediction (NWP) system. The atmospheric General Circulation Model (GCM) is coupled with the Land Surface Model (LSM), and both use a 4-Dimensional Data Assimilation (4DDA) process to integrate past forecasts with observations to improve performance*

12.1.2 Land Data Assimilation

Data assimilation is a numerical scientific tool that can improve land, water and energy budget model estimations through the incorporation of observational constraints. This leads not only to better overall predictions, but also helps to diagnose model weaknesses and can suggest where better parameterizations are needed. The fusion of operational model and observation data via data assimilation requires access to large near-real time surface atmospheric and hydrologic information volumes. Data processing, modelling and data assimilation require large computational and data storage resources that are becoming more achievable with evolving computer technology.

Additionally, new remotely sensed land surface observations are becoming available that will provide the additional information necessary to constrain land surface predictions at multiple time and space scales. These constraints can be imposed two ways (Figure 12.1). First, by forcing the land surface primarily by observations (e.g. precipitation and radiation), the often-severe atmospheric numerical weather prediction biases can be avoided. Second, by employing innovative land surface data assimilation techniques, land surface storage observations such as soil temperature and moisture can be used to constrain unrealistic simulated storages. Land data assimilation techniques also have the ability to maximize the utility of limited land surface observations by propagating their information throughout the land system to times and locations which lack observational data. The primary thrust of this chapter is to highlight land, energy and water data assimilation. Specifically we will do so through the 'Land Data Assimilation Systems' (LDAS) framework.

12.2 Land Data Assimilation Systems (LDAS)

Significant land-surface observation and modelling progress has been made at a wide range of spatial and temporal scales. Projects such as the International Satellite Land Surface Climatology Project (ISLSCP) (Hall *et al.*, 2002), the Global Soil Wetness Project

(GSWP) (Koster and Milly, 1997), and the Global Energy and Water Experiment (GEWEX) Continental-Scale International Project (GCIP), among others have paved the way for operational Land Data Assimilation System (LDAS) development. The LDAS development serves as an integrating linkage between a variety of Earth science disciplines and geographical locations. The LDAS are forced with real time output from numerical prediction models, satellite data, and radar precipitation measurements. Many model parameters are derived from high-resolution satellite-based vegetation coverage. But most importantly, LDAS integrates state-of-the-art modelling and observation on an operational basis to provide timely and consistent high quality land states to be used in real-time applications such as coupled land-atmosphere models and mesoscale climate models. A primary LDAS goal is to provide a broad range of information useful for applications, policy making and scientific research. We are currently associated with three LDAS projects: (1) the North American LDAS; (2) the Global LDAS; and (3) the Land Information System (LIS). The Global and North American LDAS provide real-time and selected retrospective simulations and the LIS team is currently developing the high-performance computation capability to perform a relatively high spatial resolution (1 km) global land prediction.

12.2.1 The North American LDAS

The North American LDAS (NLDAS) was initiated in 1998 primarily to derive land surface modelling with observation and model-based forcing fields (e.g. radiation and precipitation) to avoid biases from atmospheric models (Mitchell *et al.*, 1999). NLDAS consists of a number of land surface models that use remote sensing and *in situ* observations gridded to 1/8 degree. NASA, NOAA National Center for Environmental Prediction (NCEP), Princeton University, Rutgers University, the University of Maryland and the University of Washington implement NLDAS in near real time using existing Land Surface Models (LSM's). NLDAS has also been run for a 50-year retrospective period from 1950–2000 (Maurer *et al.*, 2002). NLDAS uses NCEP 'Eta' model analysis fields, along with observed precipitation and radiation fields to force several different land surface models in an uncoupled modelling system.

12.2.2 The Global LDAS

The Global Land Data Assimilation System (GLDAS) enlarges upon NLDAS to the global scale using many of the same algorithms, but necessarily requiring global forcing and parameter fields (Rodell *et al.*, 2002). Whereas a primary NLDAS emphasis is to improve local weather forecasts, a primary GLDAS emphasis is to provide initialization of global coupled weather and climate prediction models. In addition to improved weather prediction and climate modelling, both systems have a broad range of applications in studies of the terrestrial energy and water cycles. Examples include flood and drought assessment, water usage for crops, snowpack and snowmelt for water availability, water and energy data for ecosystem modelling and net primary productivity.

Both NLDAS and GLDAS are able to drive multiple land surface models under one system. A summary of the key system features is given in Table 12.1. Furthermore, input and output fields for NLDAS and GLDAS are summarized in Table 12.2. Outputs from the LDAS simulations are freely available (see http://ldas.gsfc.nasa.gov) and the LDAS

Table 12.1 *Typical programme execution options in LDAS (from Rodell et al., 2002)*

Attribute/Option	Description
Spatial resolution	0.125° (NLDAS); 0.25° to 2.5° (GLDAS)
Land surface model	MOSAIC; CLM; NOAH; catchment (in preparation)
Forcing	Various model and satellite-derived products
Initialization	None (constant value); restart file; forcing data
Subgrid variability	1–13 tiles per grid cell
Elevation adjustment	Temperature; pressure; humidity; long-wave radiation
Data assimilation	Surface temperature; snow cover; soil moisture
Soil classification	Look-up table; Reynolds *et al.* (1999)
Leaf area index	Look-up table; AVHRR/model derived
Inland water tiles	CLM lakes option

Reproduced by permission of the American Meteorological Society

website includes a real-time image generator and data subsetting tool that permits viewing and acquisition of recent LDAS results.

12.2.3 Land Information System

Increases in GLDAS resolution to 1 km are planned and will improve land-atmosphere process understanding. However, to process a year of 1 km global data using conventional computers would require an unavailable amount of computer runtime. A new Land Information System (LIS), building on the same land surface modelling and observation fields as GLDAS, is being constructed (http://lis.gsfc.nasa.gov). The LIS uses a high performance, massively parallel computer including a 192-node Beuwolf computer cluster to support throughput demands of a near real-time global 1 km land prediction and data assimilation system. The system will have a web-based user interface designed to facilitate broad and efficient usage. Input and output are based on developing an Earth System Modelling Framework (ESMF, http://www.esmf.ucar.edu) that demonstrates the interoperability of disparate model components and enables the use of remotely sensed data in coupled Earth system models.

12.3 LDAS Components

Land data assimilation typically refers to the incorporation of observational (*in situ* and remote sensing) data into an LSM. Land surface modelling provides the spatial and temporal predictions, whereas observations are typically used as input data or 'correction' data for the numerical simulation. The three central land surface data assimilation system components are the following: (1) land surface simulation; (2) land surface observations; and (3) land surface data assimilation.

12.3.1 Land Surface Simulation

Recent advances in understanding soil-water dynamics, plant physiology, micrometeorology, and biosphere-atmosphere interactions have spurred LSM developments that seek to realistically represent mass, energy, and momentum transfer between a vegetated surface and the atmosphere (Dickinson *et al.*, 1993; Sellers *et al.*, 1986). LSM predictions are regular in time and space, but are influenced by model structure, input forcing and model parameter errors, and inadequate sub-grid-scale spatial variability treatment. Consequently, LSM hydrology and energy prediction will likely be much improved by using assimilation strategies to remove biases and better constrain boundary conditions.

There are many different approaches to land surface prediction, which has led to great diversity in LSMs. Four LSMs either used or soon to be used in LDAS are presented here. These are the: (1) Mosaic LSM of Koster and Suarez (1992); (2) the Catchment LSM of Koster *et al.* (1998); (3) the **N**ational Centers for Environmental Prediction (NCEP), **O**regon State University (OSU), United States **A**ir Force (USAF) and Office of **H**ydrology (OH), LSM, called the **NOAH** LSM (Mitchell, 2002); and (4) the recently emerging Community Land Model (CLM) (Bonan, 1998 and Dickinson *et al.*, 2000).

The Mosaic LSM addresses the sub-grid heterogeneity issue by subdividing each GCM grid cell into a user-specified *mosaic* of tiles (after Avissar and Pielke, 1989), with each tile having different vegetation characteristics and hence water and energy balance. Surface flux calculations for each tile are similar to those described by Sellers *et al.* (1986). Like the plethora of LSMs that have been developed over the past decade, tiles do not directly interact with each other, but influence each other indirectly by their collective influence on the coupled overlying atmosphere (Henderson-Sellers *et al.*, 1993). Vukovich *et al.* (1997) report using tiles reduced the average error for three test sites by 11% for sensible heat and 20% for latent heat fluxes. Toll *et al.* (2001) report much of the error is in geographic regions with contrasting cover such as forests, grasses, and crops when using only one land cover class per grid cell. Although the Mosaic LSM is well suited to modelling the vertical exchange of mass, energy and momentum between the land surface and the overlying atmosphere, Mosaic includes no lateral moisture movement representation, which can significantly impact soil water, surface energy fluxes and runoff variation at some scales.

Recognizing this weakness, Koster *et al.* (1998) developed a new, *catchment*-based LSM that includes a more realistic hydrological process representation, including soil water lateral transport through the subsurface. The catchment-based land surface model uses a topographic land atmosphere transfer scheme (TOPLATS) that relies heavily on the concepts originally put forth by Famiglietti and Wood (1994) and Peters-Lidard *et al.* (1997) (i.e., the TOPLATS model). It represents a major advance in LSMs for the following two reasons. First, TOPMODEL's topographically based framework (Beven and Kirkby, 1979) will result in improved runoff prediction, and consequently, a more realistic catchment-scale water balance. Second, the downslope moisture movement within the watershed will yield sub-catchment-scale surface and unsaturated-zone moisture variations, which will result in more realistic prediction of intra-catchment surface flux variations. Ultimately, improved runoff simulation will result in more realistic continental-scale stream flow estimates from the land to the oceans, and similarly, the intra-catchment surface flux variation will improve catchment-average exchanges with the atmosphere.

The NOAH LSM simulates soil moisture (both liquid and frozen), soil temperature, skin temperature, snowpack water equivalent, snowpack density, canopy water content, and the traditional energy flux and water flux terms of the surface energy and surface water balance (Mitchell, 2002). NOAHLSM uses global satellite-derived monthly climatological vegetation greenness fraction, albedo values for different surfaces, and accounts for the snow albedo. Recently, the NOAH has significantly improved both cold season process physics and bare soil evaporation. This model has been used in: (1) the NCEP-OH submission to the 'PILPS-2d' tests for the Valdai, Russia site; (2) the emerging, real-time, North American Land Data Assimilation System (NLDAS); (3) the coupled NCEP mesoscale model and the Eta model's companion 4-D Data Assimilation System (EDAS); and (4) the coupled NCEP global Medium-Range Forecast model (MRF) and its companion 4-D Global Data Assimilation System (GDAS).

The Community Land Model (CLM) is under development by a grass-roots collaboration of scientists who have an interest in making a general land model available for public use. The CLM development philosophy is to use only proven and well-tested physical parameterizations and numerical schemes. The current CLM version includes superior components from each of several contributing models (Bonan, 1998 and Dickinson *et al.*, 1993). The CLM code is managed in an open source style, in that updates from multiple groups will be included in future model versions. Also, the land model was run for a test case suite including many of the Projects for the Intercomparison of Land Parameterization Schemes (PILPS) case studies (Koster and Milly, 1997).

There are strong justifications for using an uncoupled LSM modelling system for LDAS. Although coupling the LSM to an atmospheric model permits the study of the interaction and feedbacks between the atmosphere and land surface, coupled modelling also imposes strong Numerical Weather Prediction land surface forcing biases on the LSM (Mitchell *et al.*, 1999). These precipitation and radiation biases can overwhelm the behaviour of LSM physics. In fact, several numerical weather prediction centres must 'correctively nudge' their LSM soil moisture estimates towards climatological values to eliminate soil moisture 'drift' (Mitchell *et al.*, 1999). By using an uncoupled LSM, LDAS users can better constrain land surface forcing via observations, use fewer computational resources, and still address all of the relevant LDAS goals. The physical understanding and modelling insights gained from implementing distributed, uncoupled land-surface schemes with observation-based forcing have been demonstrated in recent off-line land surface modelling projects such as the GEWEX Global Soil Wetness Project (Koster and Milly, 1997).

12.3.2 Observations and Observation-Based Data

Atmospheric model simulations and observation-based data are used as a baseline to provide forcing fields as input for LDAS and LIS (Table 12.2). Forcing fields for both NLDAS and GLDAS may be obtained through the LDAS website (http://ldas.gsfc.nasa.gov) and from NCEP (ftp://ftp.ncep.noaa.gov/pub/gcp/ldas/noaaoutput). The LDAS is forced with real-time output from numerical prediction models, satellite data and surface radar precipitation measurements. Many LDAS vegetation parameters are derived from high-resolution AVHRR and MODIS observations. See Mitchell *et al.* (1999) for a further description of NLDAS forcing and see Rodell *et al.* (2002) for GLDAS forcing.

Atmospheric Forcing Data. GLDAS provides a choice between three meteorological model inputs. First, the NASA/Goddard Earth Observing System (GEOS) data assimilation system supports Level-4 data products from the NASA TERRA satellite. The data

Table 12.2 LDAS Model output summary using the mosaic LSM

Atmospheric	Land Surface and Subsurface	
Net shortwave radiation (W m^{-2})	Surface runoff (Kg mv)	Top 1 m soil moisture (Kg m^{-2})
Net longwave radiation (W m^{-2})	Subsurface runoff (Kg mv)	Snowmelt (Kg m^{-2})
Downward solar radiation flux (W m^{-2})	Average surface temperature (K)	Layer 2 soil moisture (Kg m^{-2})
Downward longwave radiation flux (W m^{-2})	Surface Albedo paul (%)	Layer 3 soil moisture (Kg m^{-2})
Snowfall, frozen (Kg m^{-2}) precipitation (Kg m^{-2})	Total soil column wetness (%)	Snowpack water equivalent (Kg m^{-2})
Rainfall, unfrozen (Kg m^{-2})	Snow depth (m)	Root zone wetness (%)
Surface pressure (Pa)	Snow cover (%)	Canopy surface water evaporation (W m^{-2})
Air temperature, 2 m (K)	Plant canopy surface water storage (kg mv)	Canopy transpiration (W m^{-2})
Specific humidity, 2 m (Kg/Kg)	Deep soil temperature (K)	Aerodynamic conductance (m s^{-1})
U wind component (m s^{-2})	Total column soil moisture (Kg m^{-2})	Canopy conductance (m s^{-1})
V wind component (m s^{-2})	Root zone soil moisture (Kg m^{-2})	Sensible heat flux (W m^{-2})
Convective precipitation (Kg m^{-2})	Vegetation greenness (%)	Latent heat flux (W m^{-2})
	Leaf area index (dimensionless)	Ground heat flux (W m^{-2})

are produced on a 1-degree global grid in 3-hour assimilation datasets. GEOS incorporates Physical-Space Statistical Analysis System (PSAS) data assimilation techniques that combine current boundary conditions (e.g. sea surface temperature) with updated observations and error statistics. Second, the Global Data Assimilation System (GDAS) operational weather forecast model of National Centers for Environmental Prediction (NCEP) is available (Derber *et al.*, 1991). GDAS data at a native resolution of 0.7° is mapped to a 1-degree global grid in the GLDAS project. Third, the European Centre for Medium-Range Weather Forecasts (ECMWF) products are also available. ECMWF produces 6-hour forecasts at approximately 39-km spatial resolution. The analysis includes *in situ* conventional and satellite-derived data using a four dimensional multivariate data assimilation technique. Rodell *et al.* (2002) report the GEOS data provides the best comparison to observation data and was chosen as the primary forcing data for GLDAS.

When possible, observation-based data from satellites, precipitation gauges and Doppler radar is used. For example, in GLDAS and NLDAS observational data may be used to replace or update modelled data that is spatially and temporally contiguous. Observational data are typically used when validation and quality control efforts indicate satisfactory accuracy. With increases in computer power, observational data are now used more frequently in data assimilation (Cohn *et al.*, 1998; Mitchell, 2002).

Land Surface State Data. Operational and accurate global land surface hydrologic predictions require the assimilation of spatially distributed remote sensing-derived observations. Observations of interest include temperature, soil moisture (surface moisture content, surface saturation, total water storage), lake/river height and flow, snow areal extent and snow water equivalent, land cover, leaf area index and albedo.

Surface temperature remote sensing is a relatively mature technology. The land surface emits thermal infrared radiation at an intensity directly related to its emissivity and temperature. The absorption of this radiation by atmospheric constituents is smallest in the 3–5 and 8–14 μm wavelength ranges, making them the best windows for sensing land surface temperature. Some errors due to atmospheric absorption and improperly specified surface emissivity are possible, and the presence of clouds can obscure the signal. Generally, surface temperature remote sensing can be considered an operational technology (Ma *et al.*, 2002), with many spaceborne sensors making regular observations (i.e., Landsat Thematic Mapper, NOAA Advanced Very High Resolution Radiometer (AVHRR), Terra and Aqua Moderate Resolution Imaging Spectroradiometer (MODIS) and Terra Advanced Spaceborne Thermal Emission and Reflection Radiometer (ASTER)). The land surface temperature evolution is linked to all other land surface processes through physically based relationships, which makes its assimilation possible.

Soil moisture remote sensing is a developing technology, although the theory and methods are well established (Eley, 1992). Long-wave passive microwave remote sensing is ideal for soil moisture observation, but there are technical challenges in correcting for the vegetation and roughness effects. Soil moisture remote sensing has previously been limited to aircraft campaigns (e.g. Jackson, 1997a), or Defense Meteorological Satellite Program (DMSP) Special Sensor Microwave Imager (SSM/I) analysis (Jackson, 1997b) data. SSM/I data was also successfully employed to monitor surface saturation/inundation (Basist and Grody, 1997). The Advanced Microwave canning Radiometer (AMSR) instrument provides additional C-band microwave observations that may be useful for soil moisture determination. The Tropical Rainfall Measuring Mission's (TRMM) Microwave Imager (TMI), which is very similar to AMSR radiometrically, is much better suited to soil moisture measurement (because of its 10 Mhz channels) than SSM/I, and is also currently available. All of these sensors have adequate spatial resolution for land surface applications, but have a very limited quantitative measurement capacity, especially over dense vegetation. However, Sipple *et al.*, (1994) demonstrated that it is possible to detect saturated areas through dense vegetation using Scanning Multichannel Microwave Radiometer (SMMR), which can greatly aid land surface predictions. The SMMR has similar radiometric characteristics to AMSR. Because of the large remotely sensed microwave soil moisture observation error, there is a real need to maximize its information content by using algorithms, such as data assimilation, that can account for measurement error and extend satellite information in time and space.

There is a potential to monitor total water storage variations (ground water, soil water), surface waters (lakes, wetlands, rivers), water stored in vegetation, and snow and ice using time variable gravity field satellite observations. The Gravity Recovery and Climate Experiment (GRACE), an Earth System Science Pathfinder mission, will provide highly accurate terrestrial water storage change estimates in large watersheds. Wahr *et al.* (1998) note that GRACE will provide water storage variation estimates to within 5 mm on a monthly basis. Rodell and Famiglietti (1999) have demonstrated the potential utility of these data for hydrologic application is aimed more at large watersheds (>150,000 km^2). They further discuss the potential power of GRACE to constrain land surface modelled water storage when combined with surface soil moisture and altimetery observations. Birkett (1998) demonstrated the potential of satellite radar altimeters to monitor height variations over inland waters, including climatically sensitive lakes and large rivers and wetlands. Such altimeters are currently operational on the ERS-2, TOPEX/POSEIDON, ENVISAT and JASON-1 satellites.

Key snow variables of interest to land data assimilation include areal coverage and snow water equivalent. While snow water equivalent estimation by satellite is currently in research mode, snow areal extent can be routinely monitored by many operational platforms (Tait *et al.*, 2000), including AVHRR, GOES, MODIS and SSM/I. Recent algorithm developments even permit snow cover fraction determination within Landsat-TM pixels (Rosenthal and Dozier, 1996). Cline *et al.* (1998) describe an approach for retrieving snow water equivalent from the jointly using remote sensing and energy balance modelling.

Other key variables conducive to remote sensing are surface albedo (Toll *et al.*, 1997), land cover and leaf area index (Justice *et al.*, 1998). They are each key variables to global climate, ecology, hydrology and biogeochemical models that may help describe the energy, mass (e.g. water and CO_2) and momentum exchanges between the land and atmosphere. In addition, they are available as data products globally from satellite sensors such as the Terra and Aqua MODIS (Justice *et al.*, 1998).

12.3.3 Data Assimilation

The presence of model and observation error causes the study of highly interactive large-scale land hydrology and energy budgets to be a complex task. A combination of information, including LSM data, remote sensing observations and *in situ* surface data is used for study. There have been recent data assimilation theory advances that have provided quantitative methods for merging the various information types to provide more accurate estimates (Errico, 1999). Lorenc (1995) defines assimilation as the process of finding the model representation that is most consistent with the observations. In essence, data assimilation merges a range of diverse data fields with a model prediction to provide that model with the best estimate of the current state of the natural environment so that it can then make more accurate predictions. Most data assimilation techniques can be applied to almost any dynamic geoscience problem, but are often limited by computational feasibility. Earth scientists now face the challenge to apply true data assimilation techniques to all problems where the incorporation of observations can provide new insights. However, this is a difficult task due to the highly nonlinear nature of land surface processes, the problem size, and the lack of data and experience to determine error statistics accurately. Consequently, data assimilation implementation always requires trade-offs between resolution, complexity, computational effort, and data availability.

Data assimilation techniques are used extensively in meteorology (Daley, 1991) and oceanography (Wunsch, 1996). For example, in meteorology data assimilation is routinely used to improve weather forecasting. Currently most operational weather forecast centres use optimal interpolation type schemes (Daley, 1991). NASA's Data Assimilation Office has recently improved this technique by developing the Physical-Space Statistical Analysis

System (PSAS) (Cohn et al., 1998). PSAS operates in the spatial domain and improves the complicated and time-dependent error covariance estimation.

Hydrologic data assimilation, especially at large scales, is in its early stages (Reichle, 2000; McLaughlin, 1995). One formidable problem is that many hydrological processes are non-linear. Currently, there are only a few studies that use distributed watershed models to assimilate field data. Reichle (2000) provides a short Earth sciences data assimilation review with a focus on hydrology. Surface hydrology data assimilation is based primarily on soil moisture information from surface observations or remote sensing. There are several soil moisture estimation data assimilation techniques that use a one-dimensional optimal estimation approach, including studies by Milly (1986), Katul et al. (1993), Parlange et al., (1993), Entekhabi et al. (1994), Galantowicz et al. (1999), Calvet et al. (1998) and Castelli et al. (1999). Several additional studies have used low-level atmospheric observations to infer soil moisture using one-dimensional optimal variation assimilation approaches (Mahfouf, 1991; Bouttier et al., 1993; Hu et al., 1999; Callies et al., 1998; Rhodin, et al., 1999). In these approaches the calculation of soil moisture is a 'parametric approach' and not physically based. Reichle et al. (2002) used four-dimensional variational data assimilation with improved physics. They used large-scale soil moisture profiles along with other soil and vegetation parameters from passive microwave measurements in their data assimilation.

Unlike previous efforts to test assimilation of soil moisture using synthetic data, two recent studies assimilated soil moisture into a LSM with actual remote sensing data. Both Crow and Wood (2003) and Margulis et al. (2002) demonstrated the utility of an extended Kalman filter that assimilated airborne microwave brightness data into a LSM. Both studies used ESTAR (Electronically Scanned Thinned Array Radiometer) brightness temperature data (1.4 GHz frequency, L-band), from the 1997 Southern Great Plains Experiment (SGP97), sensitive to soil moisture variations to 5 cm. The data assimilation was reported to be a computationally efficient and more accurate approach than modelling or remote sensing alone. The extended Kalman filter data assimilation was able to derive spatial and temporal trends of soil moisture in the root zone, significantly below the sensitivity of L-band data (to 5 cm). A summary of recent data assimilation soil moisture papers related to LDAS is given next.

Houser et al. (1998) demonstrated the feasibility of synthesizing distributed soil moisture fields by the novel application of four-dimensional data assimilation in a hydrological model. Six Push Broom Microwave Radiometer (PBMR) images gathered over the USDA-ARS Walnut Gulch Experimental Watershed in southeast Arizona were assimilated into the TOPLATS hydrological model (Peters-Lidard et al., 1997) using several alternative assimilation procedures. Modification of traditional assimilation methods was required to use the high-density PBMR observations. Information on surface soil moisture was assimilated into the subsurface using surface-subsurface correlation knowledge. Newtonian nudging assimilation procedures were found to be preferable to other techniques because generally they preserve the observed patterns within the sampled region, but also yield plausible patterns in unmeasured regions, and allow information to be advected in time.

Reichle et al. (2002) used the ensemble Kalman filter to estimate soil moisture by assimilating microwave L-band (1.4 GHz) brightness temperatures into a land surface model.

They concluded the ensemble Kalman filter specifically reduced soil moisture estimation errors in comparison to results without data assimilation. They also concluded that assimilation schemes that use 'static' forecast error covariances such as in statistical interpolation produce less accurate estimates than the ensemble Kalman filter. In addition, they found that the ensemble Kalman approach is very flexible and is applicable over a broad model error range.

Sun *et al.* (2002) used an extended Kalman filter to assimilate observed snow water equivalent, available through using satellite remote sensing for snow depth, snow temperature and snow water equivalent retrieval. They used a NASA Seasonal-to-Interannual Prediction Project (NSIPP) land surface model and derived 'true' snow states with European Centre for Medium-Range Weather Forecasting (ECMWF) atmospheric forcing data. The data were degraded and then assimilated for comparisons to the 'true' snow states. Because of snow's high albedo, thermal properties, feedback to the atmosphere and as medium-term water storage, improved snow state estimation has the potential to greatly increase the climatological and hydrological prediction accuracy.

Walker and Houser (2001) found the one-dimensional extended Kalman filter effective in assimilating near-surface soil moisture into a land surface model. They degraded a simulation by setting the initial soil moisture prognostic variables to arbitrarily wet values throughout North America. A 'true' land surface simulation was run using International Satellite Land Surface Climatology Project (ISLSCP) forcing data. The study showed assimilation of near-surface soil moisture observations from a remote sensing satellite reduced soil moisture storage error to 3% after a one-month assimilation and to 1% after a 12 month assimilation. They concluded that data assimilation of remotely sensed data may provide accurate initial conditions to GCMs and that these models need not rely on initial conditions from a spun-up LSM simulation.

12.4 LDAS Applications

Real-time North American LDAS and Global LDAS (see http://ldas.gsfc.nasa.gov) that use atmospheric forcing fields to initialize land surface models are currently in place. Figure 12.2 illustrates for July 1996 a continental U.S. 1/8° precipitation image (upper) from assimilated modelling and observational data that were input to LDAS to derive the average surface soil moisture (lower). The precipitation data were assimilated using a NCEP Eta model, NEXRAD Doppler radar and Higgins gauge precipitation data (Mitchell *et al.*, 1999) combination. The precipitation data along with radiation data are key input parameters (see Table 12.2) to LDAS. Inspection of Figure 12.2 helps illustrate the relationship between precipitation and soil moisture that is influenced by other LDAS output parameters (Table 12.2) such as soil properties, LAI and radiation supply.

Figure 12.3 shows precipitation data over North America available for use as forcing data for LDAS. Comparisons between the precipitation plots show the large variation in estimates. The DAO GEOS and NOAA NCEP-EDAS data are the primary baseline forcing fields and provide spatially and temporally contiguous data. In addition, there are other observational-based datasets that may offer an improved precipitation dataset. The U.S. Naval Research Laboratory (NRL) provided near-real time satellite precipitation data from

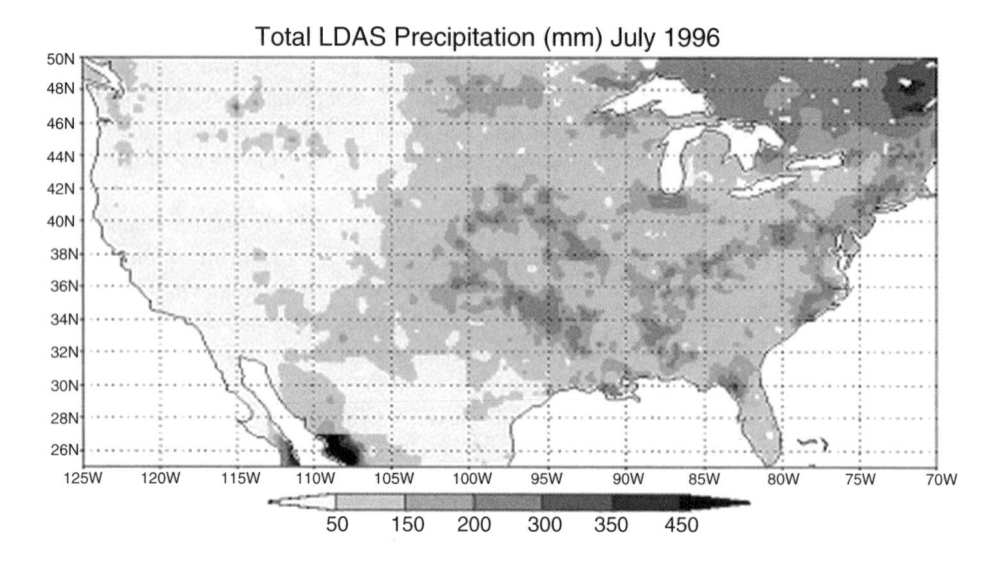

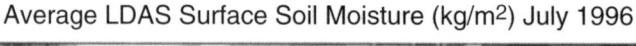

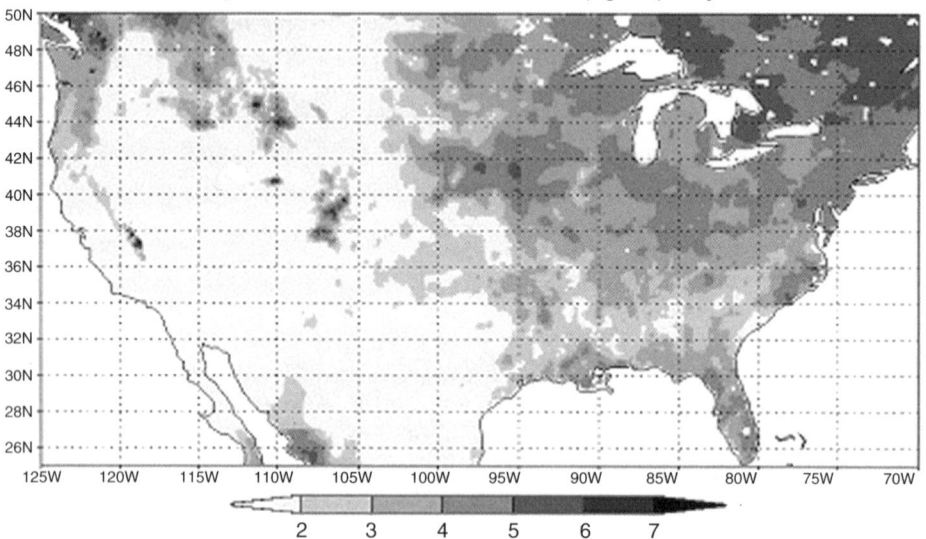

Figure 12.2 A continental United States 1/8° LDAS merged Eta model, NEXRAD Doppler radar and gauge precipitation field (a) and soil moisture field (b) (courtesy B. Cosgrove)

both infrared data and microwave sensors (Turk *et al.*, 2000). The microwave data are from TRMM and SSM/I satellites. In addition, Figure 12.3 shows other precipitation data sources which include the University of Arizona, Precipitation Estimation from Remotely Sensed Information using Artificial Neural Networks (PERSIANN) (Hsu *et al.*, 1997) and the Higgins interpolated gauge data from NOAA Cooperative Institute for Research in Environmental Sciences (CIRES) Climate Diagnostic Center (CDC). Since much of this

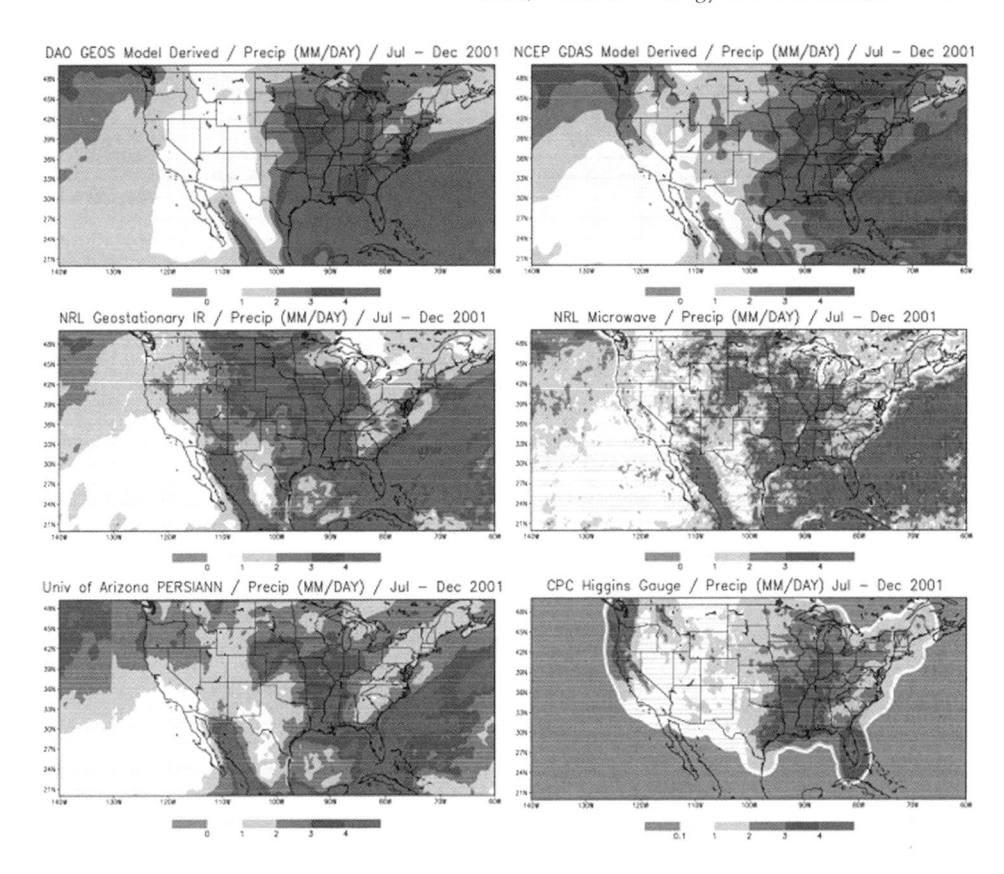

Figure 12.3 Comparison of partial North American precipitation estimates for July–December 2001. Precipitation is a key forcing parameter for LDAS modelling and data assimilation (courtesy J. Gottschalck)

data is spatially and temporally limited, LDAS also provides options to merge the sparse observational data with the baseline modelled data.

Plate 12 illustrates SMMR-derived surface soil moisture used to constrain a land surface model prediction using a one-dimensional extended Kalman filter (Walker and Houser, 2001). The SMMR has similar frequencies to the recently launched AMSR sensor on the EOS Aquas satellite. The plots in the top row show SMMR-derived soil moisture with LDAS forcing (input) data of snow depth and precipitation. The middle row shows LDAS-simulated soil moisture at the surface (left), root zone (centre) and over the soil layer profile (right). The assimilated surface moisture plots (bottom row) exhibit a soil moisture increase from the modelled data, especially in the northern Great Plains and south central US. Inspection of the Plate demonstrates how sparse observational data (upper-left) may be used in conjunction with more temporally and spatially covered model data. Approximately 15% of the area is observational data used in data assimilation with the contiguous modelling data. In the Kalman filtering technique the error covariances are used to provide an assimilated output that differs from both the observational and modelled data (Plate 12). This is the first known use of actual satellite-derived soil moisture within an assimilation framework.

12.5 Future Directions

The LDAS projects will continue to incorporate new model and observation information to improve land surface knowledge. The fourth LDAS model to be implemented soon is the Catchment Land Surface Model (CLSM) (Koster *et al.*, 1998). The catchment model signifies a major advance for the following two reasons. First, the modelling framework will result in improved runoff prediction, and consequently, more realistic catchment-scale water balance. Second, the downward slope movement of moisture within the watershed will yield sub-catchment-scale surface and unsaturated-zone moisture variations, resulting in more realistic intra-catchment surface flux variation prediction. Improved runoff simulation will ultimately yield a more realistic continental stream flow from the land to the oceans, and similarly, the within-catchment surface flux variations will result in more representative catchment-average exchanges with the atmosphere.

The significant increase in satellite observations is providing a global supply of atmosphere and land surface information available to improve land surface simulation and data assimilation. For example, higher spatial resolution data (to 1 km) will become available from MODIS and will include leaf area index (LAI), land surface temperature and surface albedo products. Also, improved global land cover datasets (to 1 km) will be updated every three months. New microwave sensors such as AMSR will permit improved precipitation estimation critical to LSM water and energy budgets. Moreover, the planned Global Precipitation Mission (GPM) will involve a satellite constellation that will enhance precipitation estimates to 3 hours temporally and to 4 km horizontal and 250 m vertical global cells. GPM should greatly improve land surface water and energy budget data, and the improved accuracy will benefit data assimilation.

Enhanced modelling and observational data with expanded computer power will improve data assimilation in hydrological sciences. Previous developments of data assimilation in hydrological applications have lagged behind in comparison to atmospheric and ocean applications. However, the wealth of emerging Earth science data coupled with hydrological land surface model physical improvements and computer capability will permit data assimilation to be more routinely implemented. Massively parallel computing techniques such as those developed by the Land Information System will support the near real-time data assimilation throughput demands. In this mode, users will have a web-based interface designed to facilitate broad and efficient information use for land, water and energy data assimilation.

Acknowledgements

Special thanks to Brian Cosgrove, Jon Gottschalck, Rolf Reichle, Jonathan Triggs and Matt Rodell for their inputs and paper review. We also thank Rolf Reichle, Chaojiao Sun and Jeffery Walker for inputs on data assimilation. We express gratitude to other members of the LDAS team for their support: Christa Peters-Lidard, Jared Entin, Jon Radakovich, Mike

Bosilovich, Urszula Jambor, Kristi Arsenault, Jesse Meng, Aaron Berg, Guiling Wang, Ken Mitchell, John Schaake, Eric Wood, Dennis Lettenmaier, Alan Robock and Jeff Basara.

References

Avissar, R. and Pielke, R., 1989, A parameterization of heterogeneous land surfaces for atmospheric numerical models and its impact on regional meteorology, *Monthly Weather Review*, **117**, 2113–2136.

Basist, A. and Grody, N., 1997, Surface wetness and snow cover, AMS Annual Meeting, *Proceedings of the 13th Conference on Hydrology*, 190–193.

Beven, K. and Kirkby, M., 1979, A physically-based variable contributing area model of basin hydrology, *Hydrological Sciences Journal*, **24**, 43–69.

Birkett, C.M., 1998, Contribution of the TOPEX NASA radar altimeter to the global monitoring of large rivers and wetlands, *Water Resources Research*, **34**, 1223–1239.

Bonan, G.B., 1998, The land surface climatology of the NCAR land surface model coupled to the NCAR community climate mode, *Journal of Climate*, **11**, 1307–1326.

Bouttier, F., Mahfouf, J.-F. and Noilhan, J., 1993, Sequential assimilation of soil moisture from atmospheric low-level parameters. Part I: sensitivity and calibration studies, *Journal of Applied Meteorology*, **32**, 1335–1351.

Callies, U., Rhodin, A. and Eppel, D.P., 1998, A case study on variational soil moisture analysis from atmospheric observations, *Journal of Hydrology*, **212–213**, 95–108.

Calvet, J.-C., Noilhan, J. and Bessemoulin, P., 1998, Retrieving the root-zone soil moisture from surface soil moisture or temperature estimates: a feasibility study based on field measurements, *Journal of Applied Meteorology*, **37**, 371–386.

Castelli, F., Entekhabi, D. and Caporali, E., 1999, Estimation of surface heat flux and an index of soil moisture using adjoint-state surface energy balance, *Water Resources Research*, **35**, 3115–3125.

Cline, D.W., Bales, R.C. and Dozier, J., 1998, Estimating the spatial distribution of snow in mountain basins using remote sensing and energy balance modeling, *Water Resources Research*, **34**, 1275–1285.

Cohn, S.E., Da Silva, A., Guo, J., Sienkiewicz, M. and Lamich, D., 1998, Assessing the effects of data selection with the DAO physical-space statistical analysis system, *Monthly Weather Review*, **126**, 2913–2926.

Crow, W.T. and Wood, E.F., 2003, The assimilation of remotely sensed soil brightness temperature imagery into a land-surface model using Ensemble Kalman filtering: a case study based on ESTAR measurements during SGP97, *Advances in Water Resources*, **26**, 137–159.

Daley, R., 1991, *Atmospheric Data Analysis* (New York: Cambridge University Press).

Derber, J.C., Parrish, D.F. and Lord, S.J., 1991, The new global operational analysis system at the National Meteorological Center, *Weather Forecasting*, **6**, 538–547.

Dickinson, R.E., Henderson-Sellers, A. and Kennedy, P.J., 1993, Biosphere-Atmosphere Transfer Scheme (BATS) Version 1e as coupled to the NCAR community climate model, *NCAR Technical Note*, NCAR/TN-387+STR, National Center for Atmospheric Research (NCAR).

Eley, J., 1992, Summary of Workshop, Soil Moisture Modeling, *Proceedings of the NHRC Workshop* held March 9–10, 1992, *NHRI Symposium No. 9 Proceedings, Canada*.

Entekhabi, D., Nakamura, H. and Njoku, E.G., 1994, Solving the inverse problem for soil moisture and temperature profiles by sequential assimilation of multifrequency remotely sensed observations, *IEEE Transactions on Geoscience and Remote Sensing*, **32**, 438–448.

Errico, R.M., 1999, Workshop on assimilation of satellite data, *Bulletin of the American Meteorological Society*, **80**, 463–471.

Famiglietti, J.S. and Wood, E.F., 1994, Multi-scale modeling of spatially-variable water and energy balance processes, *Water Resources Research*, **30**, 3061–3078.

Galantowicz, J.F., Entekhabi, D. and Njoku, E.G., 1999, Tests of sequential data assimilation for retrieving profile soil moisture and temperature from observed L-band radiobrightness, *IEEE Transactions on Geoscience and Remote Sensing*, **37**, 1860.

Gornitz, V., Rosenzweig, C. and Hillel, D., 1997, Effects of anthropogenic intervention in the land hydrologic cycle on global sea level rise, *Global Planetary Change*, **14**, 147–161.

Hall, F.G., Meeson, B., Los, S., Steyaert, L., de Colstoun, E. and Landis, D., 2002, ISLSCP Initiative II, DVD/CD-ROM, NASA/Goddard Space Flight Center.

Henderson-Sellers, A., Yang Z.-L. and Dickinson, R.E., 1993, The project for intercomparison of land-surface parameterization schemes, *Bulletin of the American Meteorological Society*, **74**, 1335–1349.

Houser, P., Gupta, H., Shuttleworth, W.J. and Famiglietti, J.S., 2001, Multiobjective calibration and sensitivity of a distributed land surface water and energy balance model, *Journal of Geophysical Research*, **106**, 421–433.

Hsu, K., Gao, X., Sorooshian, S. and Gupta, H.V., 1997, Precipitation estimation from remotely sensed information using artificial neural networks, *Journal of Applied Meteorology*, **36**, 1176–1190.

Hu, Y., Gao, X., Shuttleworth, W.J., Gupta, H., Mahfouf, J. and Viterbo, P., 1999, Soil-moisture nudging experiments with a single-column version of the ECMWF model, *Quarterly Journal of the Royal Meteorological Society*, **125**, 1879–1902.

Jackson, T.J., (1997a) Southern Great Plains, 1997, (SGP97) Hydrology Experiment Plan, United States Department of Agriculture, (http://hydrolab.arsusda.gov/sgp97).

Jackson, T.J., 1997b, Soil moisture estimation using special satellite microwave/imager satellite data over a grassland region, *Water Resources Research*, **33**, 1475–1484.

Justice, C., Vermote, E., Townshend, J.R.G., Defries, R., Roy, D.P., Hall, D.K., Salomonson, V.V., Privette, J., Riggs, G., Strahler, A., Lucht, W., Myneni, R., Knjazihhin, Y., Running, S., Nemani, R., Wan, Z., Huete, A., van Leeuwen, W., Wolfe, R., Giglio, L., Muller, J-P., Lewis, P. and Barnsley, M., 1998, The moderate resolution imaging spectroradiometer (MODIS) land remote sensing for global change, *IEEE Transactions on Geoscience and Remote Sensing*, **36**, 1228–1249.

Katul, G.G., Wendroth, O., Parlange, M.B., Puente, C.E., Folegatti, M.V. and Nielsen, D.R., 1993, Estimation of in situ hydraulic conductivity function from nonlinear filtering theory, *Water Resources Research*, **29**, 1063–1070.

Koster, R.D. and Milly, P.C.D., 1997, The interplay between transpiration and runoff formulations in land surface schemes used with atmospheric models, *Journal of Climate*, **10**, 1578–1591.

Koster, R.D. and Suarez, M.J., 1992, Modeling the land surface boundary in climate models as a composite of independent vegetation stands, *Journal of Geophysical Research*, **97**, 2697–2715.

Koster, R.D., Suarez, M.J., Ducharne, A., Kumar, P. and Stieglitz, M., 1998, A catchment-based approach to modeling land surface processes in a GCM. Part 1: model structure, *Journal of Geophysical Research*, **99**, 14,415–14,428.

Lorenc, A.C., 1995, *Atmospheric Data Assimilation*, Meteorological Office Forecasting Research Division Scientific Paper No. 34, UK Meteorological Office.

Ma, X.-L., Wan, Z., Moeller, C.C., Menzel, W.P. and Gumley, L.E., 2002, Simultaneous retrieval of atmospheric profiles, land-surface temperature and emissivity from moderate resolution imaging spectroradiometer thermal infrared data: extension of a two-step physical algorithm, *Applied Optics*, **41**, 909–924.

Mahfouf, J.-F., 1991, Analysis of soil moisture from near-surface parameters: a feasibility study, *Journal of Applied Meteorology*, **30**, 1534–1547.

Margulis, S.A., McLaughlin, D., Entekhabi, D. and Dunne, S., 2002, Land data assimilation and soil moisture estimation using measurements from the Southern Great Plains 1997 field experiment, *Water Resources Research* (in press).

Maurer, E.P., Wood, A.W., Adam, J.C., Lettenmaier, D.P. and Nussen, B., 2002, A long-term hydrologically based dataset of land surface fluxes and states for the conterminous United States, *Journal of Climate*, **15**, 3237–3251.

McLaughlin, D., 1995, Recent developments in hydrologic data assimilation. *Reviews of Geophysics*, Supplement, 977–984.

Milly, P.C.D., 1986, Integrated remote sensing modeling of soil moisture: sampling frequency, response time, and accuracy of estimates, *Integrated Design of Hydrological Networks* (Proceedings of the Budapest Symposium, July 1986). IHAS Publ. no. 158, pp. 201–211.

Mitchell, K., 2002, The Community NOAH Land-Surface Model (LSM) User's Guide, Version 2.5.2. (ftp://ftp.ncep.noaa.gov/pub/gcp/ldas/noahlsm/ver_2.5.2/NOAH_LSM_USERGUIDE_2.5.2.doc).

Mitchell, K., Houser, P., Wood, E., Schaake, J., Tarply, D., Lettenmaier, D., Higgins, W., Marshall, C., Lohman, D., Ek, M., Cosgrove, B., Entin J., Duan, Q., Pinker, R., Robuck, A., Habelts, F. and Vinnikov, K., 1999, GCIP Land Data Assimilation System (LDAS) project now underway, GEWEX News, 9(4),3–6, GEWEX Project Office, Silver Spring, MD.

Parlange, M.B., Katul, G.G., Folegatti, M.V. and Nielsen, D.R., 1993, Evaporation and the field scale soil water diffusivity function, *Water Resources Research*, **29**, 1279–1286.

Peters-Lidard, C.O., Pan, F. and Wood, E.F., 2001, A re-examination of modeled and measured soil moisture spatial variability and its implications for land surface modeling, *Advances in Water Resources*, **24**, 1069–1083.

Peters-Lidard, C.O., Zion, M.S. and Wood, E.F., 1997, A soil-vegetation-atmosphere transfer scheme for modeling spatially variable water and energy balance processes, *Journal of Geophysical Research*, **102**, 4304–4324.

Reichle, R.H., 2000, Variational assimilation of remote sensing data for land surface hydrologic applications, Ph.D. dissertation, Massachusetts Institute of Technology.

Reichle, R.H., McLaughlin, D.B. and Entekhabi, D., 2002, Hydrologic data assimilation with the Ensemble Kalman filter, *Monthly Weather Review*, **130**, 103–114.

Reynolds, C.A., Jackson, T.J. and Rawls, W.J., 1999, Estimating available water content by linking the FAO soil map of the world with global soil profile databases and pedo-transfer functions, *American Geophysical Union, Fall Meeting, EOS Transactions*, AGU, 80.

Rhodin, A., Kucharski, F., Callies, U., Eppel, D.P. and Wergen, W., 1999, Variational analysis of effective soil moisture from screen-level atmospheric parameters: application to a short-range weather forecast model, *Quarterly Journal of the Royal Meteorological Society*, **125**, 2427–2448.

Rodell, M. and Famiglietti, J.S., 1999, Detectability of variations in continental water storage from satellite observations of the time-variable gravity field, *Water Resources Research*, **35**, 2705–2723.

Rodell, M., Houser, P.R., Jambor, U., Gotschalck, J., Meng, C.-J., Arsenault, K., Cosgrove B., Radakovich, J., Bosilovich, M., Entin, J.K., Walker, J., Toll, D. and Mitchell, K., 2004, The Global Land Data Assimilation System (GLDAS), *Bulletin of the American Meteorological Society*, in press.

Rosenthal, W. and Dozier, J., 1996, Automated mapping of montane snow cover at subpixel resolution from the Landsat Thematic Mapper, *Water Resources Research*, **32**, 115–130.

Sellers, P.J., Mintz, Y. and Dalcher, A., 1986, A simple biosphere model (SiB) for use within general circulation models, *Journal of Atmospheric Science*, **43**, 505–531.

Sipple, S.J., Hamilton, S.K., Melak, J.M. and Choudhury, B.J., 1994, Determination of inundation area in the Amazon river flood plain using SMMR 37 GHz polarization difference, *Remote Sensing Environment*, **48**, 70–76.

Sun, C., Walker, J.P. and Houser, P.R., 2002, Snow assimilation in a catchment-based land surface model using the extended Kalman filter. Extended abstract and poster. *Symposium on Observations, Data Assimilation and Probabilistic Prediction, AMS 82nd Annual Meeting, Orlando, Florida*.

Tait, A.B., Hall, D.K., Foster, J.L. and Armstrong, R.L., 2000, Utilizing multiple datasets for snow-cover mapping, *Remote Sensing Environment*, **72**, 111–126.

Toll, D.L., Entin, J. and Houser, P., 2001, Land surface heterogeneity on surface energy and water fluxes, in M. Owe and G. D'Urso (eds), *Remote Sensing for Agriculture, Ecosystems, and Hydrology III, Proceedings of SPIE, Toulouse France*, 17–19 September 2002, Volume 4542, 267–271.

Toll, D.L., Shirey, D. and Kimes, D.S., 1997, NOAA AVHRR surface albedo algorithm development, *International Journal of Remote Sensing*, **18**, 3761–3796.

Turk, F.J., Rohaly, G., Hawkins, J.D., Smith, E.A., Grose, A., Marzano, F.S., Mugnai, A. and Levizzani, V., 2000, Analysis and assimilation of rainfall from blended SSM/I, TRMM and geostationary satellite data, *AMS 10th Conference on Satellite Meteorology and Ocean.*, Long Beach, CA, 9–14 January, 66–69.

Vukovich, F.M., Wayland, R. and Toll, D.L., 1997, The surface heat flux as a function of ground cover for climate models, *Monthly Weather Review*, **125**, 573–586.

Wahr, J., Molenaar, M. and Bryan, F., 1998, Time-variability of the Earth's gravity field: hydrological and oceanic effects and their possible detection using GRACE, *Journal of Geophysical Research*, **103**, 30,205–30,229.

Walker, J.P. and Houser, P., 2001, A method for initializing soil moisture in a global climate model: assimilation of near-surface soil moisture observations, *Journal of Geophysical Research*, **106**, 11,761–11,774.

Wunsch, C., 1996, *The Ocean Circulation Inverse Problem* (New York: Cambridge University Press).

13
Spatial Modelling of the Terrestrial Environment: Outlook

Richard E.J. Kelly, Nicholas A. Drake and Stuart L. Barr

13.1 The Importance of Spatial Modelling

Spatial modelling of the terrestrial environment encompasses a wide variety of research activities in the Earth and environmental sciences. In this volume we have collected examples from a small subset of those activities which serve to illustrate some important challenges that modellers currently face. Some of these issues are terrestrial-specific and apply mostly to land-based activities while other issues are more generally applicable. However, the issues are not merely abstract in importance since they often have a direct influence on policy or social activity. For example, in the chapter by Devereux *et al.* (Chapter 11) there is a clear demonstration of how models can be used in environmental planning applications. They conclude: "(a GIS) can provide the basis for testing and learning about the likely outcomes of policy measures before they are implemented and before their undesirable effects become too difficult to reverse". This is a very attractive feature for policy makers who are keenly aware of the consequences of policy decisions that are founded on a poor understanding of complex issues. It also reflects the fact that we are now capable of sophisticated computer simulations of the kind described by Casti (1997) and referred to in Chapter 1.

13.2 Key Research Issues in Spatial Modelling of the Terrestrial Environment

Throughout the book, different combinations of remote sensing, numerical models and GIS have been discussed in a variety of contexts. These general science activities are important components of many spatial modelling frameworks. Furthermore, from the eleven contributions, three key research issues recur throughout the book as important components that affect spatial modelling activities and require further research or technological advancement for the modelling science to develop significantly. First, the development of improved digital elevation models, second, the accurate representation of vegetation in models and third, the need to better understand the impact of spatial resolution on model prediction. We expand on each of these three issues in the next three sub-sections.

13.2.1 Digital Elevation Models (DEM): Improved Accuracy and Error Characterization

The need for accurate DEMs in spatial modelling of terrestrial processes is particularly important in the natural environmental sciences. In Bamber's chapter, accurate topography information is shown to be critical for ice sheet modelling and he argues that there is not enough accurate data to enable us to answer some of the important science questions relating to ice sheet dynamics and ultimately to global climate change. In this particular example, elevation data are required to test and verify several types of ice sheet models so it is important that accurate DEMs are available. With planned new satellite technologies such as NASA's ICESat or ESA's Cryosat missions, the estimation accuracy of topographic variables (elevation, slope, aspect, etc.) should be increased which, in turn, will improve our ability to characterize the complex behaviour of ice sheets. At a smaller domain scale, Bates *et al.* describe an application to estimate flood inundation using fine spatial resolution DEMs from satellite and aircraft remote sensing data. These data are used to parameterize hydraulic models of flood-water flow. In regions where these fine spatial resolution data exist, flood inundation modelling should result in smaller estimation errors. By testing and validating the models on well-defined river catchment flood plains, future Earth observation instruments, with finer spatial resolution, will offer the potential to extend the flood inundation mapping to more remote regions and at finer spatial scales than can currently be achieved. With the release of NASA's Shuttle Radar Topographic Mission data, which provides elevation information of a large portion of the Earth's surface in great detail and accuracy, there is a good potential for increased accuracy of estimates from spatial models of land processes that use DEMs.

In using DEMs for spatial modelling, it is also apparent that error propagation needs to be carefully managed. The chapter by Lane *et al.* demonstrates how relatively small elevation errors in a DEM that is used in a process erosion model can have a significant impact on the prediction accuracy of the model. Thus, while increased accuracy of DEMs are a general requirement by the terrestrial environmental science community, it is also necessary to better account for the errors that can arise from the derivatives of DEMs when used in process models.

13.2.2 Vegetation Cover: Improved Characterization

The characterization of vegetation is very important for many spatial models of terrestrial processes. This is because vegetation often constitutes an important spatially and temporally dynamic boundary layer in many terrestrial environments. For example, in their chapter on soil erosion, Drake *et al.* note that vegetation is a stabilizing factor in the process of soil erosion and Burke *et al.* show how vegetation cover not only affects soil moisture status but also must be accounted for when soil moisture retrievals are conducted. For estimates of snow pack properties (such as snow depth), Kelly *et al.* note that forest cover influences the accuracy of retrievals from passive microwave instruments. In all of these examples, small errors in vegetation parameter estimation lead to significant errors in the retrieved or the modelled variable.

Vegetation characterization can be achieved in many ways. The most usual remote sensing tool is the multispectral scanning instrument such as Landsat or the Advanced Very High Resolution Radiometer (AVHRR). Enhanced instruments have more recently been used for vegetation mapping and potentially enable the improved characterization of different vegetation parameters. For example, NASA's Moderate Resolution Imaging Spectroradiometer on board the Terra and Aqua platforms has at least twice the fineness of spatial resolution of the AVHRR with a near daily repeat and 32 (mostly narrow) spectral bands in the visible and infra-red part of the electromagnetic spectrum. This is ideal for classical vegetation classification and leaf area index mapping. With better specified instruments for particular science applications, and with the increased understanding of instrument synergisms for particular applications (for example, combining optical reflectance signals with radar canopy geometry models to produce vegetation canopy type condition and physical geometry), more diverse and accurate vegetation parameters will undoubtedly be available for modellers.

13.2.3 Spatial Resolution: Scales of Variation and Size of Support

Spatial resolution is a cross-cutting issue running throughout many of the book contributions (especially in the chapters by Okin and Gillette, Kelly *et al.*, Drake *et al.*, Wooster *et al.*, Barr and Barnsley). It is a multi-dimensional issue that is important for several different reasons. Spatial resolution refers to the spatial resolving power of the desired activity. In the case of remote sensing it usually refers to the size of instantaneous field of view of the sensor (which generally translates to a pixel) but in spatial modelling it could easily refer to the cell size used for the modelling framework. Both definitions should be determined for the modeller by the scale(s) of variation of the environmental variable under consideration. However, spatial resolution of remote sensing instruments is technologically determined and until recently, spatial resolution of the cell grid in numerical modelling has been determined by the computational hardware available, i.e. processor speed, memory and storage. With advancements in both computer and remote sensing technologies, the spatial resolution issue is becoming controlled more by limitations in our understanding of the scale of variation of an environmental variable than by technological limitations. This is an important development because the scale of variation of an environmental variable should control the way that modellers discretize space. In the past, the technological constraints have determined the size of support (or spatial framework) used by the model and have resulted in sometimes very inappropriate spatial models. We are now at a stage of technological development when scientists can often define (at least theoretically) levels of spatial resolution that can reflect the natural scale of variation of an environmental variable. However, this statement assumes that scientists can accurately define the spatial scale of variation of a desired variable but in many cases this is only possible with a high degree of uncertainty. For example, in the chapter by Kelly *et al.*, the scales of variation of snow depth observed from space and point measurements on the ground are compared. While there are similarities evident in the computed variograms, we are still uncertain of the spatial variation characteristics of snow depth beyond those that can be measured with a sparse network of points on the ground or from coarse resolution satellite observations. As Burke *et al.* point out, downscaling methods using statistical techniques or land surface modelling techniques can address this scaling issue but they are not straightforward.

An example of a science community attempting to improve our understanding of the spatial variability of environmental parameters is the Cold Lands Processes Experiment (Cline *et al.*, 2001). This was a two-year field experiment in Colorado, USA, conducted during the winters of 2002 and 2003 and designed to improve our fundamental understanding of the hydrology, meteorology, and ecology of the terrestrial cryosphere. It is a process-oriented experiment that seeks to address several key questions relating to mass and energy dynamics in the cryosphere and the spatial variability of these dynamics. The ultimate goal of the project is to provide a comprehensive benchmark data set that can help refine models of cryospheric processes and define the nature and scale of observations required from remote sensing instruments that could be used effectively to quantify key cryospheric processes (snow cover, frozen ground, etc.). The emphasis is firmly on spatial variability of processes and how processes scale up or down. This information should then be used to inform modellers and technologists alike to which spatial scales their models and instruments should be calibrated.

13.3 Outlook

The future of spatial modelling of the terrestrial environment, therefore, appears to look very promising. The spatial resolution issue is related to our understanding of the scaling of processes and different research communities have shown great coherence in addressing this issue (e.g. the cryosphere, soil moisture, rainfall and vegetation monitoring communities). The production of increasingly accurate DEMs (both in the vertical and horizontal dimensions) is ongoing and the global characterization of vegetation from enhanced satellite instruments is constantly being improved. These advances are the result of coherence of the research communities at the international research community level and will help pave the way for the important enhancement of improved land surface models, which will enable us to better parameterize and constrain land surface conditions for global environmental models. If we are to be able to predict future environmental trends in our world with increased accuracy, spatial models of land surface processes (in both natural and human environments) are a vital part of that effort.

References

Casti, J.L., 1997, *Would-be Worlds: How Computer Simulation Is Changing the Frontiers of Science* (Chichester: John Wiley and Sons).

Cline, D., Armstrong, R., Davis, R.E., Elder, K. and Liston, G., 2001, NASA Cold Land Processes Field Experiment plan 2002–2004, *NASA Earth Science Enterprise: Land Surface Hydrology Program*, NASA: Washington, DC.

Index

Table in **Bold**, figure in *Italic*.

ablation, estimates of mass loss by 31
ablation rates 20
Advanced Microwave Scanning Radiometer EOS
    (AMSR-E) 43, **44**, 54, 253, 259
Advanced Spaceborne Thermal Emission and
    Reflection Radiometer (ASTER) 185, 186
Advanced Very High Resolution Radiometer
    (AVHRR) 158, 266
  cloud-free images, Lake Tanganyika 164–5, 168
air pollution, fire-related 177
air quality modelling 233
air quality monitoring 228–9
airborne laser altimetry (LiDAR) *see* LiDAR
airborne stereo-photogrammetry 86
Airborne Visible/IR Imaging Spectrometer (AVIRIS)
    186
albedo-ice positive feedback 14
Along Track Scanning Radiometer (ATSR-2) 158,
    164–5, 168
Antarctic Ice Sheet 14, 20–1
  balance velocities of grounded portion 26
  DEM(s)
    satellite-derived 31
    for whole ice sheet 21, *23*
  east Antarctica, ice rheology 30
  iceberg fluxes 25–6
  mass budget 13

anthropogenic disturbance 138
assimilation techniques 70
Assimilation Value Index 69
ASTER *see* Advanced Spaceborne Thermal Emission
    and Reflection Radiometer (ASTER)
atmospheric dust 137, 138–9
atmospheric forcing data 251–2
  European Centre for Medium-Range Weather
    Forecasts (ECMWF) products 252
  Global Data Assimilation System (GDAS) weather
    forecast model 251
  NASA/Goddard Earth Observing System (GEOS)
    251–2
automated data correction methods 133–4
automated data generation 114
automatic mesh generation *95*, *96*
automation, in error identification 121
AVHRR *see* Advanced Very High Resolution
    Radiometer (AVHRR)

balance velocities 20–1, 26
  Antarctica and Greenland 26, *27*, *28*
basal sliding 20
best estimates 11
Bi-spectral InfraRed Detection (BIRD) satellite, Hot
    Spot Recognition System (HSRS) 187
  comparison with MODIS *188*, 189–91

Spatial Modelling of the Terrestrial Environment. Edited by R. Kelly, N. Drake, S. Barr.
© 2004 John Wiley & Sons, Ltd. ISBN: 0-470-84348-9.

Bi-spectral InfraRed Detection (BIRD) (*cont.*)
  usefulness of higher spatial resolution 189–90, *190*
bias, in statistical analysis 116–17, 120
biodiversity, Lake Tanganyika catchment 158
biodiversity monitoring, of remote locations 5
biomass burning 177
biomass combustion estimates 182, 184, *184*
boundary friction 79
built-form connectivity models 5, 204–11
  Kruger's original model 204–8
  recognition of built-form constellations 211–18
  a region-based, graph-topological implementation 209–11
    built-form spatial structure 210–11
    built-form connectivity model 209
    representation of regional morphology/spatial structure 209–10
    pre-processing 211
built-form constellation structure 218–23
  built-form unit area 220, *221*
    mixed land-use categories 220
  built-form unit compactness 220–3
  built-form unit packing and density 218–19
built-form constellations 211–18
  containment relation
    analysis of for corresponding node set in XRAG 211–12, *213*
    summary information on for scene as a whole 213–15, **214**
  depth-first graph-searching algorithm, use of 215–16
  relations encapsulating hierarchical containment patterns 216–18

Cambridgeshire, UK
  building the Cambridgeshire model 235–7
    calibration data sources 235
    land use model 236
    transport model 236–7, *236*
  examination of planning strategy (1996–2000) 229
  formulating and testing a sustainable policy package 237–9
    environmental problems needing solutions 238
    planning problems of rapid growth 237–8
  modelling emissions impact in 235–41
    emissions impact of the policy scenarios 239–41
    reference case and policy case and scenarios 235, 238–9
  wide area estimates of emissions concentrations required 234–5
Canadian Meteorological Center 37
catchment-based LSM 250, 259
river channel routing 93
Chavenet principle 121

climate change studies, and snow 35
climate prediction and snow extent/volume 37
climate system, impacts of ice sheets on 13–14
cloud masking technique 165, 159–60
coastal zone colour scanner 110
Cold Lands Processes Experiment (CLPX) 54, 267
combustion
  chemical equation for 178
  combustion efficiency 180
  process in a spreading fire 179
  in wildfires 178–9
Community Land Model (CLM) 251
coupled land surface and microwave emission models 60–3
  MICRO-SWEAT 61–2
    emission component based on Wilheit coherent model 61–2
    Dobson *et al* model 62
    time series of modelled and measured brightness temperatures 62–3, *62*
Cryosat 14, 32, 265
cryosphere, study of, primary objective for satellite missions, Cryosat and ICESat 14, 31–2

Darcy-Weisbach friction factor 87, 94
data
  accuracy, reliability and precision in terms of errors 116
  distinction between accuracy and bias 116–17
data assimilation 11, 246, 247, 254–6
  data assimilation theory 254
  hydrologic data assimilation 255
  in meterology and oceanography 253–4
  see also Land Data Assimilation Systems
  Physical-Space Statistical Analysis System (PSAS) 252, 254–5
  soil moisture estimation 255
data quality
  determination of the SDE important 117–18
  local vs. global measurements of 118–19, 135
    description of surface quality needs careful thought 118–19
    three main issues 118
  measured by the RMSE (Root Mean Square Error) 117
deforestation, Lake Tanganyika's 158
DEM quality 113–14
DEMs 111
  coarse resolution 124–6, *125*, *126*
  accuracy and error characterization 167, 265
    NASA Shuttle Radar Topographic Mission data 265
  InSAR-derived 19, *24*
  error in 116–21
  SRA-derived 30, 31

dense media radiative transfer (DMRT) model 50–1, 52
desert dust 139–40
digital elevation models *see* DEMs
distributed process models 10

Earth System Modeling Framework (ESMF) 4
economic multiplier effect 232
eddy viscosity parameter 86
emissions model 233–5
　DRMB 233–4
　validation 233
environmental modelling 3, 4
environmental problems, Cambridgeshire, needing solutions 238
equifinality 80, 98, 99–100, *99*
erosion
　modelling in large catchments 158–9
　controlling parameters derived from remotely sensed imagery 110–11
error correction 121–6, 134–5
error identification 121–7
　localized error 121–6
　systematic surface error: banding 126–7
error management 111
　DEMs 114
　　aerial photography with photo-control points 114–15
　　assessment of DEM quality 115–16
　　banding 120
　　DEMs of study reach produced 115, 119–20, *119*
　　error 119–21
error propagation
　spatial modelling 265
　DEM-derived geo morphological and hydrological parameters 113
　analytical approaches 54
error(s) 5
　associated with remote sensing products 11
　LiDAR data 92
　in DEMs 116–18
　　systematic error, blunders (gross errors) and random errors 116, 118, 134
　explanation of 127–33
　localized
　　causes 128–9
　　identification 121–6
　　and severe, may be masked by global indicator of precision 118
　in surface temperature remote sensing 253
　in traditional hydraulic investigation 80
　Waimakariri study
　　associated with proximity to wet areas 128
　　significantly greater along channel margins 128, **128**
　　systematic error 119–21, *119*, 129, 135
　see also misclassification errors/problems
ERS-1 and ERS-2 17, 83, 84
eucalyptus
　high fire intensities 181
　high heat yield 180
European Centre for Medium-Range Weather Forecasts (ECMWF) 252, 256
European Space Agency (ESA), Soil Moisture Ocean Salinity (SMOS) mission *see* SMOS mission
exhaust emissions 228
extinction coefficient 65

Famine Early Warning System (FEWS) 158
filters, local topography-based 121–3
fire ecology 177–8
fire intensity 179, 180–2
　fireline intensities 181–2, 190–1
　wild-land fires 180–1
fire modelling 111–12
fire power *see* fire intensity
fire propagation modelling
　potential use of fire radiative energy (FRE) in 192–3
　useful measures from 193
fire radiative energy (FRE) 178, 182
　derivation of from MIR spectral radiance 186–92
　　geostationary satellite imagery 191–2
　　polar-orbiting satellite imagery 186–91
　fire propagation modelling 192
　remote sensing of 183–6
fires
　heat lost to convection/ conductive transfer 191
　heat generation in 179–80
　　heat yield parameter 180, 181
　　fuel moisture content 179–80
　effects across multiple scales in time and space 177
　excessive release of particulates and gases 177
　self-sustaining 179
　spreading, combustion process in 179
　wild-land 176
flood envelope
　inundation extent prediction 80–1
　fuzzy maps 81, 98
flood inundation modelling 79–106
　development of spatial fields for 81–92
　integration of spatial data with hydraulic models 92–100, *101*
　research needs 102
　value of spatial data 102–3
flood inundation models
　integration in a GIS 93
　raster-based 93

floodplain maps, UK 92
flow velocity 85
form drag 86–7
friction data, spatially distributed 94–5
  benchmark validation dataset 94–5
friction, in hydraulic models 86–92
  skin friction for in-channel flows 87
  vegetation biophysical attributes 88–92
  vegetation classification 87–8

gauging stations, national spacing defined by flood warning role 81–2
Geosat 17
Geostationary Operational Environmental Satellites (GOES:US)
  GOES Precipitation Index (GPI) 162
  remotely sensed FRE from 191, *192*
GEWEX Global Soil Wetness Project 251
GIS 3, 4, 264
  integrated, and policy measures 241–2
  and remote sensing, in fire modelling 112
  see also SHIRE 2000 GIS
Glen's flow law 15, 20
  and diagnostic velocities 26
Global Data Assimilation System (GDAS), weather forecast model 251
GOES Precipitation Index (GPI) 162
gravitational driving force, and ice sheet dynamics 14–15, 30
Gravity Recovery and Climate Experiment (GRACE) 253–4
Greenland Ice Sheet 14, 20–1, 31
  balance velocities calculated for grounded portion 26
  DEMs for *16*, 21, *22*, 31–2
    InSAR-derived, north-east Greenland ice stream *24*
  ice divides 23–5, *25*
  uncertainty in mass budget 13

heat yield parameter 180, 181
Helsinki University of Technology (HUT) snow emission model 51–2, *51*, 52
hydraulic model calibration and validation, practical consequences for flood envelope estimation and flood risk maps 80–1
hydraulic modellers, predictive uncertainty a significant problem 80
hydraulic models
  friction 86–7
  integration of spatial data with 92–100, *101*
  inundation extent 82
  physically-based parameterization using LiDAR data 94, *95*
  of reach scale flood inundation 81–2

parameterization 86
  of friction 94
  trials against consistent inundation datasets 96
  use of LiDAR and airborne stereo-photogrammetry for automated broad-area mapping 86
hydraulic resistance 79–80
  a lumped term 86
hydrological cycle fluxes 246
hydrological models, attempt to represent explicitly mass or energy transfers 10
hydrology
  land surface modelling at large regional scales 5–6
  spatial modelling in 9–12
    models constantly evolving 10
    physically-based models 9
HYDROS (HYDROspheric States Mission) (NASA) 66, 71, 73

ice deformation 27
ice divides 23–4
  for Greenland and Antarctica 24–5, *25*
ice flow models 11
ice mass geometry 20
ice sheet dynamics
  derived datasets 23–6
  fast-flow features
    ice sheet interior 26
    mechanisms 20, *27*
  force-budget approach 30
  numerical models 19–21
  relationship between thickness and rheology 15, 30
  thermo-mechanical models, *in situ* rheology of the ice 28, 30
  validation of models 26–30
    Greenland Ice Sheet, balance velocities cf. diagnostic velocities 26–9
    representation of thermodynamics 28–9
    isothermal cases 27–8
    limitations in model resolution 29
    shallow ice approximation 29
    use of accurate surface DEMs for 30, 31
ice sheet models 19–21
  key feedbacks 20
ice sheet topography 4, 31
  derived from satellite radar altimetry 16–18
  from InSAR 18–19, *19*, 22–3, *24*
  from SRA 21–2
  parameterization in numerical modelling 14–15
ice sheets
  impacts on the climate system 13–14
  surface profile, dependence on ice rheology 30
  thickness of 15
iceberg fluxes 25–6

Index

ICESat
  estimation of topographic variables 265
  study of cryosphere a primary objective 14, 31–2
IFOV (instantaneous field of view) 43, 44
image banding
  causes of 129–33
    simulation of random error 129, *130*, 131
  main problem, how to deal with 131–3
    adding additional tie points, main effects 131–2, **131**
    method based on using ground control points 132–3, **133**
    refinement of stitching process 132, *132*
  in systematic surface error 126–7
interferometric Synthetic Aperture Radar (InSAR) 18–19
  topography from 22–3, *24*
inundation extent 82–5
  remote sensing systems 82–5, **83**
    air photo datasets 82
    airborne optical platforms 82
    satellite optical platforms 82
    Synthetic Aperture Radar (SAR) 83–5
  hydraulic models 82
  validation 96

Jornada del Muerto Basin, New Mexico
  wind erosion 138–9, *139*
  Chihuahuan Desert 140–1, *140*
  mesquite dunelands 143, 152
  wind erosion and dust flux 152, 153
  SWEMO **145**, 146–53
    model, operation 148–53
    soil data **145**, 146, *146*
    vegetation data 146–7, *147*, **147**
    wind data 147–8, *148*

Kalman filter 70
  ensemble Kalman filter 255–6
  extended Kalman filter 255, 256, 258–9
kernel-based reclassification techniques 203

L-band radar, use in canopy penetration 84
L-band (satellite) missions, potential 60, 66
land cover, and land use 202
land data assimilation systems (LDAS) 59–60, 69–70, 247–9, *247*
  multiple land surface models under one system 248–9, **249**, **252**
  applications 256–9
    precipitation forcing data 256–8, *258*
    SMMR-derived surface soil moisture data 258–9
    atmospheric forcing fields 256, *257*
  components of 249–56, **252**
  data assimilation 254–6
  land surface simulations 249–51
  observations and observation-based data 251–4
  uncoupled vs. coupled modelling systems 251
  future directions 259
  global LDAS (GLDAS) 248–9, 251–2
  land information system 249
  North American LDAS (NLDAS) 248, 251
  use of existing Land Surface Models 248
  primary goal 248
land form change, application of remote sensing to 110
land information system (LIS) 249, 251
land surface models 12
  assimilation of soil moisture remote sensing data into 255
  coupled with microwave emission models 60–3
  to be used in LDAS 248, 249, 250–1
    catchment-based LSM 250, 259
    Community Land Model (CLM) 251
    Mosaic LSM 250
    NOAH LSM 250–1
  use of probability density function (PDF) with 71
land surface state data 253–4
  snow variables 254
  soil moisture remote sensing 253
  surface temperature remote sensing 253
land use, and atmospheric mineral dust 139
land use and transport modelling framework 231–3
  demand elements 232
  important features for monitoring environmental impact of traffic emissions 233
  land use system 231–2
  transport model, based on trade pattern between zones 232–3
  modal split procedure 233
Landsat Thematic Mapper (TM) 82, 203, 229
large catchments, problems of modelling erosion and deposition in 158
LDAS *see* land data assimilation systems (LDAS)
LiDAR
  high spatial-density airborne LiDAR 201
  topographical mapping 86
  r.m.s. errors 92
LiDAR data, variogram analysis of 92–3
LiDAR and SAR data
  coupled to a numerical flood model 5
  used in flood inundation modelling 12, 79–106
LiDAR segmentation system 88, *89*, 90–1, *90*, 96
  advantage of segmentation 91
LISFLOOD-FP inundation model 93, 96, *97*
  application of 98–100, *99*
  model uncertainty 100

Long-Term Ecological Research (LTER) Network, National Science Foundation 138, 141
loss of lock (SRA) 17
LSM *see* land surface models

METEOSAT 138
METEOSAT-based rainfall estimates 158
  warm cloud precipitation estimation 162–3
  derivation of overland flow from curve number maps 162–3
MICRO-SWEAT 61–2
  microwave component used in exploratory SMOS retrieval algorithm 71–2, *72*
  prediction of surface soil-deeper soil moisture relationship 69–70, *69*
Microwave Doppler radar 252
  for determining water velocities 85
microwave emission model of layered snowpacks (MEMLS) 52
  SNTHERM and Crocus 50
microwave emission models 50, 52, 53
  coupled with land surface models 60–3
  from soil, using coherent or non-coherent models 61
  when soil surface is rough, use of Mo and Schmugge model 62
microwave remote sensing algorithms, to 49, 50
microwave sensors
  useful in polar regions 15
    interferometric synthetic aperture radar (InSAR) 18–19
    satellite radar altimetry 16–18
misclassification errors/problems 83–4
  national land cover map 87–8
Moderate Resolution Imaging Spectroradiometer (MODIS) Airborne Simulator 182, 186, 266
Moran's *I* 210, 220
  values of for built-form compactness *222*, 223

NASA/Goddard Earth Observing System (GEOS) 251–2
  incorporates Physical-Space Statistical Analysis System (PSAS) data assimilation techniques 252
national censuses, suitability of for studying/modelling urban systems 200
national land cover map 87–8
NOAH LSM 250–1
Normalized Difference Vegetation Index (NDVI) 67
  Lake Tanganyika catchment 160–1
north-east Greenland ice stream (NEGIS) 26
  InSAR-derived DEM 22–3, *24*
Numerical Weather Prediction land surface forcing biases, in coupled modelling 251

optical satellite images 201
  potential for urban land-use mapping 203–4
  pattern recognition techniques 204, *205*
Orpington, Bromley (London)
  identification of built-form constellations 217–18, *217*
  land-cover *containment* hierarchy 215–16
  used to evaluate the built-form connectivity approach 211

passive microwave remote sensing
  of soil moisture 45–46, *45*, 72–3
  of snow volume 36
  validation 53
photogrammetric data collection, some post processing with local variance based filter 121–3
  bilinear interpolation 126, *127*
  effects of different DEM resolutions and elevation tolerance values 125–6, *126*
  main parameters: radius of search area 123
  main parameters: SD tolerance 123–4
  tolerance levels tested 123–4
Waimakariri River case study 114–15
  discrepancies between check data and photogrammetric data 128–9
  errors and the stero-matching process 129
  correction methods 133–4, **133**
planning policy development 5
policy scenarios, Cambridgeshire, emissions impact 239–41
  index of emissions impact for Reference and Policy cases 240–1, *240*, 241
precision, theoretical 134
probability density function (PDF) 71
Program for Integrated Earth System Modelling (PRISM) 4
pyrogenic energy emissions
  combustion 178–9
  fire intensity 180–2
  heat generation 179–80
pyrolysis 179

radar data, images of flooding 84
radar interferometry 18
RADARSAT 83, 84–5
radiative transfer models (emission models) 50, *51*
radiative transfer theory 63–6
  discrete, simple and extended Wilheit models 63–4
    discrete model 64
    extended Wilheit model 65–6
    simple model 64–5
random function and regionalized variable 39
range window, satellite radar altimeter 17, *17*

raster grid representation 10
reaction intensity/reaction intensity curve 193
Region Search Map (RSM) 209
remote sensing 264
　coupled with spatial hydrological models 11–12
　estimates of snow depth/SWE
　　recent approaches and limits to accuracy 43–7
　　spatial representivity of SSM/I snow depth estimates: example 47–9
　of fire radiative energy (FRE) 183–6
　　spectral-matching technique to satellite imagery 184–5
　　small-scale experiment using spectroradiometer 183–4
　　shortwave IR wavebands on ASTER 185–6
　of floodplain environments 79
　measuring fireline intensity 182
　methods to retrieve snow depth/SWE 44–7
　　passive microwave estimates cf. visible/infrared global snow maps 44–6, 45
　of near surface sediment concentrations, Lake Tanganyika 164–5, 172
　　atmospheric correction and cloud masking essential 165
　passive microwave systems 43–4
　total water storage variations 253–4
　coupled to hydrological models 10–11
　of soil moisture 253
　spatial resolution 266
regional scale application 48, 53
　of surface temperature 253
　of SWE and snow areal coverage 254
　of erosion 110
　see also satellite remote sensing
remote sensing data 2, 3, 4, 5
reproductive biology, of species in fire-prone systems 177
river channel research 5
　remotely sensed topographic data for 113–36
river reaches, real, traditional methods of hydraulic investigation 79–80

saltation, negative physical consequences for vegetation 138
saltation equation 141–2
SAR see Synthetic Aperture Radar (SAR)
satellite radar altimetry (SRA) 16–18
　corrections to 17–18
　topography from 21–2
satellite remote sensing 14
　development of new applications 110–11
　see also remote sensing
scale, and spatial modelling 3
Scanning Multichannel Microwave Radiometer (SMMR) 43, **44**, 253

SCS (Soil Conservation Service) model
　estimating overland flow using FEWS rainfall data 161–3
　problem implementing with FEWS rainfall data 163
Seasat 16–17
sediment plumes, Lake Tanganyika
　buoyancy of 164
　detection by remote sensing 164, 168, *169–71*
　the Malagarasi River plume 167–8
　mapping 158
　Ruzizi river mouth plume 168
sediment transport and routing, Lake Tanganyika 163–4, 167–8, 172
　alternative method to model deposition 163–4
　routed delivery ratio 164
　sediment transport capacity 163
seedbank depletion 139
Severn Basin, and LiDAR segmentation system 88, *89*, 90–1, *90*
Severn River
　finite-element hydraulic model, computational mesh 94–5, *95*
　floods (1998 and 2000) 84, 84–5
　observed and simulated inundated areas 93
shallow ice approximation 29
Shallow Water models 94
SHIRE 2000 GIS 229–30, **231**
　creation of framework for modelling emissions impact *230*, 231
　emissions model 230
　'hot spots' of traffic emissions 239
　key assumptions 229
　land use and transport models 230
　visualization of traffic emissions concentrations-settlement pattern relationship 239
Shuffled Complex Evolution optimization procedure 71
slope-induced error 18, *18*
smoothing functions, and error in DEMs 127
SMOS mission 66, 73
　downscaled estimates of soil moisture 71
　opacity coefficient 67–8, *68*
　estimates of optical depth 67–8
　use of MICRO-SWEAT microwave emission component for exploratory retrieval algorithm 71–2, *72*
snow representation in climate models 35–6
snow cover
　and climate 35
　distribution 37
　global, quantitative maps using satellite-derived snow cover estimates 37
　McKay and Gray classification 37, 38–9
snow cover area, qualitative maps of 36–7

snow depth measurement networks 37–8, *38*
  U.S. COOP station network 38, *38*, 40, *40*, 41
  FSU network 38, *38*, *40*, 41
  WMO GTS network 37, 38, *38*, *40*, 41
snow depth/SWE 11
  estimation using pysically based models 42–3, 50
  improving estimates of at all scales
    combining models and observations 49–52
  validation frameworks 53
  methodological approaches to 50, *51*
  modelling spatial variations of using *in situ* snow measurements 36–42
  remote sensing estimates
    recent approaches and limits to accuracy 43–7
    spatial representivity of SSM/I snow depth estimates 47–9
    retrieval schemes based on empirical formulations 46–7
  spatial dependency of 37
  spatial variability investigated using variograms 39–42
  use of terrain and meteorological variables in spatial modelling 42
snow hydrology models 42–3, 53
  combined with microwave emission models 53
snow maps, uses of 37
snow pack energy balance models 49
snow packs
  layering 46
  interaction with vegetation cover 266
snow variables and land data assimilation 254
snow volume
  global, retrieval of 4
  satellite passive microwave estimates 36
snow water equivalent (SWE)
  general estimation approach 53–4
  changes at hemispheric level 35
  regionally calibrated approaches 36
SNTHERM model, coupled with dense media radiative transfer (DMRT) model 50–1, 54
soil erosion 109
  required parameters 111
  models of erosion by water 109–10
  regional scale, near real-time modelling of 157–73
    Sediment flux 168, 172
    source areas of erosion 168
soil erosion model, applied to Lake Tanganyika 158, 159–63
  overland flow estimated using SCS model with FEWS data 161–3
  soil erodibility computed from soil properties maps 161, **162**

  vegetation cover estimated using LAC AVHRR 159–61
  relationship between vegetation cover and NDVI 160–1, *160*
  scaling problem, overcome by use of Polya function 161
soil moisture, improving accuracy of in models 59
soil moisture estimates 11
  downscaling
    four-dimensional assimilation algorithm 71
    modified fractal interpolation technique 71
  extending estimates from deeper within the profile 69–70
  use of statistical methods 69–70
  using assimilation techniques 70
  from ground-based and aircraft radiometer systems 4–5
  remotely sensed observations using L-band passive microwave radiometer 60
  subpixel heterogeneity 70–2
Soil Moisture Ocean Salinity (SMOS) mission *see* SMOS mission
soil moisture retrieval algorithms
  effects of vegetation in 66–9
    ancillary information to estimate optical depth 66–7
    use of NDVI 67
    quantifying errors due to assumptions about the vegetation 68–9
space syntax theory 204
spatial autocorrelation 223
  of snow depth 39
spatial data
  incorporation into environmental models 102–3
  integration with hydraulic models 92–100, *101*
    automatic mesh generation 96
    high resolution topographic data 92–4
    model calibration and validation studies 96–8
    spatially distributed friction data 94–5
    uncertainty estimation using spatial data and distributed mapping 98–100, *101*
spatial models/modelling 2–4
  the future 267
  in hydrology 9–12
  importance of 264
  of the terrestrial environment, key research issues 264–7
    DEMs: improved accuracy and error characterization 265
    spatial resolution: scales of variation and size of support 266–7
    vegetation cover: improved characterization 265–6
spatial reclassification techniques 203
spatial resolution 267

higher, usefulness of 189–90, *190*
passive microwave systems 43–4
scales of variation and size of support 266–7
spatial variability, of environmental parameters 267
spatially explicit wind erosion and dust flux model 143–6
  issues in integration of parameters for 144–6, **145**
  processing stream for *145*
  predicting wind erosion and dust flux at the Jornada Basin 146–53
    data sources and model inputs **145**, 146–8, *146*, *147*, **147**, *148*
  relations between model parameters and vegetation/soil parameters 144, **144**
Special Sensor Microwave Imager (SSM/I) 43, **44**, 253
SPOT-HVR XS images 203
SSM/I snow depth estimates, spatial representivity of, an example 47–9
standard deviation (SD), relationship with error *122*, 123–4, *124*
Stefan's Law 187
stereo-photogrammetry, limitations over ice sheets 15
Structural Analysis and Mapping System (SAMS) 210, 215
  and built-form constellations 209
surface mass balance model 19–2020
surface quality
  description 118–19
  quantification 116–18
surface temperature remote sensing 253
suspended sediment concentration, used to map flow patterns 82
sustainable planning policy 229
sustainable urban development 200
SWE *see* snow water equivalent (SWE)
SWEMO *see* spatially explicit wind erosion and dust flux model
Synthetic Aperture Radar (SAR)
  airborne 84, 97–8
  dynamic flooding processes 83
  River Severn floods (1998 and 2000) 84–5
  polarimetric or multi-frequency and accurate flood mapping 84
  new sensors for measurement of inundation extent 83–4
  statistical active contour methods (snakes) 83–4

Tanganyika, Lake 5, 157–8
  deficiency in sediment delivery model, Malagarasi River sediment 167–8
  estimation of lake sediment concentrations 164–5
    use of AVHRR and ATSR-2 imagery 164–5
  key regions prone to erosion, deforested and degraded 166, *167*

Lake Tanganyika Biodiversity Project (LTBP)
  Special Sediment Study 158
  mapping of sediment plumes 158
  remote sensing of sediment plumes 164, 168, *169–71*
    Ruzizi River plumes 168
  sediment monitoring system 158
    estimation of lake sediment concentrations 164–5
    sediment transport and routing 163–4, 167–8, 172
    soil erosion modelling 159–63
  sediment yield estimations 172
  sensitive to sedimentation problems 157, 158
  severe erosion in the northern catchment, Rwanda, Burundi and eastern Zaire 166, 167, *167*
  soil erosion model of the catchment 158
terrestrial environment 1
  connected to spatial modelling 1–2
topographic data, high resolution 92–4
  integration with standard hydraulic models 92
topography
  controlling sediment route through a catchment 164
  in hydraulic models 86
  ice sheet 4, 14–15, 31, 265
    from InSAR and SRA 21–3, *24*
  see also ice sheet topography; surface quality
Total Ozone Mapping Spectrometer (TOMS) 137–8, 139
trace gases and aerosols, emissions from fires 177, 182
tractability, in computer modelling 3
traffic congestion, Cambridgeshire 238
traffic emissions, modelling impact of on the urban environment 228–43
  integrated land use and transport modelling framework 231–3
  emissions model 233–5
  integrating models within the framework of SHIRE 2000 GIS 229–30, **231**
  modelling emmissions impact in Cambridgeshire 235–41
  emissions impact of the policy scenarios 239–41
Tropical Rainfall Measuring Mission (TRMM)
  Microwave Imager (TMI) 253

UK
  local authorities responsible for an 'Air Quality Strategy' 228–9
  National Atmospheric Emissions Inventory (NAIE) 229
uncertainty estimation
  using spatial data and distributed risk mapping 98–101
  probability mapping 100, *101*
uniform flow theory 86
universal soil loss equation (USLE) 109
urban growth 200

urban land use
  in remotely sensed images, automated analyses using multi-stage approach 202–3
    use of kernel-based re-classification techniques 203
urban land-use categories, recognised by human photo-interpreters 202
urban systems
  graph-theoretic measures of built-form connectivity (Kruger) 207, **207**, *208*, **208**
  modelled by Kruger as structured trees 206–7, *206*

variograms/semi-variograms 39–42, 52
  U.S. COOP data 40, *40*
  different sampling scales 39
  WMO dataset, range, nugget and sill variance *40*, 41
vector representations of space 10
vegetation
  accounting for effects of in retrieval algorithms 66–9
  amount of heat released during burning 180
  biophysical attibutes 88–92
  burnt and unburnt components 193
  determining drag coefficient of 87
  effect on microwave emission from the soil 63–6, 73
  use of height data to specify an 'effective', individual friction factor 94, *95*
vegetation cover
  estimated using LAC AVHRR data 159–60
  improved characterization 265–6

errors in the estimation 161
relationship with NDVI 160–1, *160*
vehicular emissions, exposure to harmful 228
von Karmann's constant 142

Waimakariri River case study 114–34
water erosion modelling 110
  Lake Tanganyika 111
waveform retracking 17
wildfire events
  close to Sydney, Australia, MODIS and HSRS MIR images compared 187, *188*, 189–91
  possible role in climate change/species extinction 176
Wilheit (1978) model 64, 65–6
  multi-layered canopy 65
  dielectric constant for the canopy 65–6
  time series of modelled and measured brightness temperature *62*, 66
wind erosion
  local averages 144–5
  wind speed 153
wind erosion and dust emission 5
  little information about occurrence of 137
wind erosion and dust flux model 141–2, 143
  see also spatially explicit wind erosion and dust flux model
wind erosion modelling 110, 111
  basic equations relating to 141–3

XRAG (eXtended Relational Attribute Graph) 209, 210, 211